MW01060860

EC&M's Electrical Calculations Handbook

EC&M's Electrical Calculations Handbook

John M. Paschal, Jr., P.E.

McGraw-Hill

New York San Francisco Washington, D.C. Auckland Bogotá
Caracas Lisbon London Madrid Mexico City Milan
Montreal New Delhi San Juan Singapore
Sydney Tokyo Toronto

Library of Congress Cataloging-in-Publication Data

Paschal, John, P.E.
EC&M's electrical calculations handbook / John M. Paschal, Jr.
p. cm.
Includes index.
ISBN 0-07-136095-6
1. Electric engineering—Mathematics—Handbooks, manuals, etc.
2. Engineering mathematics—Formulae—Handbooks, manuals, etc.
I. Title.
TK151.P355 2000 00-051115
621.3'01'51—dc21

McGraw-Hill 
A Division of The ***McGraw-Hill*** *Companies*

1 2 3 4 5 6 7 8 9 0 DOC/DOC 0 6 5 4 3 2 1 0

ISBN 0-07-136095-6

The sponsoring editor for this book was Zoe G. Foundotos, the editing supervisor was Stephen M. Smith, and the production supervisor was Sherri Souffrance. It was set in New Century Schoolbook per the MHT 5 x 8 design by Joanne Morbit of McGraw-Hill's Hightstown, N.J., Professional Book Group composition unit.

Printed and bound by R. R. Donnelley & Sons Company.

This book is printed on recycled, acid-free paper containing a minimum of 50% recycled de-inked fiber.

Contents

Preface

We all frequently need electrical reference material, and sometimes we need an explanation of how certain electrical equipment works, what dimensions are acceptable or unacceptable, or approximately which values of things such as voltage drop or wire size are reasonable. I have observed over the years that there are certain electrical engineering and design resources that I refer to more frequently than any others. In my work I have also noticed that there are certain types of calculations that are important enough to occur frequently, but not frequently enough for me to have memorized all of the dimensional or output data associated with them. In addition, making calculations without reference values to "go by" sets the stage for errors that could have been avoided if similar calculations could be referred to. Finally, there is a need for good explanatory material that can be shared with fellow engineers or designers or with owners. Such information is invaluable in helping them to make sound decisions, since most thinking individuals can make a good decision when given the correct data to consider.

It was with all of these in mind that I conceived of this electrical calculations handbook. It is intended to be a handy tool that provides in just one place much of the information that one normally seeks from reference manuals; it also provides solved "go-by" problems of the most-often-encountered types in the electrical industry to expedite solutions and make calculations easy. Instead of simply providing formulas without explanations, I took care to explain each problem type and formula, and to prepare step-by-step solutions. The problems covered in this book range from explanations of Ohm's law and generator sizing, to lighting calculations and electrical cost estimating and engineering economics calculations. I made every effort to make the book concise enough to be portable, while still including the very best graphic illustrations. I also included, following this preface, a detailed listing of problem types in alphabetical order to make finding the proper "go-by" calculation easy and fast.

I sincerely hope that you will find that keeping this "electrical calculation reference library in one book" close by will save you from having to carry several other reference books, and that it will expedite your work while making it easier and more accurate. I hope that the knowledge and insight gained from it will add even more fun to your work in our terrific electrical industry.

John M. Paschal, Jr., P.E.

List of Problems

Figure	Solve for
14-3	Working space dimensions for equipment operating at 0–150 V to ground
14-4	Working space dimensions for equipment operating at 151–600 V to ground
14-5	Working space dimensions for equipment operating at over 600 V to ground
9-14	X/R ratio of transformer from transformer impedance and full-load loss

EC&M's Electrical Calculations Handbook

Chapter

1

Basic Electrical Working Definitions and Concepts

Electricity is an invisible force that is used to transfer energy into heat, light, intelligence, or motion. Electricity is explained in terms of electrical charge, potential difference (or voltage), electrical charge flow (or current), and resistance to current flow. Figure 1-1 graphically illustrates electron flow through a conductor by comparing it with water flow through a pipe. The normal unit of current measurement is the ampere, whereas the normal unit of voltage measurement is the volt. The unit of opposition to current flow, or resistance, is the ohm.

Voltage as Potential Difference

The basic property of every operating electrical system is that different parts of the circuit contain items having different polarities. Another way of saying this is that the "negatively" charged parts contain a surplus of negatively charged electrons, whereas the "positively" charged parts contain a deficiency of electrons. When molecules

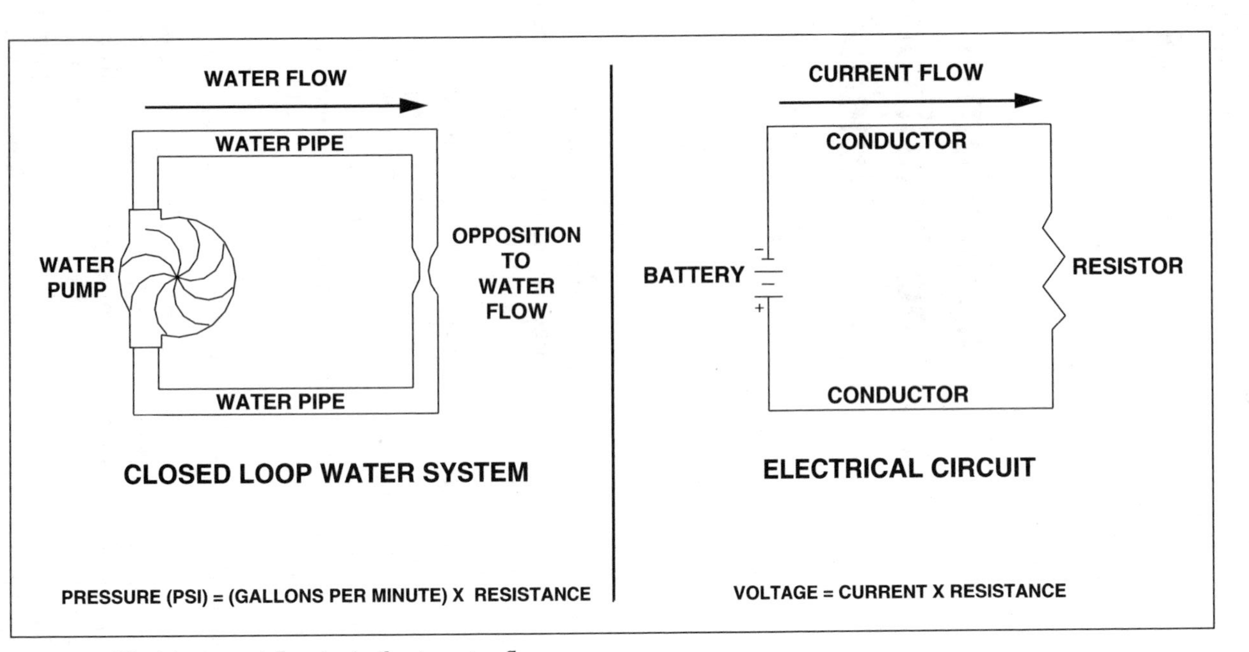

Figure 1-1 Electric current flow is similar to water flow.

contain more protons than electrons, they have a deficiency of electrons, and relatively speaking, this means that they have a "positive" overall charge. In nature, there is a natural attraction by protons for sufficient electrons to equalize the positive and negative charges of every molecule. The greater the charge between different parts of the circuit, the greater is the potential difference between them. The standard way of describing this state is to say that the circuit driving voltage, or source voltage, increases.

Surplus electrons exist on the negative (−) terminal of a battery, whereas "slots in molecular outer orbits" for missing electrons exist in molecules of the positive (+) terminal of a battery. In an electrical circuit, a conductor "makes a complete path" from the negative to the positive battery terminals, and electrons then flow from the negative terminal to the positive terminal through the conductor. Within the circuit conductor, electrons flow from one molecule to the next and then to the next.

Some molecules permit the easy movement of electrons, and the materials composed of these molecules are said to be *conductors*. When materials do not permit the easy flow of electrons, they are said to be *insulators*. The entire key to electrical systems is to "show electrons where *to flow*" by installing conductors and to "show electrons where *not to flow*" by surrounding the conductors with insulators. Practically speaking, most circuit conductors are made of either copper or aluminum. Insulators can be rigid or flexible. Everyday examples of rigid insulators are glass and plastic, and common examples of flexible insulators are rubber and air.

Current

In an attempt to provide a quantifying image of how many electrons are required to form a current flow of one ampere, one can state that 6.25×10^{18} electrons flowing past any one point in an electrical circuit constitutes one ampere of current flow.

Resistance

The voltage is the "pressure" that forces the electrons, or current, to flow through the circuit conductors, and the opposition to current flow in the circuit, or the circuit resistance, is measured in ohms. One volt can force one ampere to flow through one ohm of resistance. This is the basic relationship known as *Ohm's law:*

Voltage (volts) = current (amperes) × resistance (ohms)

A characteristic of all conductors is resistance, but some conductors offer more resistance to current flow than do other conductors. A conductor can be imagined to consist of bundles of molecules, each containing "spaces" where electrons are missing. In current flow, voltage can force electrons to flow into and out of these "spaces." To reduce the opposition to current flow, the conductor can be widened, thus effectively creating more parallel paths through which electrons can flow. To increase the opposition to current flow, the conductor can be made more narrow. The resistance value of the conductor also can be altered by lengthening or shortening the conductor. Longer conductors offer more opposition to current flow and thus contain more ohms of resistance. Note that the insertion of an infinitely large resistance into an electric circuit has the effect of creating an *open circuit,* causing all electric current flow to cease. This is what happens when a switch is placed in the "open" position, since it effectively places a very large value of resistance in the form of air into the circuit.

Direct-Current (dc) Voltage Sources

Various types of dc cells are available, most providing approximately 1.75 open-circuit volts across their output terminals. When higher dc voltages are required, additional cells are connected together in a series "string" called a *battery,* and the resulting overall voltage of the battery is equal to the sum of the voltages of the individual cells in the string.

Basic electrical symbols and abbreviations are shown in Fig. 1-2. Some of the symbols and abbreviations used most often are as follows:

SYMBOL	MEANING	UNITS
V, or E	Voltage in a DC system	Volts
v	Instantaneous voltage in an AC system	Volts
I	Current in a DC system	Amperes
i	Instantaneous current in an AC system	Amperes
R	Resistance in either an AC or DC system	Ohms
Z	Impedance in an AC system	Ohms
X	Reactance in an AC system	Ohms
X_L	Inductive Reactance in an AC system	Ohms
X_C	Capacitive Reactance in an AC system	Ohms
L	Inductance in an AC system	Henries
C	Capacitance in an AC system	Farads
W	Power in either an AC or DC system	Watts
w	Instantaneous power in an AC system	Watts
VA	Apparent power in an AC system	Volt-Amperes
va	Instantaneous apparent power in an AC system	Volt-Amperes
VAR	Reactive power in an AC system	Volt-Amperes Reactive
VAC	Reactive power in an AC system	Volt-Amperes Capacitive

Figure 1-2 These basic electrical symbols are used in equations and on electrical drawings.

The electrical symbol for the volt is v or V.

The symbols for current are a or I.

The symbol for resistance is the Greek capital letter omega, Ω.

The symbol for the voltage source is E.

The symbol for a conductor without resistance is a thin, straight line.

Direct and Alternating Current

Electron flow from a cell or battery is called *direct current* (dc) because it has only one direction. Some voltage sources periodically reverse in polarity, and these are identified as *alternating-current* (ac) sources. In terms of electron flow at each instant in time, the current always flows from the negative terminal through the circuit to the positive terminal. Thus 60-cycle ac power of the type found in most homes is an example of an ac system. In this example, the frequency of 60 cycles per second, or hertz (Hz), means that the voltage polarity and the current direction reverse 60 times per second. Figure 1-3 is a graph of an ac voltage system in which key facets are identified. In the ac system, the *effective voltage* is distinguished from the *peak-to-peak* voltage because the peak voltage is not always present, so effectively, it cannot be used accurately in mathematical solutions. The effective voltage value, however, accommodates the varying voltage values and their continually varying residence times to provide accurate electrical system calculations.

dc Voltage

Cells, batteries, and dc voltage

In a dc circuit, the most common voltage source is the chemical cell. Many different types of chemical cells are available commercially, and each exhibits unique characteristics. Some of the more common chemical cells, along with their voltage characteristics, are shown in Fig. 1-4.

When more than one cell is connected together in series, a *battery* is formed. When cells within a battery are connected together such that the polarities of the connections are (+) (−) (+) (−) (+) (−), then the battery is connected in *additive* polarity, and the overall battery voltage is equal to the arithmetic sum of the cell voltages (as demonstrated in Fig. 1-5). However, when the cells within a battery are connected together such that some of the cells are not connected in additive polarity, then the overall battery voltage is equal to the sum of the cells connected in additive polarity minus the voltages of the cells connected in *subtractive* polarity (as shown in Fig. 1-6). A common method of constructing a battery of the required voltage rating for a given load is shown in Fig. 1-7, where a series connection of two 12-volt (V) batteries is used to provide a 24-V battery for a diesel engine electrical system.

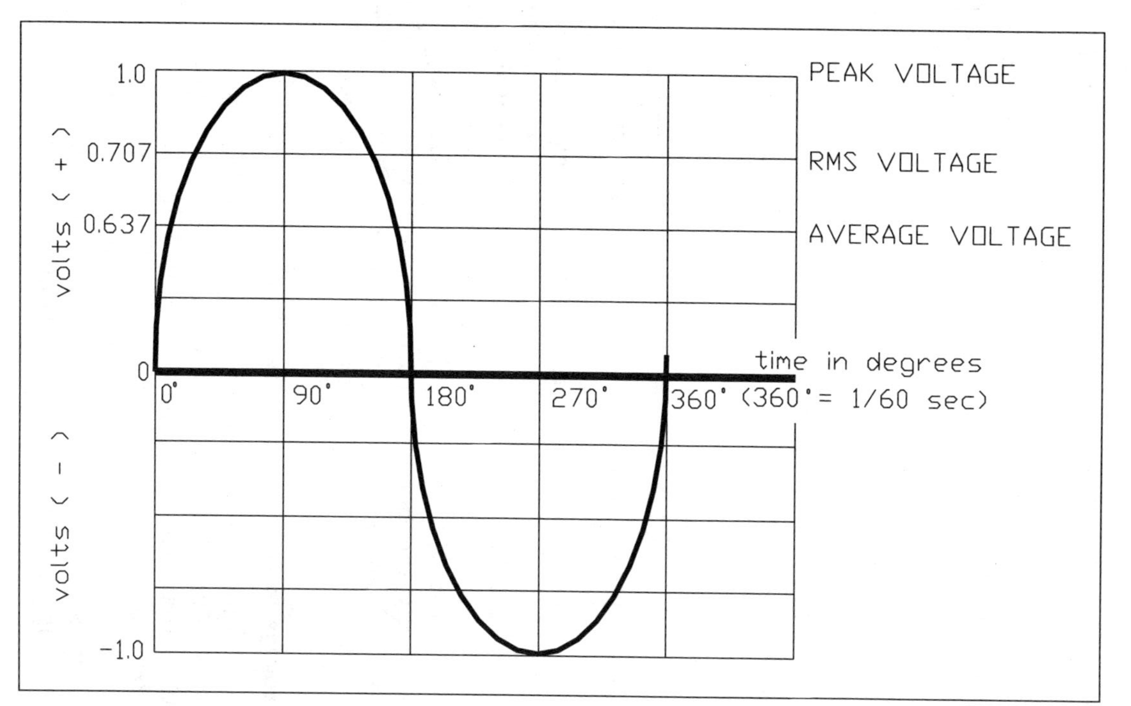

Figure 1-3 A graph of ac voltage.

TYPE OF CELL	WET OR DRY	VOLTAGE PER CELL
Carbon-Zinc	Dry	1.5
Alkaline Dry Cell	Dry	1.5
Lead Acid	Wet	2.2
Nickel-Cadmium	Wet or Dry	1.25
Mercury	Dry	1.3

Figure 1-4 Some of the more common chemical cells and their characteristics.

Battery current limitations

Sometimes individual batteries are not large enough to provide sufficient electron flow for the load to operate correctly. In such cases, additional batteries can be connected in parallel with the original batteries without changing the output voltage. All that changes when identical batteries are added in parallel is that additional electron flow is made available from the additional battery plates; the overall voltage, however, is not changed by adding batteries in parallel. Accordingly, the amount of current that flows through the resistive circuit is still simply determined by Ohm's law. See Fig. 1-8 for an illustration and an example calculation.

dc voltage source with internal resistance

Every battery is only able to deliver a finite amount of current. To understand what is actually happening within a battery that exhibits a limited current output, it is useful to draw a more detailed electrical diagram of a battery. In the more detailed diagram, the battery is shown not only to have a set of internal electron-producing and voltage-producing cells but also to incorporate an internal resistor (see Fig. 1-9). The internal resistance is an artificial manner of representing the fact that the battery has output-current limitations such that if a zero-resistance circuit path were connected from − to + at the battery terminals, current flow

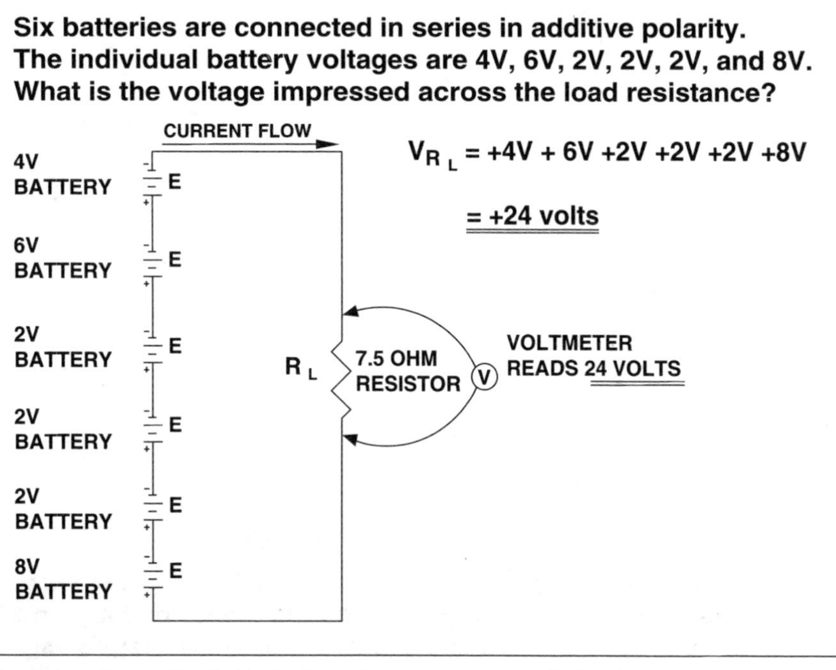

Figure 1-5 Connecting batteries in series to increase terminal voltage.

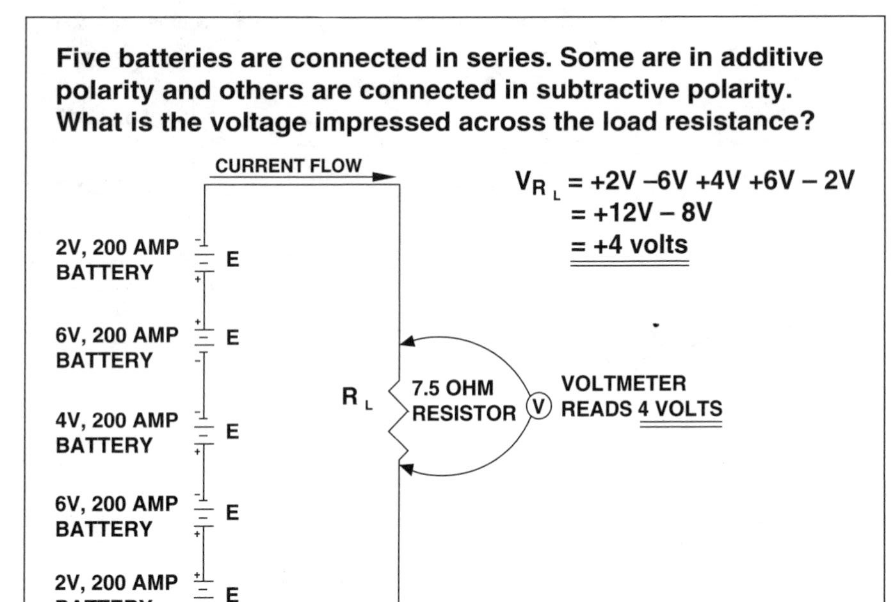

Figure 1-6 Connecting batteries in series to regulate output voltage.

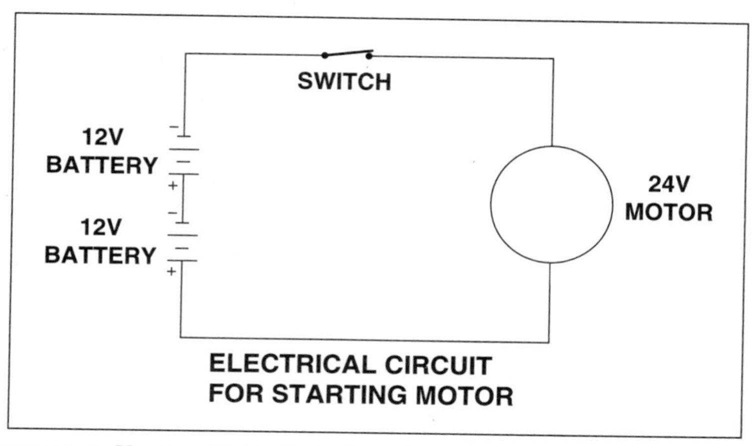

Figure 1-7 Use two 12-V batteries in series to drive a 24-V motor.

would *not* be infinite. From a practical perspective, each battery has an internal resistance, with its resistive value lessening in magnitude as the battery temperature increases. Conversely, a battery can be expected to have a very low output current during very cold ambient temperatures. For example, at a given ambient temperature of 77°F, a certain battery is nameplate-rated at 300 amperes (A) at 12 V. Following the output current versus temperature curve of Fig. 1-10, if the ambient temperature declines from +77 to +20°F, then the battery output drops to 150 A. Thus, for continued full 300-A current flow at temperatures colder than the battery nameplate rating of +77°F, an engineer or designer must oversize this battery as follows:

$$\text{Battery output at } +20°\text{F} = 150 \text{ A}$$

$$\text{Battery output at } +77°\text{F} = 300 \text{ A}$$

$$300/150 = 200 \text{ percent}$$

Thus, for continued full 300-A current flow at +20°F, a battery must be specified that has twice the 77°F name-

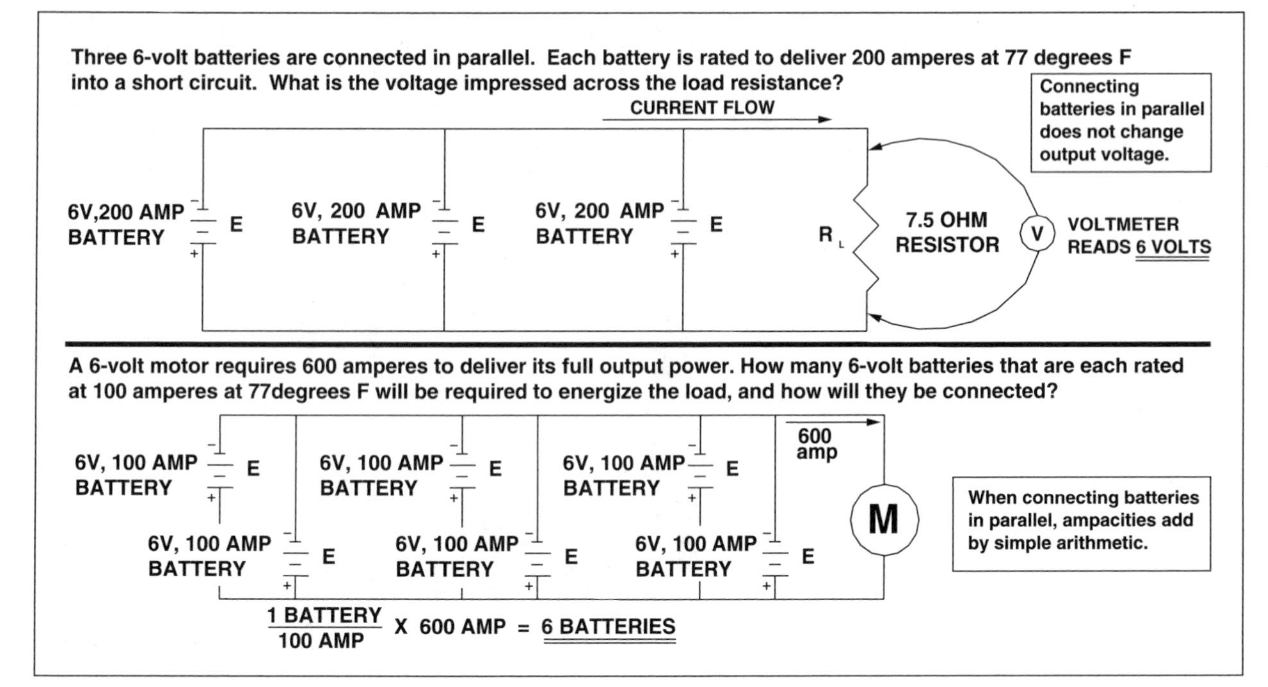

Figure 1-8 Connecting batteries in parallel for added current capacity.

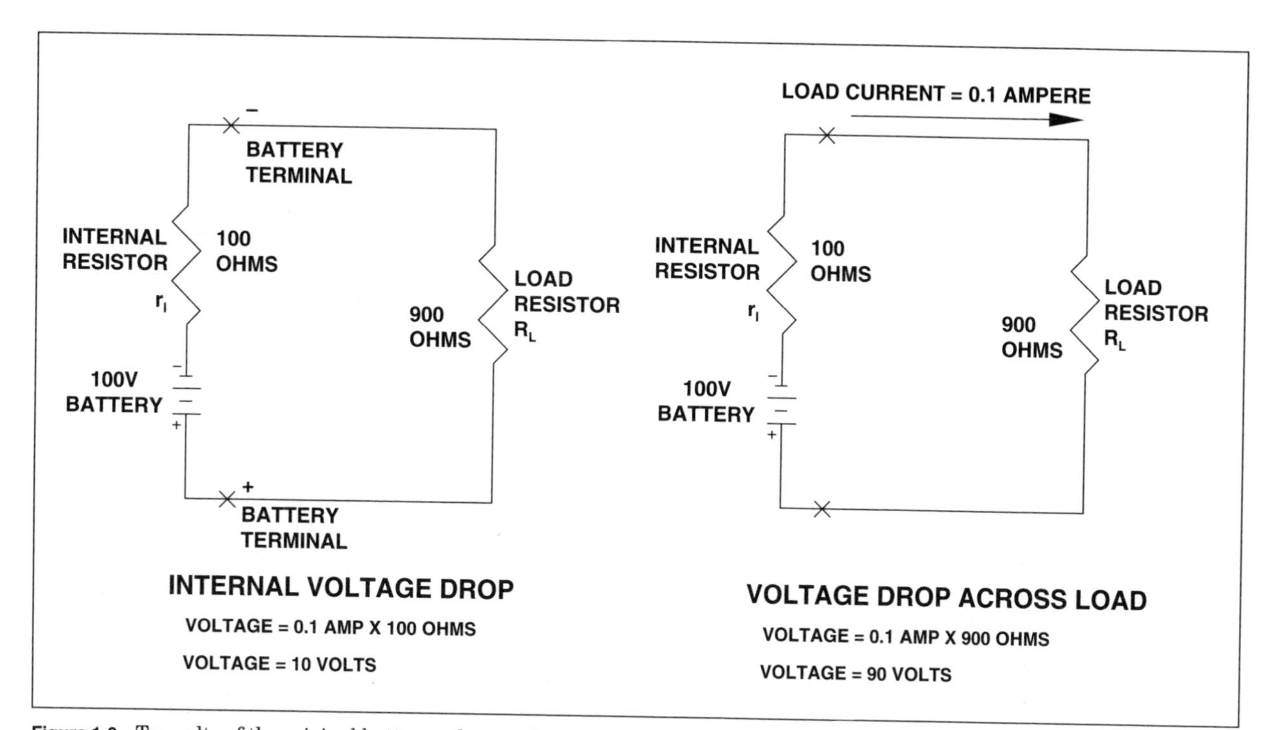

Figure 1-9 Ten volts of the original battery voltage is lost to internal resistance.

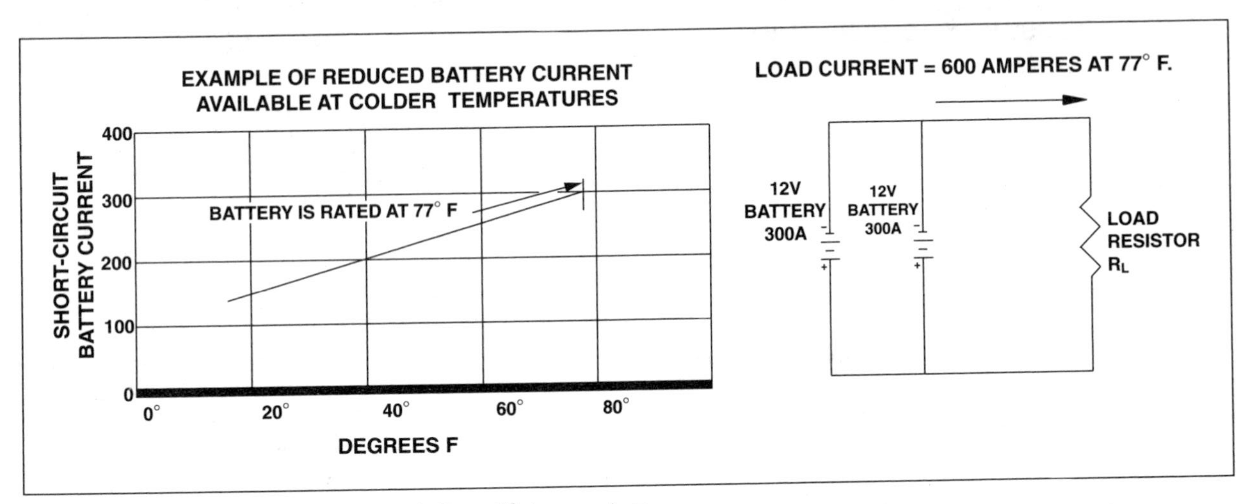

Figure 1-10 Lead acid battery output variation with temperature.

plate battery rating. Therefore, the 77°F nameplate rating of the required battery source is calculated as

$$300\text{ A} \times 2 = \mathit{600}\text{ A}$$

Different battery types exhibit slightly different temperature derating curves, but the curve shown in Fig. 1-10 provides a good approximation for them.

Current Flow in a Resistive Circuit

In a simply resistive electric circuit, as shown in Fig. 1-11, the battery source generally is shown, but its internal resistance is not shown. In addition, the resistive load normally is shown connected to the battery terminals through "perfect" conductors that are imagined to have zero resistance. That is, *all* the resistance in the circuit is represented by the load

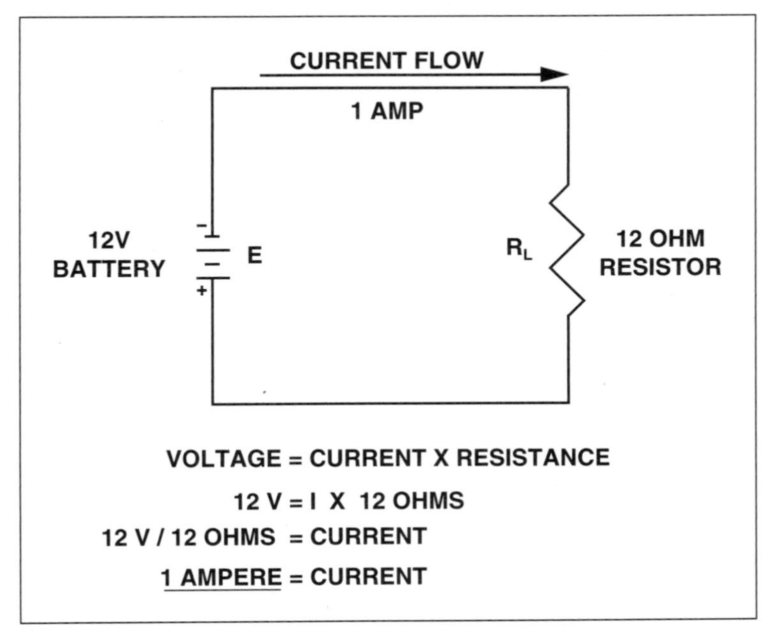

Figure 1-11 Solve for dc current in a series circuit given dc source and resistance.

resistance symbol R_L. This methodology offers many conveniences in making calculations on more complex circuits.

In a simple series circuit, a 12-V battery source (the symbol for which is E) is able to force a total of 1 A (the symbol for which is I) to flow through a 12-ohm (Ω) load resistor (the symbol for which is R). This is as would be expected from the standard Ohm's law calculation:

$$E = IR$$

$$12 = I(R)$$

$$12 = I(12)$$

$$12/12 = I = 1\text{ A}$$

Common problems that arise in electrical design include calculation of the source voltage that would be required to force a certain current magnitude to flow through a given resistance. An example of this type of problem is shown in Fig. 1-12. It shows that 10 A is to flow through a 6-Ω load resistor and asks what the source voltage must be to make this occur. The solution to the problem, using Ohm's law, is 60 V.

Similarly, often the resistance rating of a resistor is needed when the source voltage and required current flow are known. Figure 1-13 is a solved problem of this type, once again simply making use of Ohm's law.

Current Flow in a Series Resistive Circuit

When resistors are connected end to end, they are said to be in *series*. In a series circuit, all the current flows through each and every element in the circuit.

In a dc circuit, the surplus electrons that have built up on the negative battery terminal try to move to the positive battery terminal, and they will follow whatever conductive path is presented to do so. Accordingly, in Fig. 1-14, electrons leave the negative terminal of the battery and flow through resistors R_1 and R_2 on their way to the positive battery terminal. The entire source battery voltage E is "dropped" or

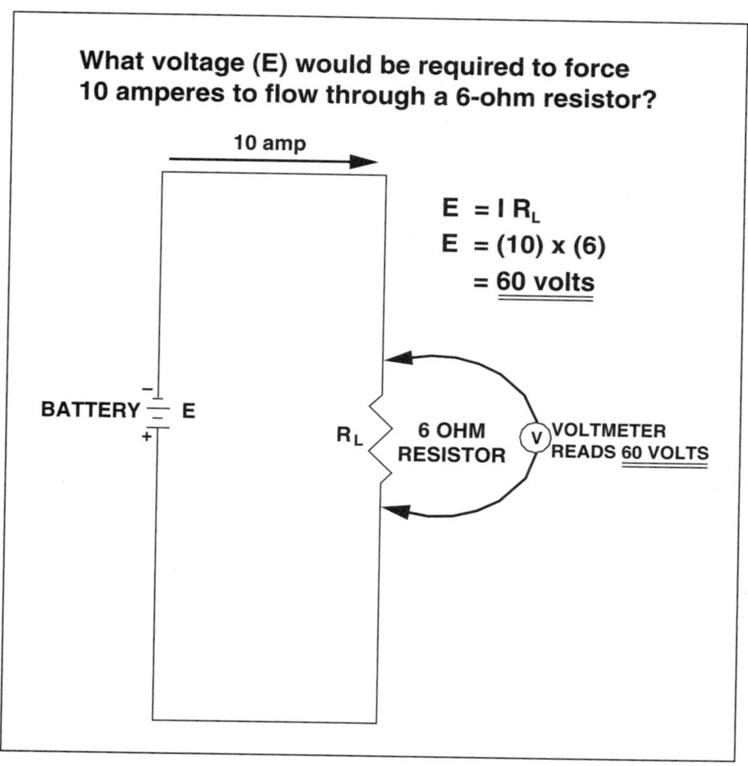

Figure 1-12 Solve for dc voltage in a series circuit given dc current and resistance.

"spread" across the entire circuit. That is, when the battery voltage is 10 V, the sum of the voltages that can be measured across the series resistor string is also 10 V.

To determine the amount of current that will flow in a series resistive circuit, it is first necessary that the resistances be resolved into one equivalent resistance, and then Ohm's law can be applied. For example, Fig. 1-15 shows the steps necessary to resolve the equivalent resistance of parallel resistances, and Fig. 1-16 shows the steps necessary to resolve the equivalent resistance of resistances that are connected in a series-parallel resistance network.

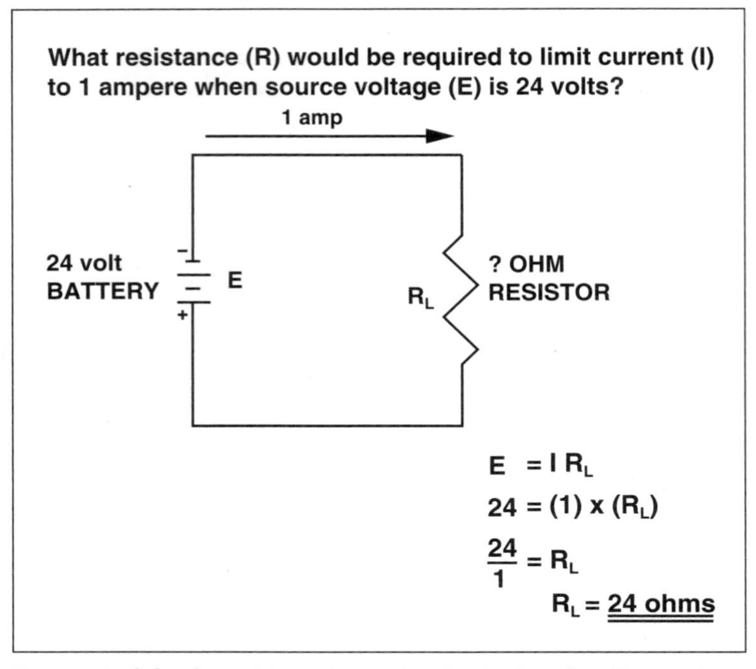

Figure 1-13 Solve for resistance in a series circuit given dc voltage and current.

Voltage Division in a Series Circuit

Both Ohm's law and the voltage divider law are useful in determining the voltage drop that can be measured across each resistor in a series resistor string. For example, Fig. 1-17 shows the calculation by each method of the voltage drops across each resistor in a series resistive circuit.

Power Rating of a Resistor

A resistor is rated not only in ohms of resistance but also in watts. The watt rating of a resistor tells the maximum quantity of watts of heat that will radiate or conduct from the resistor in operation, and it also gives an indication of how many amperes can flow through the resistor without damaging it. In

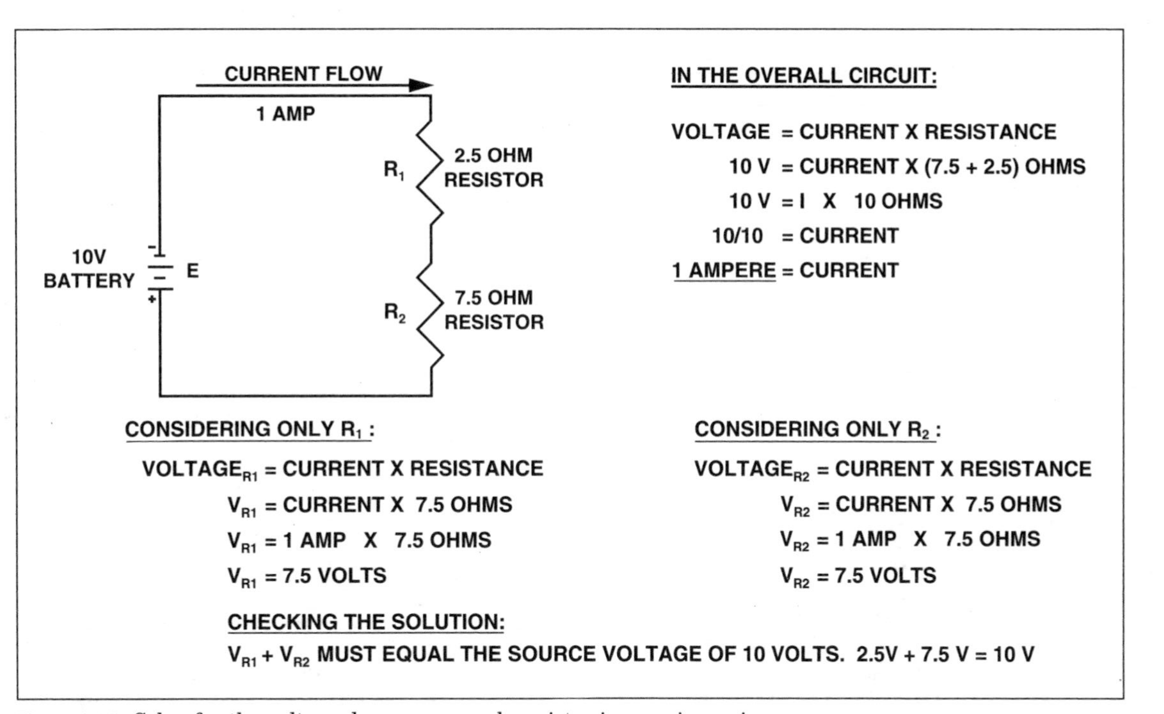

Figure 1-14 Solve for the voltage drop across each resistor in a series resistance dc circuit.

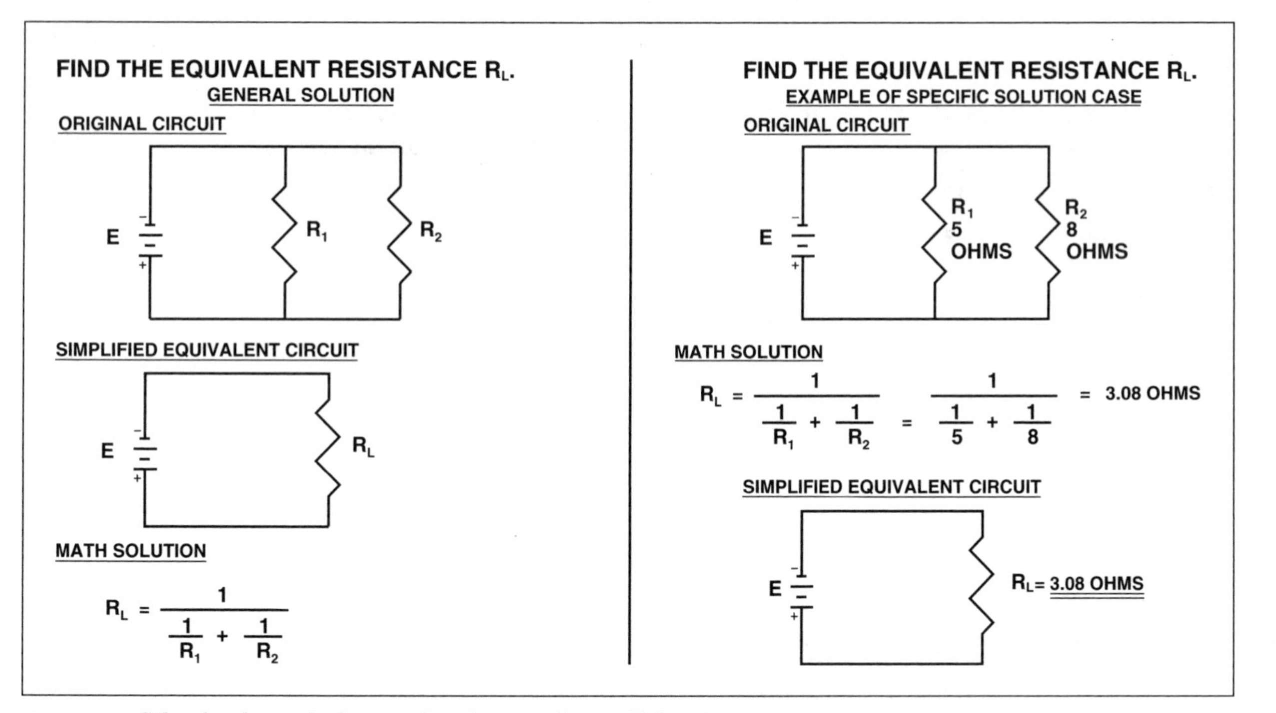

Figure 1-15 Solve for the equivalent total resistance of a parallel resistance circuit.

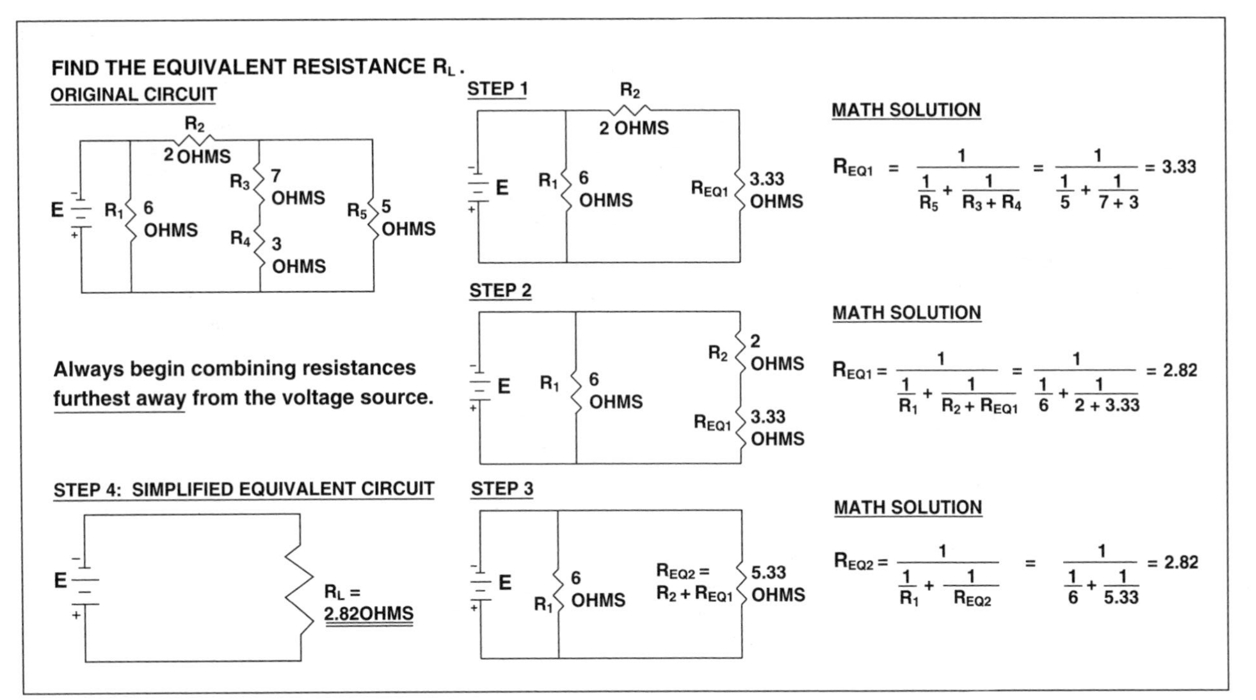

Figure 1-16 Solve for the equivalent total resistance of a series-parallel resistance network.

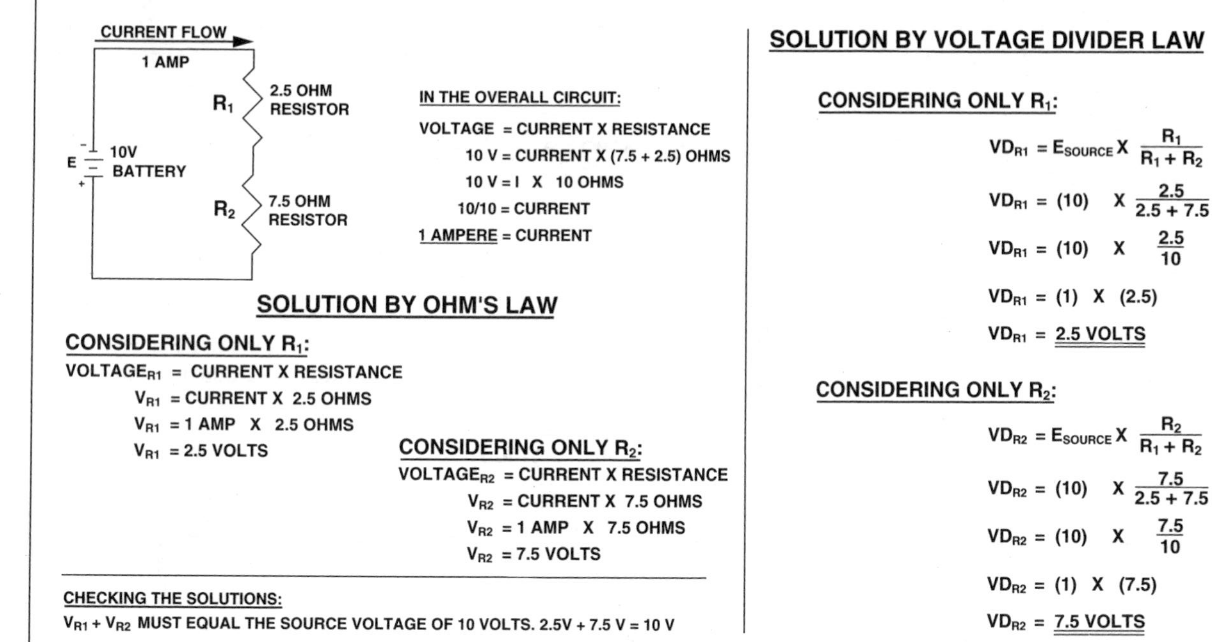

Figure 1-17 Solve for the voltage drop across each series-connected resistor by two methods.

dc systems, watts are equal to the voltage times the current ($W = E \times I$), and watts are also equal to the current squared times the resistance ($W = I^2 \times R$). The correct application of each of these formulas is shown in Fig. 1-18.

ac Voltage

The effective value of ac voltage is used in an ac resistive circuit just as dc voltage is used to calculate current in a dc circuit. Although many different voltages exist in an ac system, useful definitions of only the following three provide calculation capability for most problems:

1. Peak value
2. Average value
3. Effective, or root-mean-square (rms), value

These three voltages are illustrated graphically on the ac voltage waveform shown in Fig. 1-19.

In explanation, the *peak* value is the voltage magnitude between zero and the highest point of the voltage wave for one half-cycle. The *average* voltage value is the arithmetical average of all the values in the sine wave for one half-cycle and is equal to 63.7 percent of the peak voltage value. The rms value of the voltage is the equivalent "effective" voltage to a matching dc voltage, and the rms value is equal to 70.7 percent of the peak voltage value.

Note that these voltage values are independent of the frequency of the system; however, the system frequency causes several unique load considerations.

ac Circuits with Resistance

An ac circuit has an ac voltage source. Notice the symbol for ac voltage shown in Fig. 1-20. This voltage connected across an external load resistance produces alternating current of the same waveform and frequency as the applied voltage. Ohm's law applies to this circuit at every instant in time, as well as when considering the sine-wave voltage. When E is

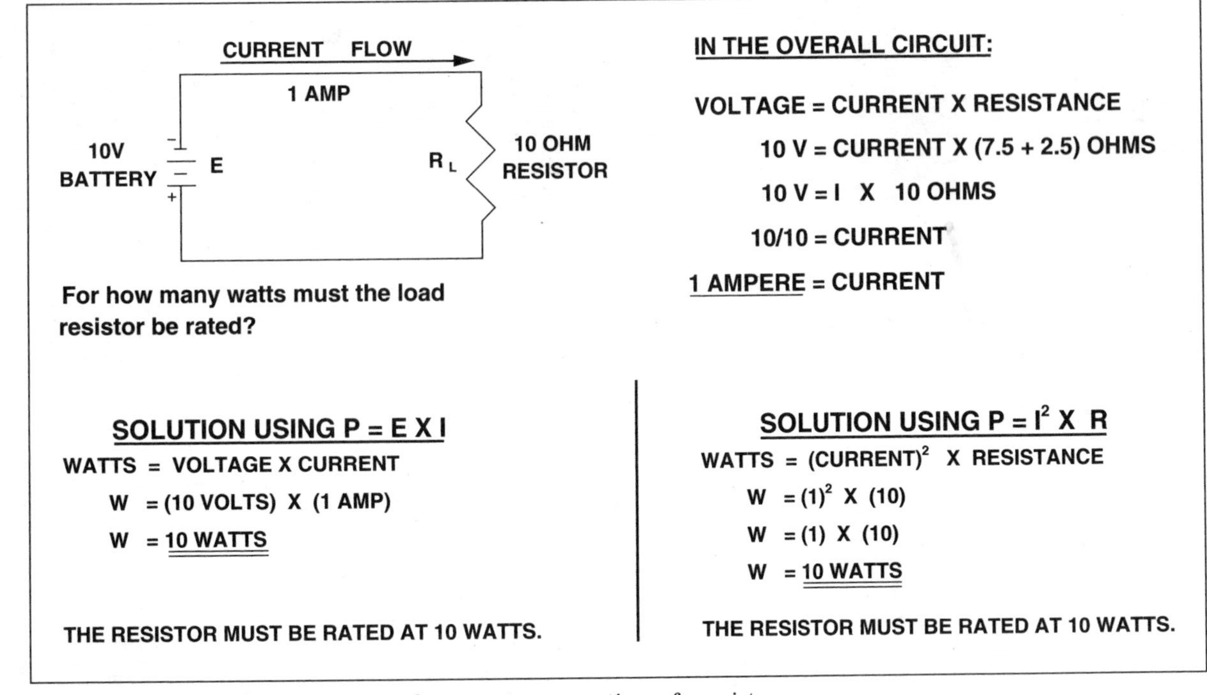

Figure 1-18 Solve for the true power and apparent power ratings of a resistor in a dc circuit.

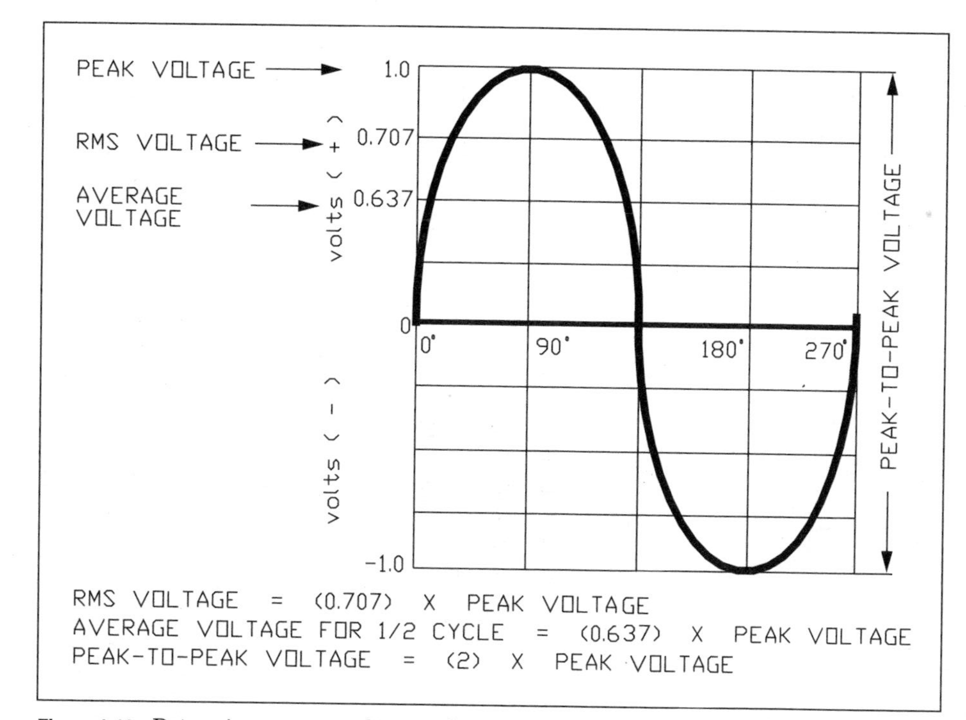

Figure 1-19 Determine average voltage and rms voltage from peak ac voltage (rms is the driving voltage).

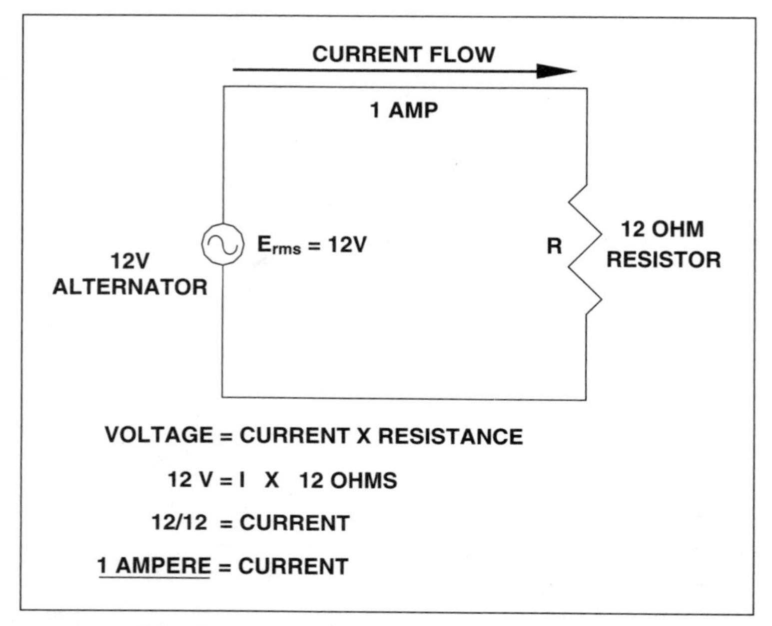

Figure 1-20 Solve for rms ac current in a resistive circuit with an ac voltage source.

an rms value, I is also an rms value, and for an instantaneous value of E during a cycle, the value of I is for the corresponding instant of time.

ac Phase Angle and Power

In an ac circuit with only resistance, the current variations i are "in phase" with the applied voltage e, as shown in Fig. 1-21. That is, at the exact instant that voltage increases in magnitude, current also increases in magnitude. As a result of this in-phase relation between e and i, an ac circuit that contains only resistance (no inductance and no capacitance) can be analyzed by exactly the same methods used in analyzing dc circuits, since there is no phase angle to consider.

On the other hand, Fig. 1-22 shows what happens to current with reference to voltage in a circuit that only contains

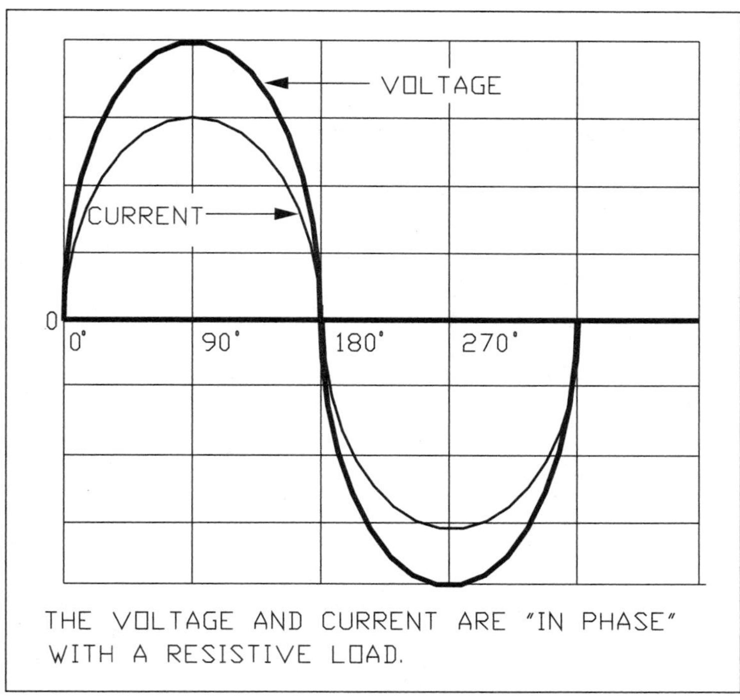

Figure 1-21 Solve for the relative phase angle between the current and voltage in an ac resistive circuit.

inductance. Specifically, the voltage changes, and 90 electrical degrees (one-quarter of one cycle, or $^1/_4$ of $^1/_{60}$ second) later, the current flows. This delay in current flow is the result of the need for time to physically build up the magnetic field that always concurrently surrounds current flowing in a conductor, and as voltage falls, a similar delay occurs as the magnetic field collapses, maintaining the current flow until the magnetic field has reduced to zero "lines of magnetic flux."

The cosine of the quantity of degrees that the current lags the voltage is known as the *ac phase angle,* and it is also known in the electrical discipline as the *power factor.*

By observation, it is possible to see in Fig. 1-22 that when voltage is positive (the voltage curve is above the "zero" x axis),

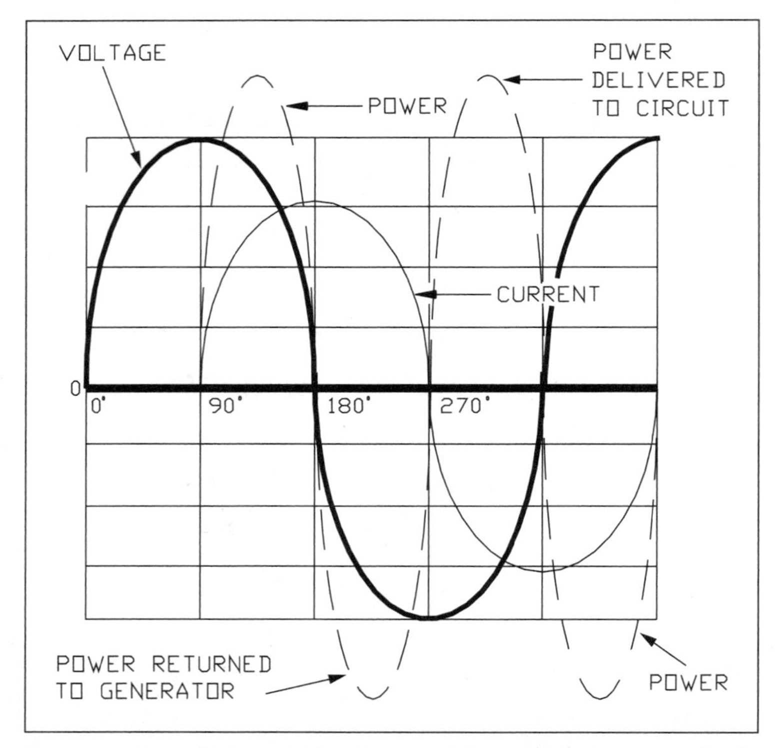

Figure 1-22 Solve for the relative phase angle between the current and voltage in an inductive circuit.

current is also positive, and when voltage is negative, current is also negative. Figure 1-23 shows that power is equal to the product of voltage times current, so in a resistive circuit power flow is always positive (from the power source to the load). That is, positive voltage times positive current equals positive power, and negative voltage times negative current equals positive power. The term *positive voltage* is merely an adopted polarity referenced in the circuit, and *positive current* is merely electron flow in a certain direction from the one referenced side of the voltage source. *Negative voltage* exists when the voltage source polarity changes, and *negative current* exists when the current flow changes direc-

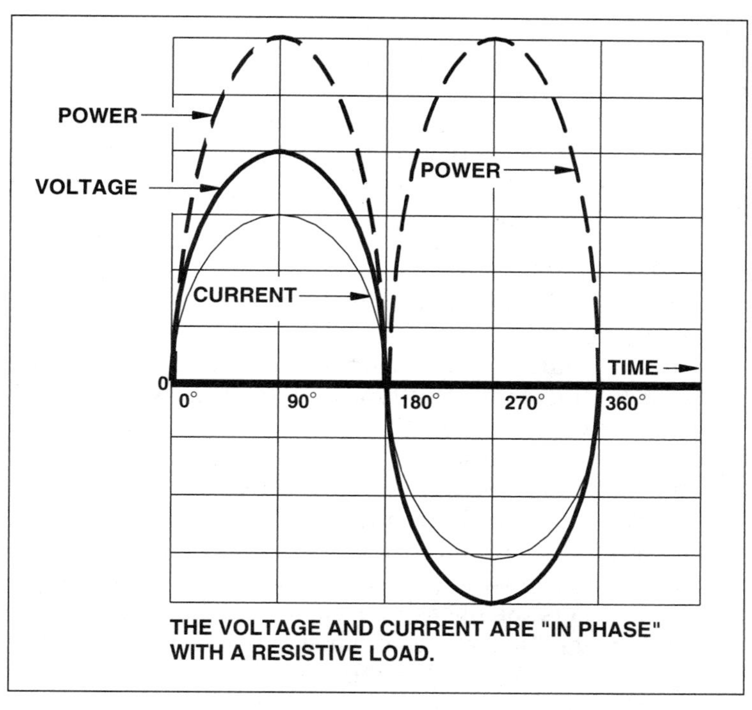

Figure 1-23 Solve for apparent power relative to current and voltage in an ac circuit.

tion from what it was during *positive current* flow. However, *positive power* means that power is physically flowing into the load portion of the circuit and *not* back into the power source.

Figure 1-22 shows that an unexpected thing happens when an inductor is the only load in the circuit. During the first part of the cycle, power (in the *positive* direction) flows from the power source to the inductor, taking energy to build up the magnetic field around the inductor. However, during the next part of the cycle, power is *negative* as the magnetic field collapses, returning the energy back to the original power source, only to have the power then cause the magnetic field to build up once again, except in the opposite magnetic polarity. These occurrences cause the power curve to be at twice the frequency of the voltage curve. A pure inductor, therefore, returns to the power source all the power that was delivered to it earlier. Therefore, 0 watts (W) of power (known as *true* power) are consumed in a purely inductive circuit, whereas many voltamperes of *apparent* power flow in the circuit. Note that 1 V times 1 A is equal to 1 voltampere (VA).

The circuit containing only an inductor as a load is only a theoretical circuit, however, because all conductors have resistance. Even the inductor itself is made of a conductor that contains resistance. Therefore, in reality, a circuit that contains an inductor also contains resistance. Since the inductance opposes current flow in a different way and at a different time than the resistance opposes current flow, it is necessary to differentiate between the two.

Figure 1-24 is a graphic explanation of the opposition to current flow in an ac circuit containing resistance and inductance. In this figure, R is the symbol for resistance, and X_L is the symbol for inductive reactance, where inductive reactance is the opposition to current flow that results from the magnetic field that must surround a conductor through which current is flowing. As was shown in Fig. 1-22, current through a pure inductance lags the voltage by 90 electrical degrees. In Fig. 1-24, this 90-degree lagging angle is depicted simply by using a familiar right triangle.

In Fig. 1-24*a,* the resistance R is shown on the x axis in a left-to-right direction. In Fig. 1-24*b,* the current through the resistance is also shown on the x axis, and in Fig. 1-24*c,* the power dissipated in the resistance (known as *true power* and measured in watts) is shown on the x axis as well.

In a similar manner, Fig. 1-24*a* depicts electric current and power through an inductance. In Fig. 1-24*a,* the inductive reactance X_L is shown on the y-axis in a vertical direction. In Fig. 1-24*b,* the current through the reactance is also shown in the vertical direction, and in Fig. 1-24*c,* the power flow through the inductive reactance (known as *apparent power* and measured in reactive voltamperes, or VARs) is shown vertically as well. Note that *no* true power, or wattage, is expended in an inductance.

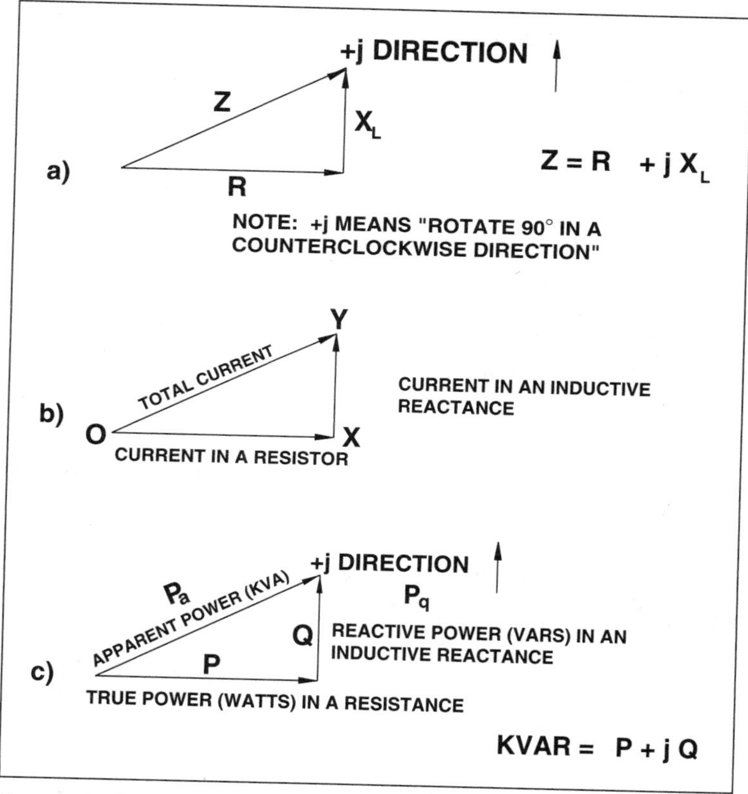

Figure 1-24 Illustrations and formulas to solve for ac current and power in resistors and in inductors.

Most real circuits consist of a combination of resistance and inductance. Vector or phasor diagrams of the type shown in Fig. 1-24 are a convenient way to represent resistance and inductive reactance, current through a resistance, and current flow through an inductance, or true power and reactive power. The reason that this representation is so expedient is that it allows the easy solution of most forms of problems with ac systems using simple basic trigonometry, as is summarized in Fig. 1-25. These are applied to electric circuits in Fig. 1-26 showing basic ac system relationships.

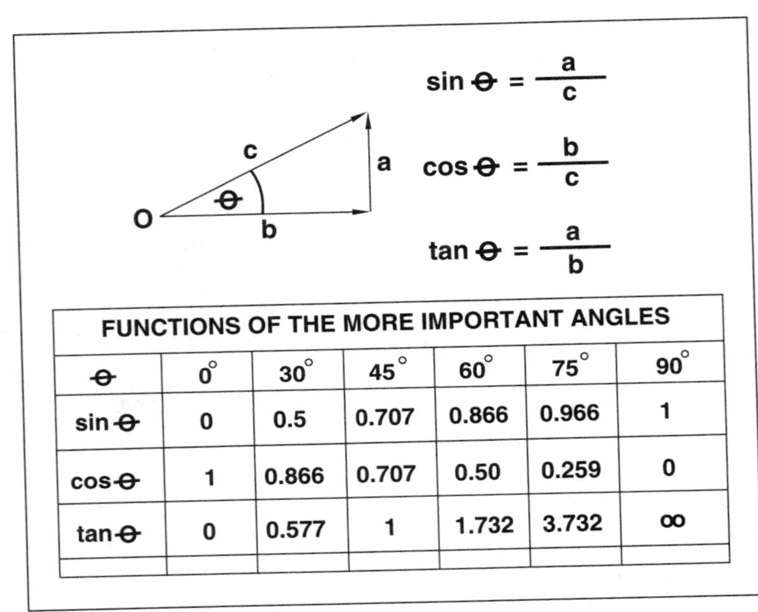

FUNCTIONS OF THE MORE IMPORTANT ANGLES

θ	0°	30°	45°	60°	75°	90°
sin θ	0	0.5	0.707	0.866	0.966	1
cos θ	1	0.866	0.707	0.50	0.259	0
tan θ	0	0.577	1	1.732	3.732	∞

Figure 1-25 Use these basic trigonometric formulas and triangles to simplify ac solutions.

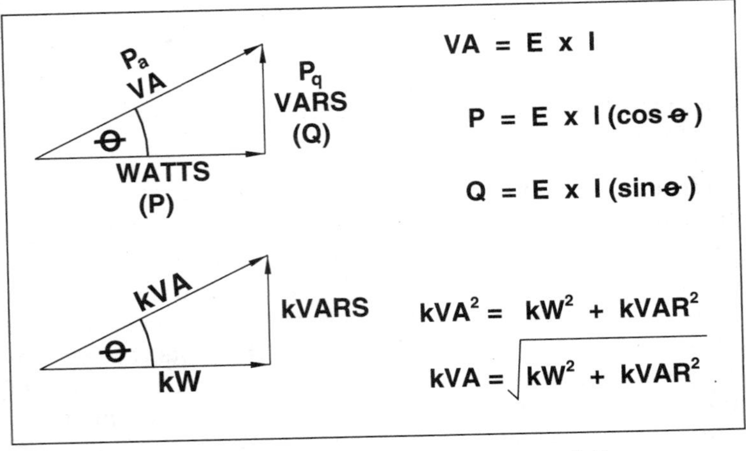

Figure 1-26 Use these basic relationships in ac system solutions.

Current and Power in a Single-Phase ac Circuit

In a single-phase ac circuit, power is related to voltage, current, and power factor, as shown in this formula:

$$P = E \times I \times \text{power factor}$$

This formula can be used to calculate current to determine wire size when power, voltage, and power factor are known, as shown in Fig. 1-27. It also can be used to calculate required voltage when power, current, and power factor are known, as shown in Fig. 1-28. Alternatively, it can be used to calculate true and apparent power when voltage, current, and power factor are known, as shown in Fig. 1-29, Moreover, it can be used to determine the power factor, as shown in Fig. 1-30.

Current and Power in a Three-Phase ac Circuit

In a three-phase ac circuit, apparent power is related to voltage, current, and power factor, as shown in this formula:

$$P_a = E \times I \times \text{power factor} \times \sqrt{3}$$

This formula can be used to calculate three-phase current to determine wire size when power, voltage, and power factor are known, as shown in Fig. 1-31. It also can be used to calculate required voltage when power power, current, and power factor are known, as shown in Fig. 1-32. Alternatively, it can be used to calculate true and apparent power when voltage, current, and power factor are known, as shown in Fig. 1-33. In addition, it can be used to determine the power factor, as shown in Fig. 1-34. These are all standard calculations of the type that electrical engineers solve frequently and are presented in a resolved step-by-step manner as reference templates that can be used in any similar problem simply by changing the numbers and maintaining the procedures.

(Text continues on p. 42.)

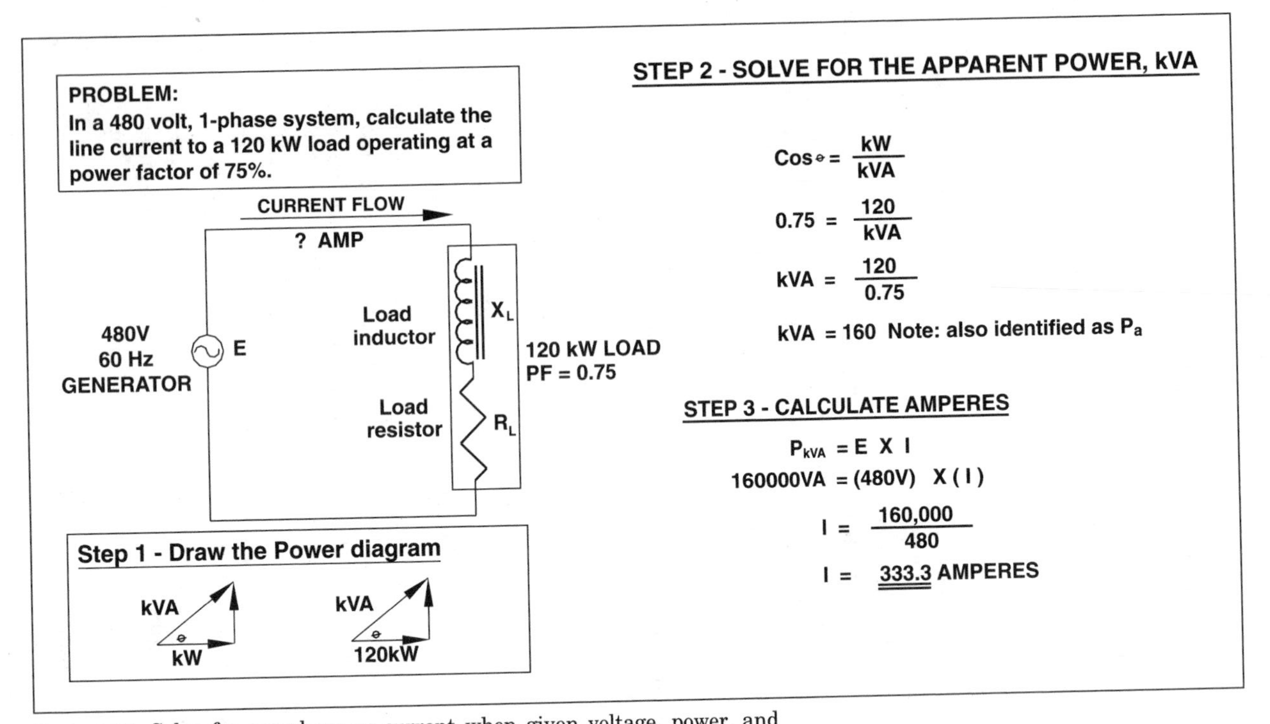

Figure 1-27 Solve for one-phase ac current when given voltage, power, and power factor.

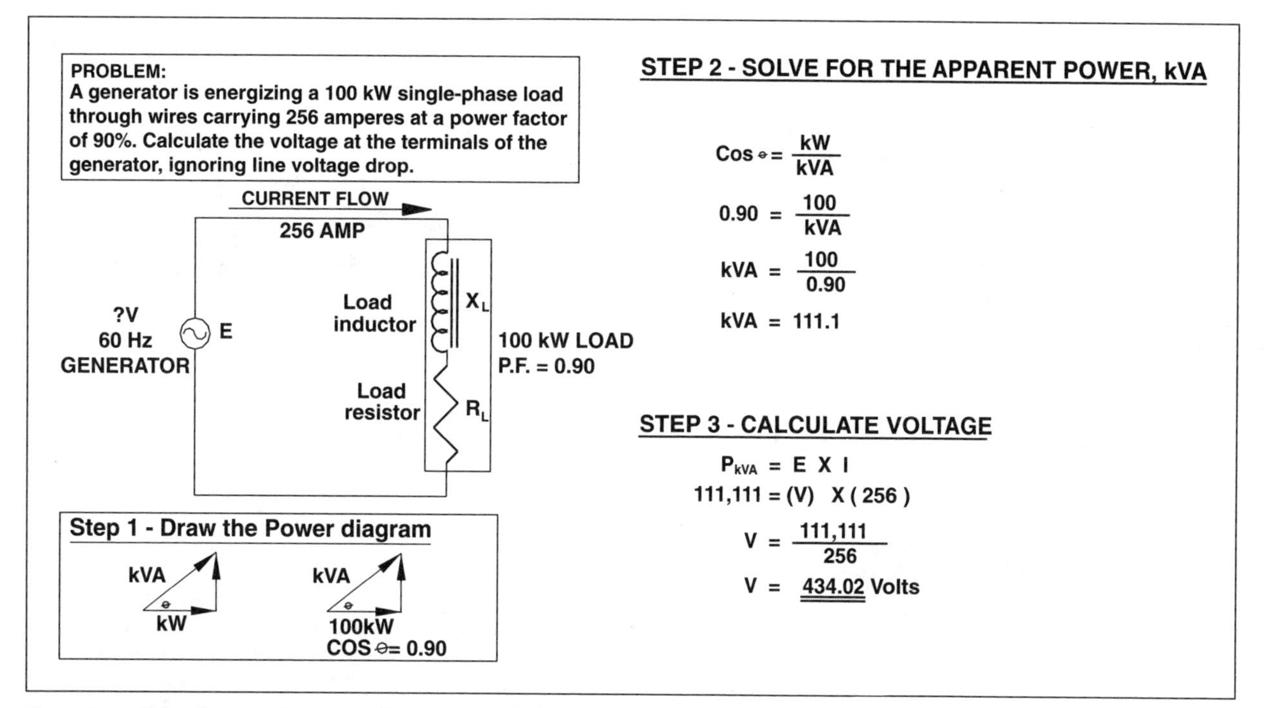

Figure 1-28 Solve for one-phase ac voltage required when current, resistance, and power factor are known.

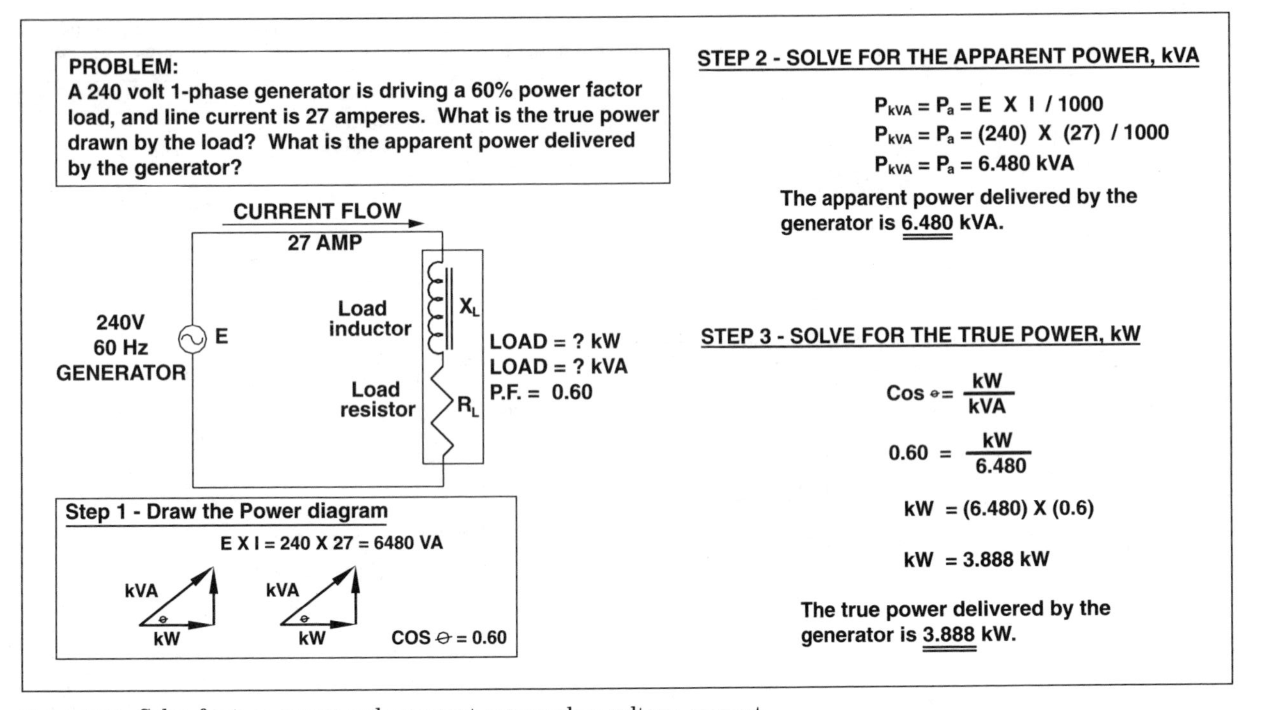

Figure 1-29 Solve for true power and apparent power when voltage, current, and power factor are known.

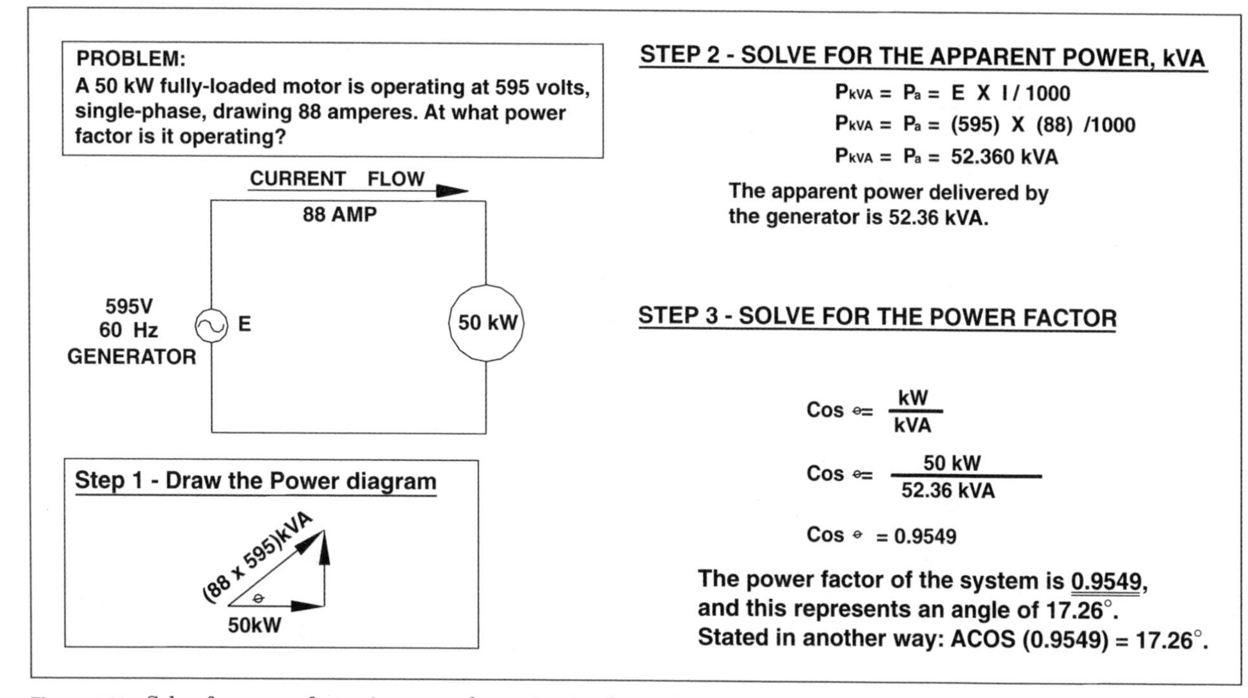

Figure 1-30 Solve for power factor in a one-phase circuit when voltage, current, and true power are known.

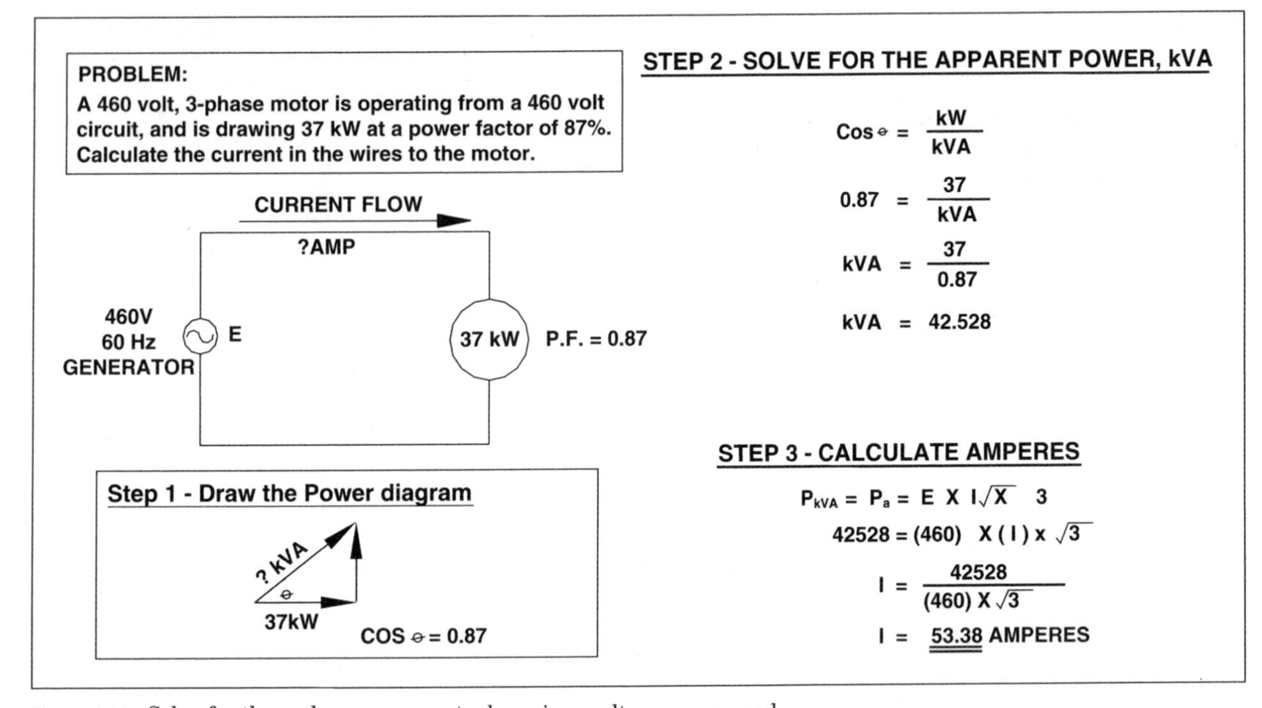

Figure 1-31 Solve for three-phase ac current when given voltage, power, and power factor.

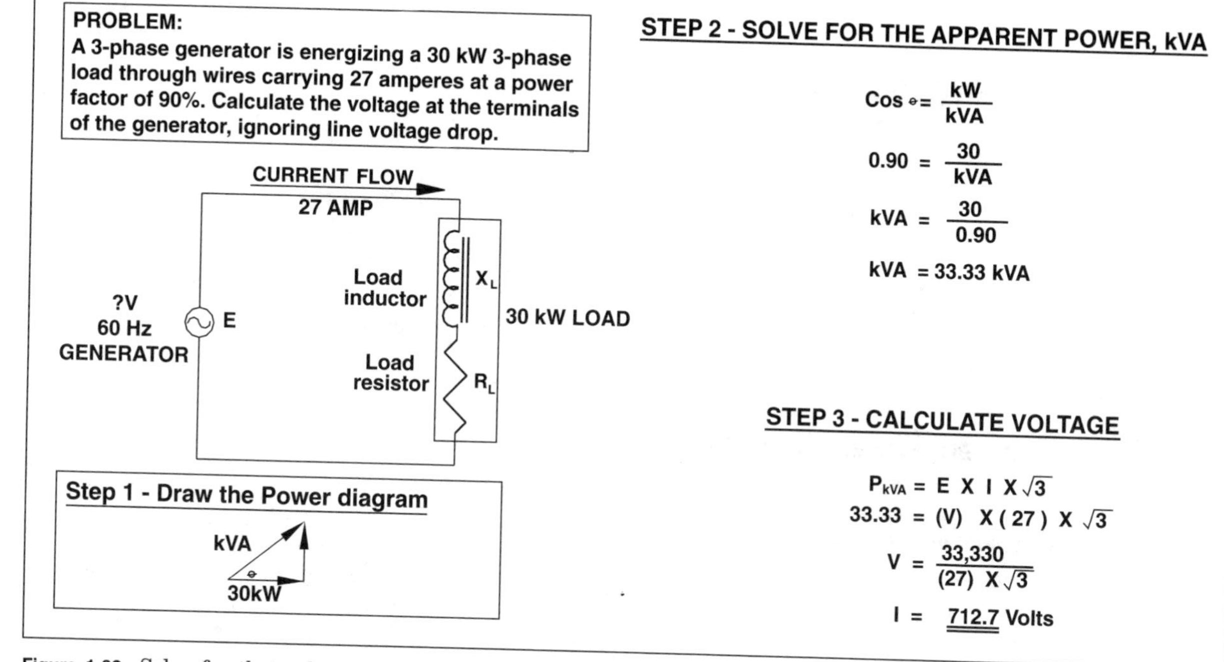

Figure 1-32 Solve for three-phase ac voltage required when current, resistance, and power factor are known.

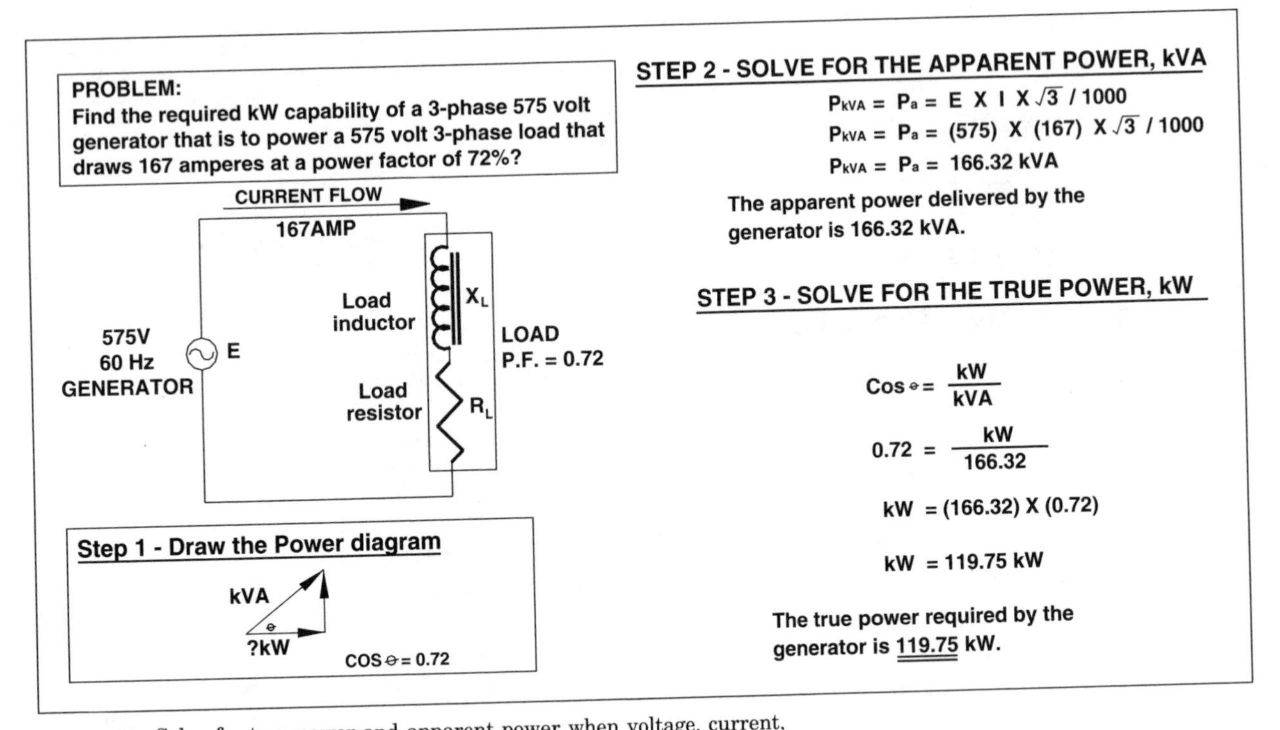

Figure 1-33 Solve for true power and apparent power when voltage, current, and power factor are known.

PROBLEM:
A system of motors and electrical heaters having an operating load of 420 kW draws 632 amperes at 480 volts, 3-phase. Calculate the power factor of the load.

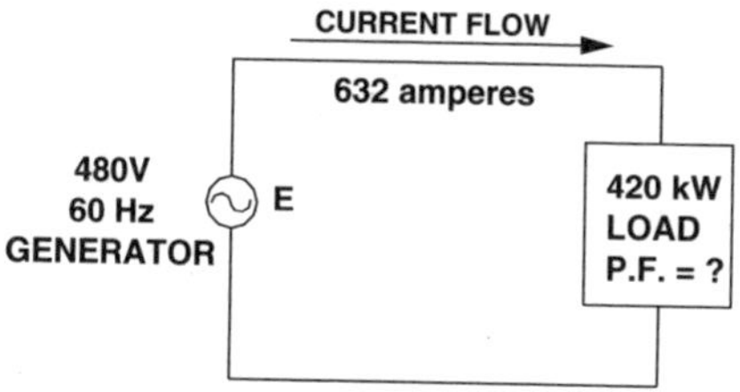

CURRENT FLOW

632 amperes

480V
60 Hz
GENERATOR

E

420 kW
LOAD
P.F. = ?

Step 1 - Draw the Power diagram

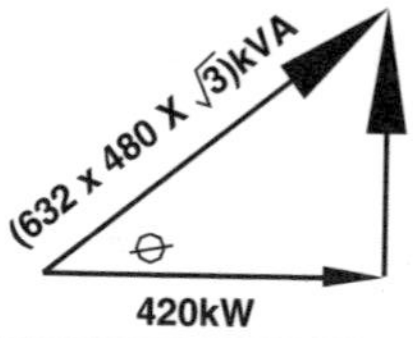

(632 x 480 X $\sqrt{3}$)kVA

θ

420kW

STEP 2 - SOLVE FOR THE APPARENT POWER, kVA

$$P_{kVA} = P_a = E \times I \times \sqrt{3} / 1000$$

$$P_{kVA} = P_a = (632) \times (480) \times \sqrt{3} / 1000$$

$$P_{kVA} = 525.435 \text{ kVA}$$

The apparent power delivered by the generator is 525.435 kVA.

STEP 3 - SOLVE FOR THE POWER FACTOR

$$\cos\theta = \frac{\text{kW}}{\text{kVA}}$$

$$\cos\theta = \frac{420 \text{ kW}}{525.435 \text{ kVA}}$$

$$\cos\theta = 0.799$$

The power factor of the system is 0.799, and this represents an angle of 36.9°.
Stated in another way: ACOS (0.799) = 36.9°.

Figure 1-34 Solve for power factor in a three-phase circuit when voltage, current, and true power are known.

Inductive Reactance

The opposition to electron flow presented by the magnetic field change surrounding a conductor through which current flow is changing is called *inductive reactance*. For this reason, an inductor in which the magnetic field is concentrated by coiling the circuit conductor is often called a *reactor*. Current flow through an inductive reactance lags the application of voltage, and the lag through a perfect inductor (with no resistive component) would be 90 electrical degrees.

Inductive reactance is measured in ohms, the same unit used to measure resistance, except that the opposition to current flow created by an inductive reactive ohm occurs 90 electrical degrees later than the opposition to current flow from a resistive ohm. Therefore, these two values cannot be added by simple algebra and instead must be added as vectors. The addition of vectors is described and illustrated in the section on adding vectors in Chap. 3.

Inductive reactance X_L can be calculated as

$$X_L = 2\pi f L$$

where $\pi = 3.14$
f = frequency, Hz
L = inductance, henries (H)

From the formula, it is apparent that a given coil can have a very low value of ohms in a 60-Hz system, whereas the same coil would have a much greater ohmic value in a 1-MHz system. A sample problem calculating the value in ohms of an inductor of a given value is shown in Fig. 1-35. A sample problem calculating the total opposition to current flow in a circuit containing a resistor in series with an inductor is shown in Fig. 1-36, and a sample problem calculating the total opposition to current flow in a circuit containing a resistor in series with an inductor is shown in Fig. 1-37. An explanation of vector addition and multiplication is given in Chap. 3. Note that the inductor contains resistance, as explained previously.

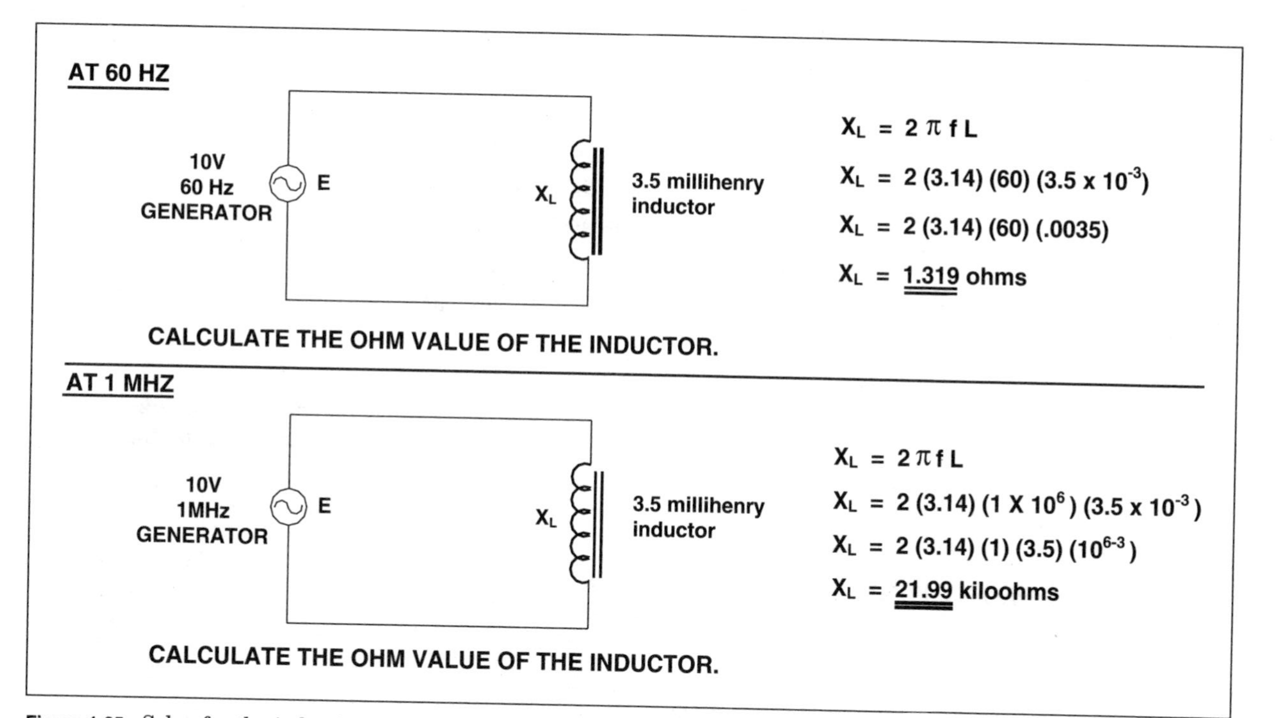

Figure 1-35 Solve for the inductive reactance of an inductor given frequency and inductance.

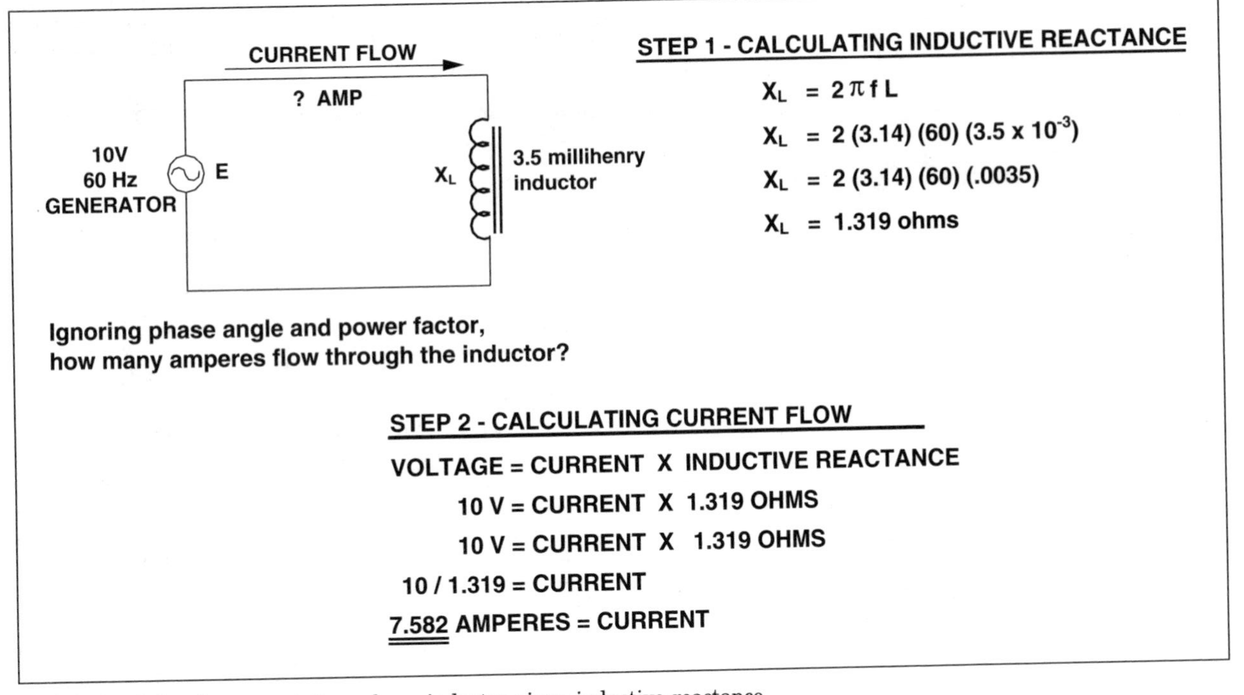

Figure 1-36 Solve for current through an inductor given inductive reactance and voltage.

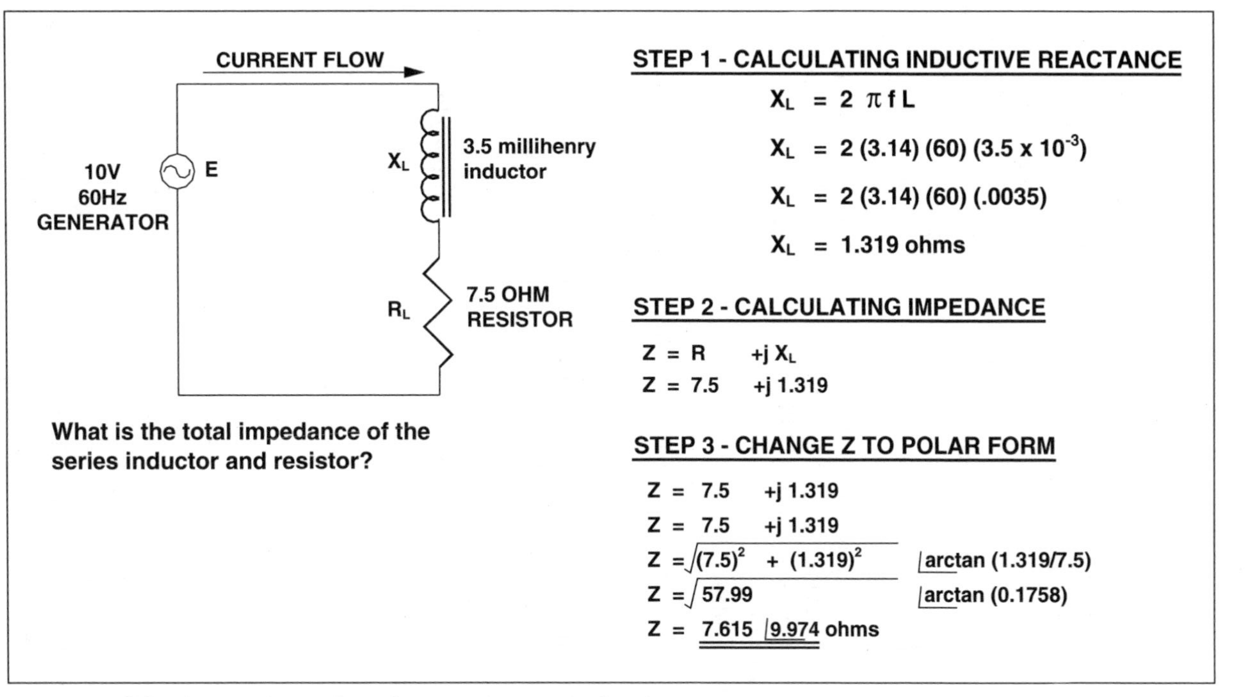

Figure 1-37 Solve for impedance of a coil given inductance and resistance.

Capacitive Reactance

A capacitor is simply a very thin insulator called a *dielectric* that is sandwiched between two conductor "plates" on which electrons "build up" under the pressure of a voltage source. No electrons actually flow through a capacitor, and in a dc circuit, a "charged" capacitor appears to be an open circuit. In an ac circuit, however, electrons build up first on one side of the dielectric and then on the other side as the source voltage polarity changes. In this way, a capacitor appears to conduct electrons in an ac circuit. The larger the capacitor plates, the greater is the quantity of electrons that can build up on them, and the lower is the apparent opposition to current flow produced by the capacitor. The measurement of opposition to current flow in a capacitor is called *capacitive reactance* X_C.

When voltage is first applied to a capacitor, the capacitor appears to be a short circuit until its plates begin to be charged with electron buildup. Contrasted with an inductor, in which current flows 90 electrical degrees *after* the application of source voltage, current flow in a capacitor actually leads the application of source voltage by 90 electrical degrees. Since both capacitive reactance and inductive reactance are "reactive" and occur 90 electrical degrees from the time of application of voltage, capacitive reactive ohms can be added to inductive reactive ohms using simple algebra.

Capacitive reactance X_C can be calculated as

$$X_C = \frac{1}{2\pi f C}$$

where $\pi = 3.14$
f = frequency, Hz
C = capacitance, farads (F)

As can be observed from this formula, the opposition to current flow at very low frequencies is quite great, whereas the ohmic value of capacitive reactance decreases at higher frequencies. This is exactly the opposite of the behavior of the ohmic value of an inductive reactance. A sample problem calculating the value in ohms of a capacitor of a given value is shown in Fig. 1-38, along with a capacitor current flow calculation.

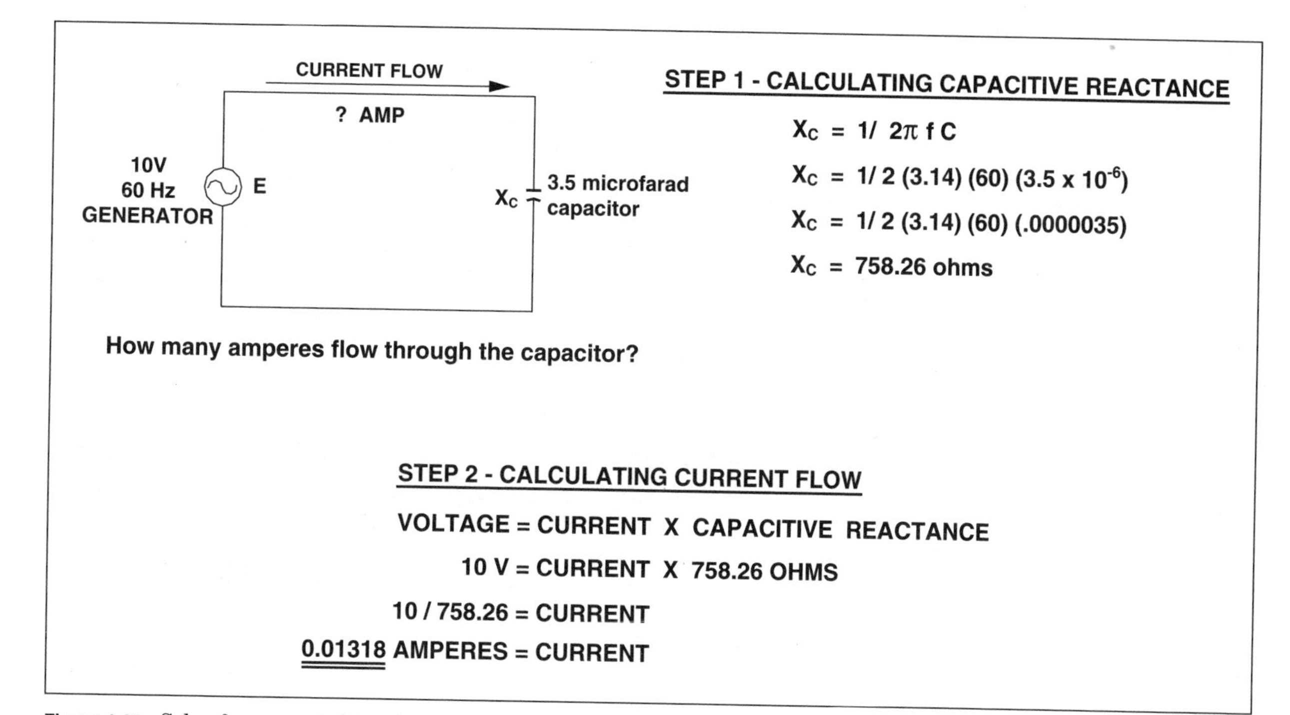

Figure 1-38 Solve for current through a capacitor given capacitance value in microfarads.

How many amperes flow through the series circuit?

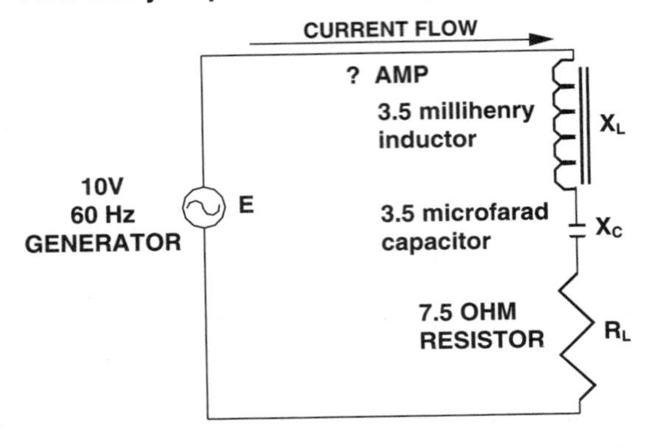

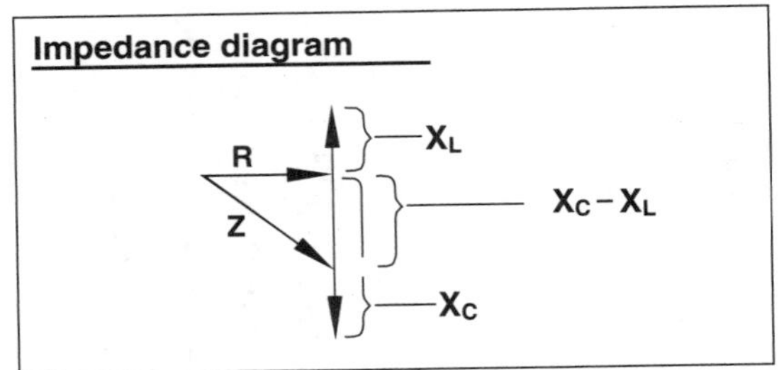

STEP 4 - CHANGE Z TO POLAR FORM

$$Z = 7.5 \quad -j \quad 756.94$$

$$Z = \sqrt{(7.5)^2 + (756.94)^2} \quad \angle \arctan (756.94/7.5)$$

$$Z = \sqrt{573014} \quad \angle \arctan (100.92)$$

$$Z = 756.97 \angle 89.43^\circ \text{ ohms}$$

THE TOTAL SERIES IMPEDANCE IS 756.97 OHMS, AND IT IS ALMOST ALL CAPACITIVE (ALMOST +90 DEGREES)

Figure 1-39 Solve for current through a series circuit of inductance, capacitance, and resistance using vectors impedance solution.

STEP 1 - CALCULATING CAPACITIVE REACTANCE

$X_C = 1/\ 2\pi\ f\ C$

$X_C = 1/\ 2\ (3.14)\ (60)\ (3.5 \times 10^{-6})$

$X_C = 1/\ 2\ (3.14)\ (60)\ (.0000035)$

$X_C = 758.26$ ohms

STEP 2 - CALCULATING INDUCTIVE REACTANCE

$X_L = 2\pi\ f\ L$

$X_L = 2\ (3.14)\ (60)\ (3.5 \times 10^{-3})$

$X_L = 2\ (3.14)\ (60)\ (.0035\ )$

$X_L = 1.319$ ohms

STEP 3 - CALCULATING TOTAL CIRCUIT IMPEDANCE

$Z = R \quad +j\ X_L\ -j\ X_C$

$Z = 7.5 \quad +j\ 1.319\ -j\ 758.26$

$Z = 7.5 \quad -j\ 756.94$

STEP 5 - CALCULATING CURRENT FLOW

VOLTAGE = CURRENT X IMPEDANCE

$10\ V\angle 0^\circ$ = CURRENT X $756.97\angle 89.43^\circ$ OHMS

$10\angle 0^\circ\ /\ 756.97\angle 89.43$ = CURRENT

$0.0132\angle 0.0^\circ - 89.43^\circ$ AMPERES = CURRENT

$0.0132\angle -89.43^\circ$ AMPERES = CURRENT

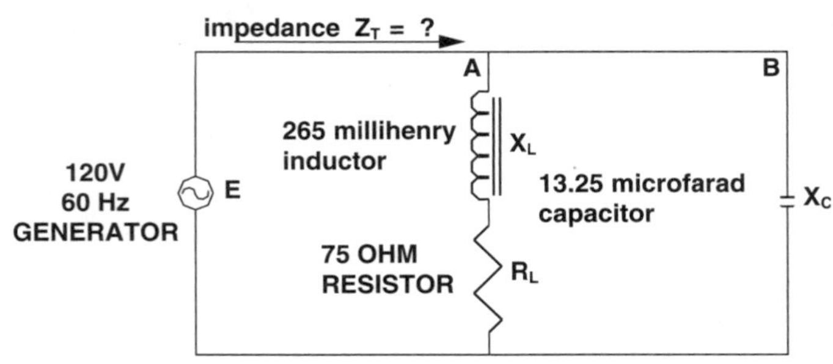

This solution is most simply done by solving for the current in each branch and summing them.

STEP 1 - CALCULATING CAPACITIVE REACTANCE

$$X_C = 1/\ 2\pi\ f\ C$$
$$X_C = 1/\ 2\ (3.14)\ (60)\ (13.25 \times 10^{-6})$$
$$X_C = 1/\ 2\ (3.14)\ (60)\ (.00001325)$$
$$X_C = 200 \text{ ohms}$$

STEP 2 - CALCULATING INDUCTIVE REACTANCE

$$X_L = 2\pi\ f\ L$$
$$X_L = 2\ (3.14)\ (60)\ (265 \times 10^{-3})$$
$$X_L = 2\ (3.14)\ (60)\ (0.265)$$
$$X_L = 100 \text{ ohms}$$

STEP 3 - CALCULATING IMPEDANCE OF BRANCH A

$$Z = R + j\ X_L$$
$$Z = 75 + j\ 100$$

STEP 4 - CHANGE Z OF BRANCH A TO POLAR FORM

$$Z = \sqrt{(75)^2 + (100)^2}\ \angle \arctan(100/75)$$
$$Z = \sqrt{15625}\ \angle \arctan(1.33)$$
$$Z = 125 \angle 53.1^\circ \text{ ohms}$$

Figure 1-40 Solve for total current in a parallel ac circuit containing inductance, resistance, and capacitance.

STEP 6 - CALCULATING CURRENT FLOW IN BRANCH A

VOLTAGE = CURRENT X IMPEDANCE

$120 \angle 0^\circ$ = CURRENT X $125 \angle 53.1^\circ$ OHMS

$120 \angle 0^\circ / 125 \angle 53.1^\circ$ = CURRENT

$0.96 \angle 0.0^\circ - 53.1^\circ$ AMPERES = CURRENT

$0.96 \angle -53.1^\circ$ AMPERES = CURRENT

STEP 7 - CALCULATING CURRENT FLOW IN BRANCH B

VOLTAGE = CURRENT X IMPEDANCE

$120 \angle 0^\circ$ = CURRENT X $200 \angle -90^\circ$ OHMS

$120 \angle 0^\circ / 200 \angle -90^\circ$ = CURRENT

$0.6 \angle 0.0^\circ - (-90)^\circ$ AMPERES = CURRENT

$0.6 \angle 90^\circ$ AMPERES = CURRENT

STEP 8 - CHANGE THE CURRENTS TO VECTOR VALUES, AND THEN SUM THE CURRENTS FROM BOTH BRANCHES

BRANCH A

$0.96 \angle -53.1^\circ$

= .96 COS -53.1 +j .96 SIN -53.1

= [(0.96)(0.6) +j (0.96)(-0.7997) = 0.576 -j 0.77

BRANCH B

$0.6 \angle 90^\circ$

= .6 COS 90 +j 0.6SIN 90

= [(.6)(0) +j (0.6)(1)] = 0 +j 0.6

= 0.576 -j 0.17

$0.576 - j\,0.17 = \sqrt{(0.576)^2 + (0.17)^2} \angle -\text{ARCTAN}(0.17/0.576)$

$\sqrt{0.3307} \angle -\text{ARCTAN}(0.2951)$

$0.6 \angle -16.4^\circ$ TOTAL AMPERES

STEP 9 - SOLVE FOR OVERALL IMPEDANCE

E = I X Z

$120 \angle 0^\circ = 0.6 \angle -16.4^\circ$ X Z

$120 \angle 0^\circ / 0.6 \angle -16.4^\circ$ = Z

$200 \angle 0^\circ - (-16.4^\circ)$ = Z

$200 \angle 16.4^\circ$ OHMS = Z_T

Impedance

Just as resistance is the measurement of opposition to current flow in a dc system, the opposition to current flow in an ac system is its *impedance Z*. Impedance is measured in ohms at a given angle of electrical displacement to describe when current flows in the circuit with respect to when the voltage is applied. Figure 1-24*a* graphically describes the vector addition of resistive ohms to inductive reactive ohms, whereas Fig. 1-39 illustrates the addition of capacitive reactive ohms to inductive reactive ohms plus resistive ohms. It shows a sample problem calculating the total opposition to current flow in an ac circuit containing a resistor in series with an inductor and a capacitor, and it also shows the calculation for current flow through the circuit. Figure 1-40 shows a sample problem calculating the total opposition to current flow in an ac circuit containing a resistor in series with an inductor, with that series string in parallel with a capacitor. An explanation of vector addition and multiplication is given in Chap. 3.

NOTES

NOTES

Chapter

2

Three-Phase Systems

Figure 2-1 shows the most common electrical system voltages for 60-hertz (Hz) systems, and Fig. 2-2 shows the most common electrical system voltages for 50-Hz systems. In general, 60-Hz systems are designed to be in compliance with Institute of Electrical and Electronics Engineers (IEEE)/American National Standards Institute (ANSI)/National Electrical Manufacturers Association (NEMA)/*National Electrical Code* (NEC) requirements, whereas, generally, 50-Hz systems are designed to be in compliance with International Electrotechnical Commission (IEC) or Australian standards. This book concentrates on 60-Hz systems but notes 50-Hz system information where it is pertinent. The immediate question arises as to how to select the most correct voltage for a system that is being designed, and the answer is equally straightforward and is shown in the flowchart in Fig. 2-3. The ultimate goal of this flowchart is to provide the load with proper current and voltage but not to exceed approximately 2500 amperes (A) at any one bus because of switchgear construction physical constraints.

In the simplest case of a single-phase circuit, an alternating-current (ac) system consists of a generator, a load, and conductors that connect them together. The generator is

SYSTEM VOLTAGE	Notes
115 Volt single-phase	Note 1
115/230 Volt single-phase	Note 2
120/208 Volt, 3-phase 4-wire wye	Note 3
240 Volt, 3-phase, 3-wire delta	Note 4
277/480 Volt, 3-phase 4-wire wye	Note 5
460 Volt, 3-phase 3-wire delta	Note 4
600/347 Volt, 3-phase, 4-wire wye	
2400 Volt, 3-phase, 3-wire delta	
2400 Coltm 3-phase, 4-wire wye	
4160/2400 Volt, 3-phase, 4-wire wye	
12470/7200 Volt, 3-phase, 4-wire wye	Note 6
24940/14400 Volt, 3-phase, 4-wire wye	Note 8
34500/19920 Volt, 3-phase, 4-wire wye	Note 9
46000	
69000	
115000	
138000	
161000	
230000	
345000	
500000	
765000	

Notes

1 Also known as 120 Volt, single-phase
2 Also known as 120/240 Volt, single-phase
3 "Professionally" referred to as 208Y/120 instead of as 120/208
4 This connection is not in frequent use any longer.
5 "Professionally" referred to as 480Y/277 or 460Y/265 instead of as 277/480
6 Actual voltage setting in this system may be from 12470 Volts to 13800 volts.
7 "Professionally" referred to as 600 Volts instead of 575 Volts.
8 "Professionally" referred to as 24940Y/14400 Volts.
9 "Professionally" referred to as 34500Y/19920 Volts

Figure 2-1 This is a listing of the most common 60-Hz ac electrical power system voltages.

simply a coil of conductors by which a magnetic field is passed repeatedly by rotating an electromagnet within the coil. The voltage output of the generator is proportional to the number of lines of magnetic flux that "cut" the coil, and the number of lines of flux is governed by the amount of current that flows through the electromagnet. Therefore, the generator output voltage is regulated simply by increasing or decreasing the "field" current through the electromagnet.

SYSTEM VOLTAGE	Notes
220 Volt single-phase	Note 1
220/380 Volt 3-phase 4-wire wye	Note 2
3300/1900 Volt, 3-phase 4-wire wye	
6600/3800 Volt, 3-phase, 4-wire wye	
11000/6350 Volt, 3-phase 4-wire wye	

Notes

1 Also known as 230 Volt, single-phase
2 Also known as 400/230 Volt or 415/240 Volt single-phase

Figure 2-2 This is a listing of the most common 50-Hz ac electrical power system voltages.

From the generator to the load are two wires so as to form a complete circuit, in addition to a "safety ground" conductor that is run with, or encloses, the circuit conductors. Chapter 7 and article 250 of the *National Electrical Code* explain when and how an ac system must be grounded.

Figure 2-4 shows a generator and a motor with three single-phase circuits that are entirely separate from one another. Note that the voltage of each of these circuits originates in a generator coil and that the load of each circuit is a motor coil that consists of resistance and inductance. All three of these coils are shown inside one generator housing within which one electromagnet is spinning, known as the *field*. The generator contains three single-phase systems, so it is called a *three-phase generator*. The one voltage regulator provides regulated field magnetic flux for all three phases simultaneously.

In a three-phase generator, the three phases are identified as phases *a*, *b*, and *c*. As the magnetic field piece rotates, it passes first by phase coil *a*, then by phase coil *b*, and then by phase coil *c*. Because of this action, the voltage is generated in coil *a* first, then in coil *b*, and finally in coil *c*, as shown graphically in Fig. 2-5. Note that the voltage of one phase is displaced from the voltage of the next phase by one-third of a 360-electrical-degree rotation, or by 120 electrical degrees. The figure also shows graphically how these voltages can be shown as vectors, and it shows the relationship of one voltage vector to the next.

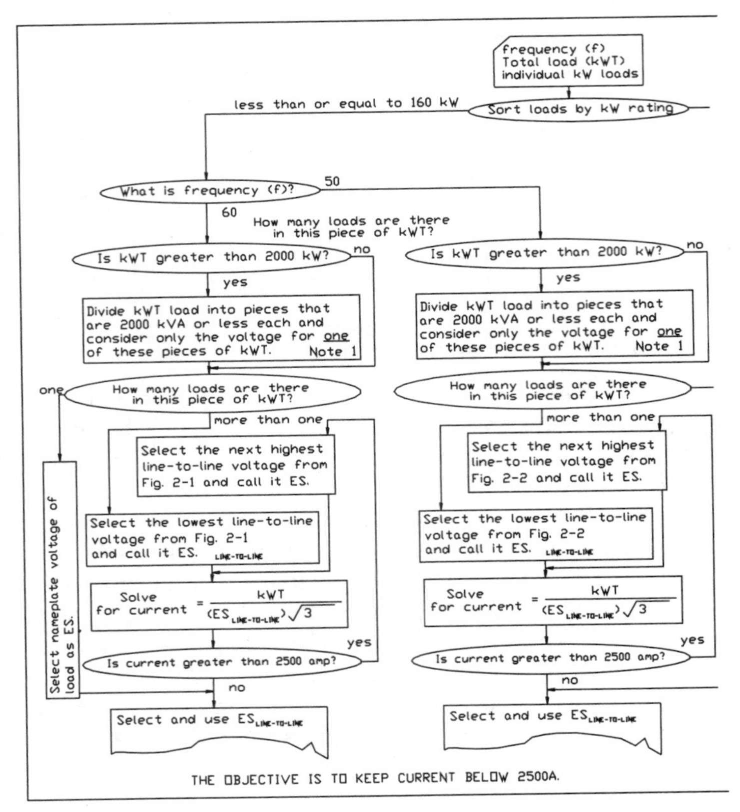

Figure 2-3 Use this logic to select the proper system voltage for an electrical load.

Wye-Connected Systems

Noting that the three vectors seem to form a semblance of the letter Y, it is apparent that all three of these voltage vectors begin at a "zero" or common point. This common point is called the *neutral point.* In actuality, generators that are connected in *wye* have one point of each of their windings connected together and to ground, and the other ends of each of the windings of phases *a, b,* and *c* are extended out to the circuit loads, as shown in Fig. 2-6.

The voltage generated in one coil of the wye-connected generator is known as the *phase-to-ground voltage,* or *line-to-*

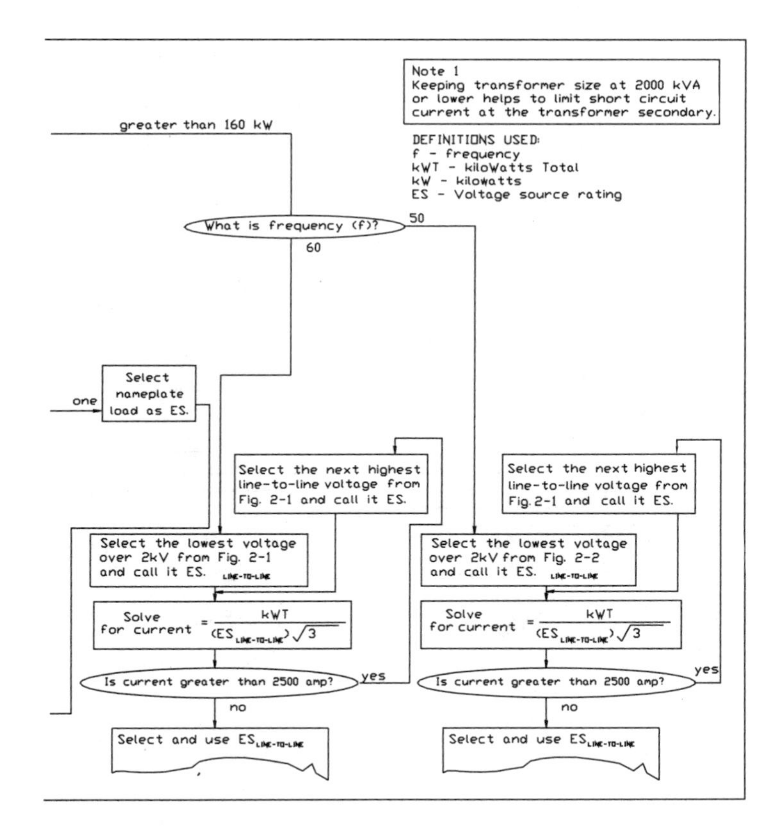

ground voltage. Since any two different phase coils within the generator are not displaced from one another by 180 electrical degrees, their voltage vectors cannot be added without considering their relative phase angle. Assuming that a generator coil voltage is 120 volts (V), Fig. 2-7 illustrates that the *phase-to-phase* (or *line-to-line*) voltage is calculated as

$$120 \angle 0° + 120 \angle 120° = 120\,(\sqrt{3}) = 120\,(1.713) = 208 \text{ V}$$

This relationship is true for all wye connections: *Phase-to-phase voltage is equal to phase-to-neutral voltage multiplied by 1.713.*

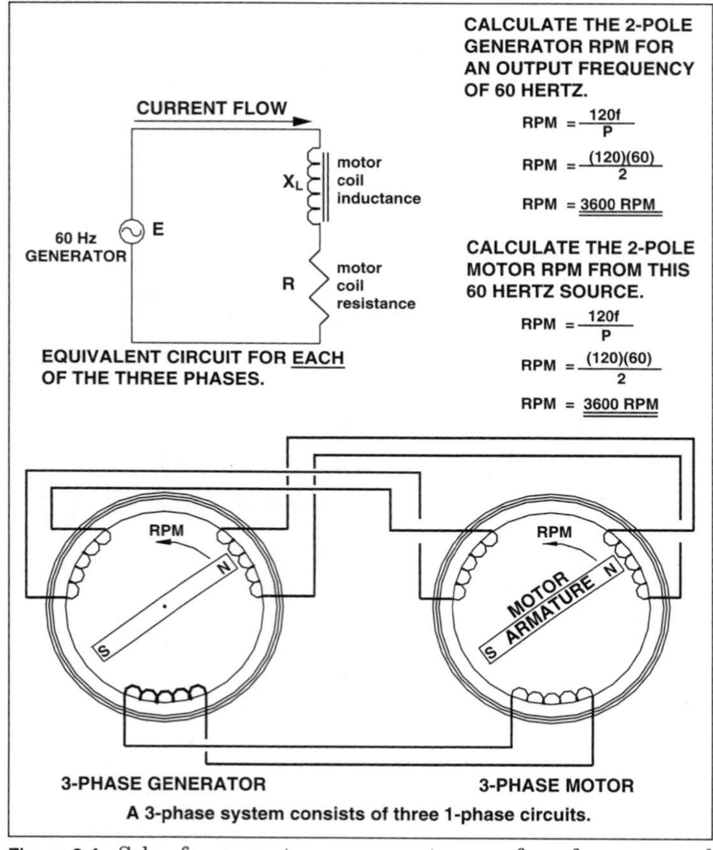

Figure 2-4 Solve for generator rpm or motor rpm from frequency and quantity of magnetic poles.

Common voltages from wye-connected systems include 120/208, 277/480, 343/595, 2400/4160, and 7200/12,470 V. Where these systems are grounded, the phase-to-neutral voltage is also the phase-to-ground voltage.

Delta-Connected Systems

An even more straightforward method of connecting the three phases together at the generator is known as the *delta* connection, as illustrated in Fig. 2-8. In this connection, the

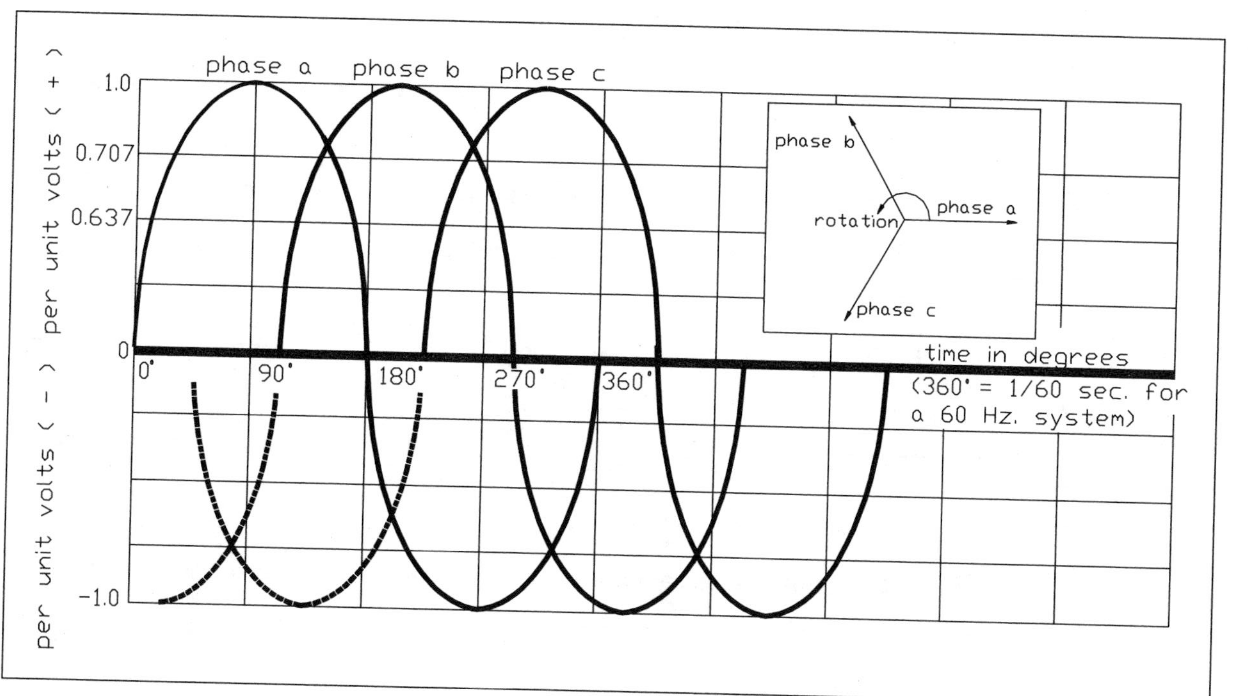

Figure 2-5 A graph of three ac voltages from a 60-Hz three-phase generator.

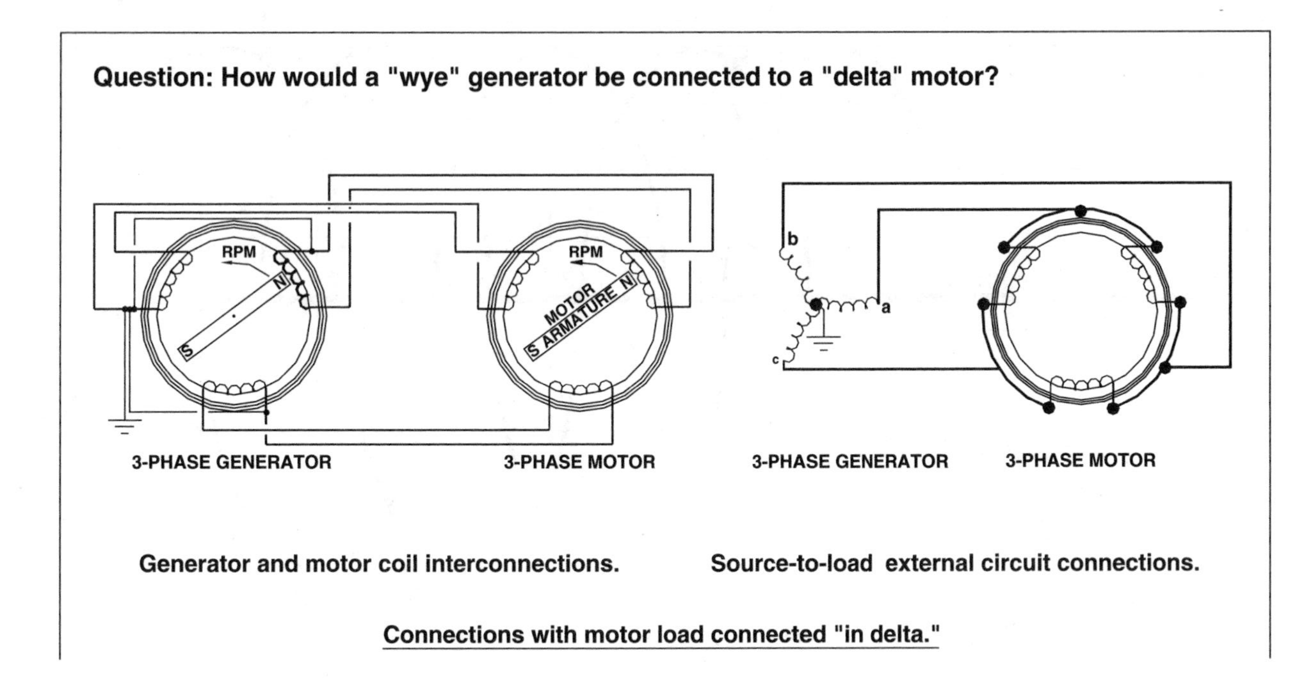
Question: How would a "wye" generator be connected to a "delta" motor?
RPM
N
S
RPM
MOTOR
ARMATURE
N
S
b
a
c
3-PHASE GENERATOR
3-PHASE MOTOR
3-PHASE GENERATOR
3-PHASE MOTOR
Generator and motor coil interconnections.
Source-to-load external circuit connections.
Connections with motor load connected "in delta."

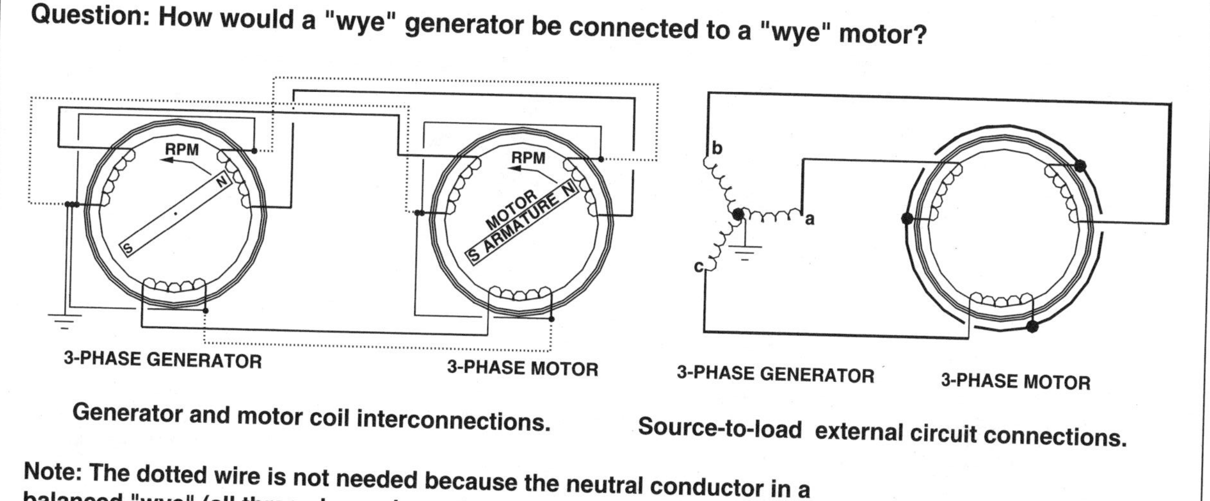

Figure 2-6 A wye three-phase system consists of three one-phase circuits connected together at a common neutral point that is normally grounded, and from the generator, connections can be made to either wye or delta loads.

Calculate the voltage across each delta-connected motor coil if wye-connected generator coil voltage is 120 volts.

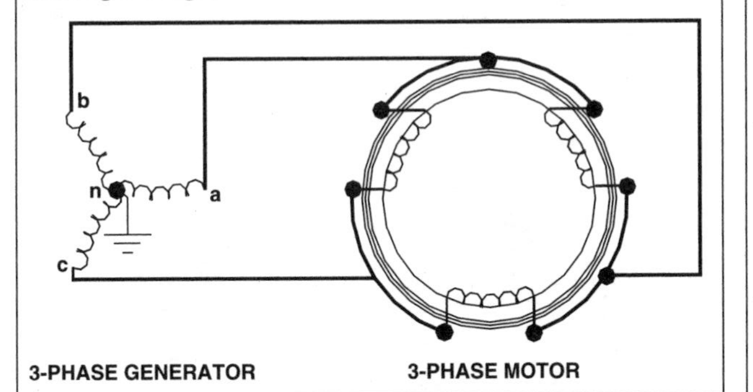

1. Diagram the system under analysis.

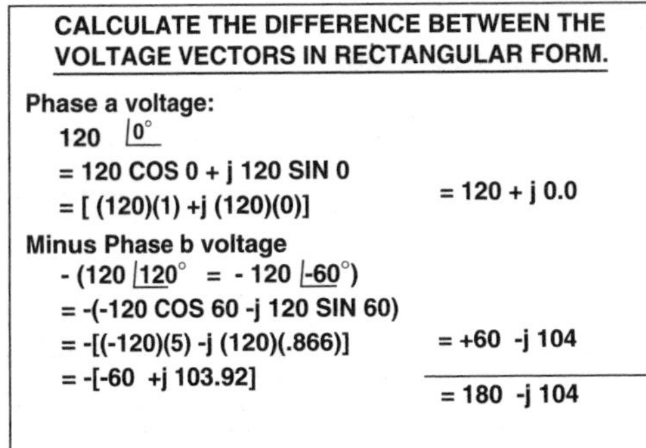

CALCULATE THE DIFFERENCE BETWEEN THE VOLTAGE VECTORS IN RECTANGULAR FORM.

Phase a voltage:

$120 \angle 0^\circ$

$= 120 \text{ COS } 0 + j\ 120 \text{ SIN } 0$

$= [\,(120)(1) + j\,(120)(0)]$ $\qquad = 120 + j\ 0.0$

Minus Phase b voltage

$-(120 \angle 120^\circ = -120 \angle -60^\circ)$

$= -(-120 \text{ COS } 60 - j\ 120 \text{ SIN } 60)$

$= -[(-120)(5) - j\,(120)(.866)]$ $\qquad = +60 - j\ 104$

$= -[-60 + j\ 103.92]$

$= 180 - j\ 104$

THEN CHANGE BACK TO POLAR FORM

$180 - j\ 104 = \sqrt{(180)^2 + (104)^2} \angle \text{ARCTAN } (-104/180)$

$= \sqrt{43199} \angle \text{ARCTAN } (-0.577)$

$= 208 \angle +30^\circ - 180^\circ = 208 \angle -150^\circ$ TOTAL VOLTS

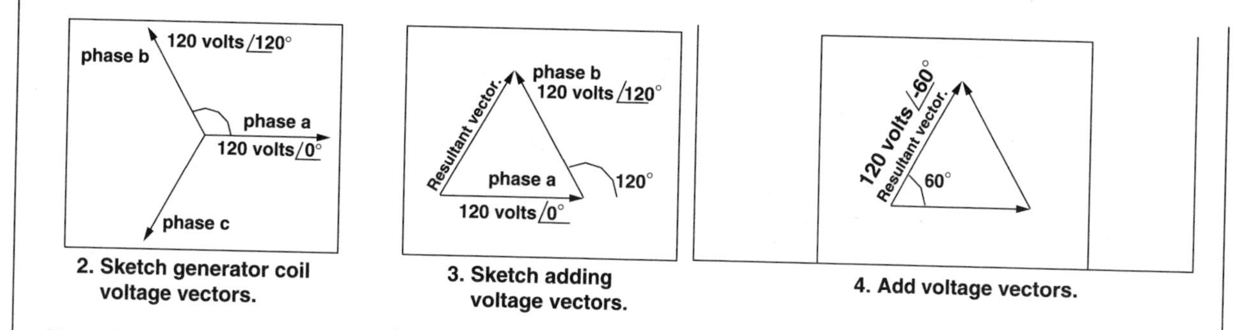

Note that the Tangent of 180° minus 120°, or 60°, also equals the $\sqrt{3}$, which is 1.732

Therefore the phase-to-phase voltage in a "wye" equals coil voltage times $\sqrt{3}$.
Stated in another way, the phase-to-phase voltage equals the phase-to-neutral voltage times $\sqrt{3}$.

Applying this general finding in the problem stated above:

Phase-to-phase voltage = delta-connected motor coil voltage = Wye coil voltage times $\sqrt{3}$

Phase-to-phase voltage = (120) (1.732) = 208 Volts

Figure 2-7 Solve for motor coil voltage using vectors given wye-connected generator coil voltage.

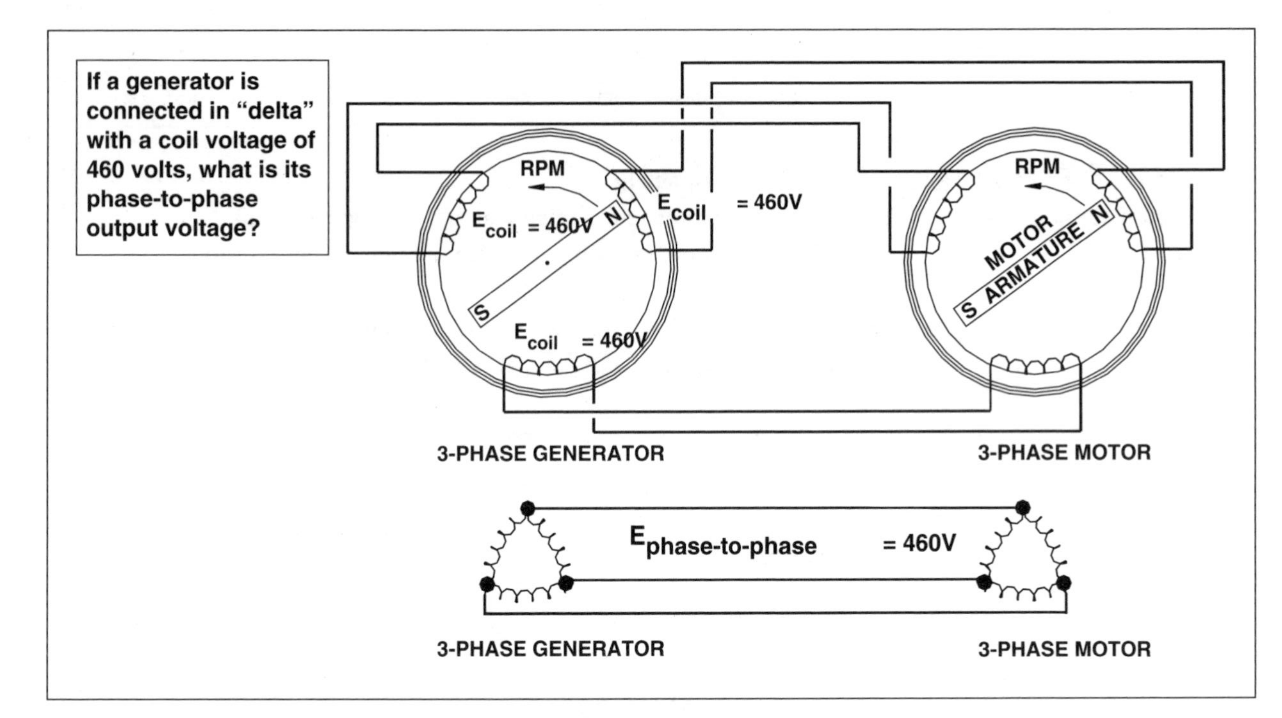

Figure 2-8 In a delta three-phase system, coil voltage is equal to phase-to-phase voltage.

end of one phase coil is connected to the end of the next phase coil, and it is connected to the other end of the first coil. The magnetic field effectively rotates within these three coils, forming voltages that are 120 electrical degrees apart, but with delta connections, the coil voltage is equal to the line-to-line voltage.

Common voltages from delta-connected systems include 240, 460, and 2400 V. Where these systems are grounded, the phase-to-ground voltages are unequal to one another, thus creating extra considerations in the load circuits.

NOTES

Chapter

3

Mathematics for Electrical Calculations, Power Factor Correction, and Harmonics

Just as the 100 pounds (lb) of force that a child exerts when pulling an object toward the east, or x direction, through a rope, cannot be added directly to the 150 lb of force that a separate child exerts simultaneously pulling the same object toward the north, or y direction, the two pulls do work together toward the goal of pulling the object in a direction that is somewhat in the x direction and somewhat in the y direction. The $+y$ direction is depicted in vector geometry as $+j$, and the $-y$ direction is depicted as $-j$. To resolve the value of the resulting force that is exerted onto the object, vector addition is used.

Changing Vectors from Rectangular to Polar Form and Back Again

Vector values are written in two different ways to represent the same values:

1. Polar coordinates: $(100 \angle 0°)$
2. Rectangular coordinates: $(+100\cos 0° \quad +j \quad 100\sin 0°)$

In polar coordinates, the 100-lb pull vector to the right in the x direction would be written as $100 \angle 0°$, and the 150-lb pull vector to the upward y direction would be written as $150 \angle +90°$.

A rectangular coordinate vector representation is simply the (x, y) location on graph paper of the tip of the vector arrow. A polar coordinate can be changed to a rectangular coordinate by the formula

$$E \angle \phi° = +E\cos\phi \quad +j \quad E\sin\phi$$

For example, the polar coordinate vector $100 \angle 0°$ can be written as

$$\begin{aligned} 100 \angle 0° &= +100\cos 0° \quad +j \quad 100\sin 0° \\ &= +100\,(1) \quad +j \quad 100\,(0) \\ &= +100 \quad +j \quad 0 \end{aligned}$$

To draw this vector on a graph, the tip of its arrow would be at $(+100, 0)$ on cartesian graph paper, and the base of the arrow would originate at $(0, 0)$.

Adding Vectors

Vectors are added most easily in rectangular coordinate form. In this form, each of the two parts of the coordinates is added together using simple algebra. For example, to sum $80 \angle +60° + 80 \angle -135°$, as is shown in the solution of Fig. 3-1, the first thing that must be done is to hand sketch the vectors to identify the angle from the x axis. In the case of $-135°$, the angle from the x axis is $180°$ minus $135°$, or $45°$.

$$\begin{aligned} 80 \angle +60° &= +80\cos 60° \quad +j \quad 80\sin 60° = +80\,(0.5) \quad +j \quad 80\,(0.866) = +40 \quad +j \quad 69.3 \\ 80 \angle -135° &= -80\cos 45° \quad -j \quad 80\sin 45° = -80\,(0.707) \quad -j \quad 80\,(0.707) = \underline{-56.6 \quad -j \quad 56.6} \\ & \qquad\qquad\qquad\qquad\qquad\qquad\qquad\qquad\qquad\qquad -16.6 \quad +j \quad 12.7 \end{aligned}$$

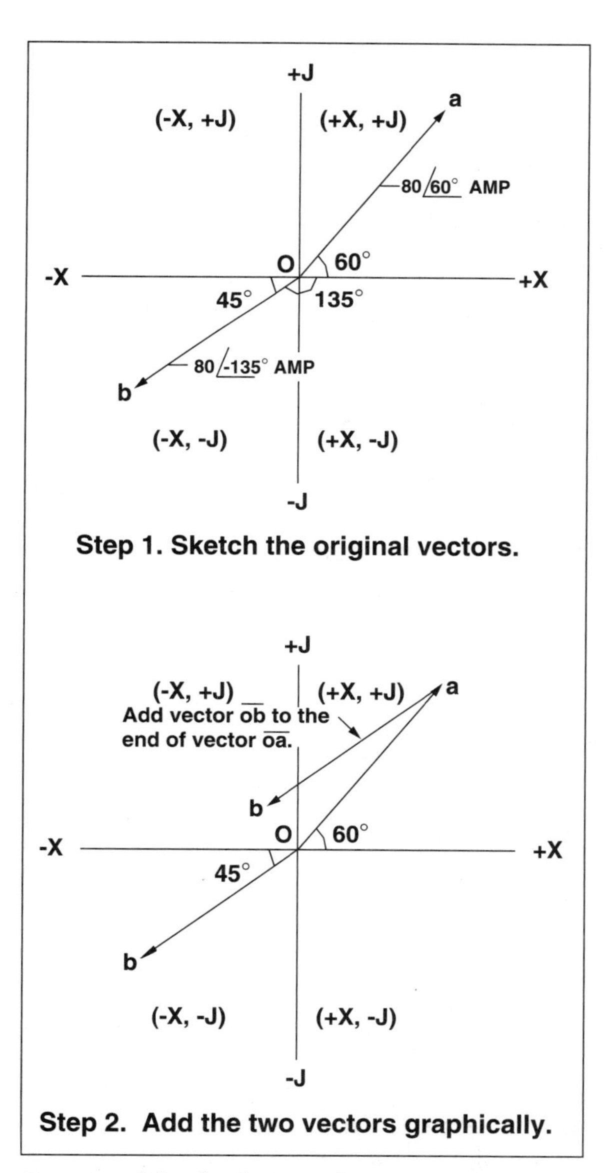

Figure 3-1 Solve for the sum of two vectors graphically given the original polar value of each vector.

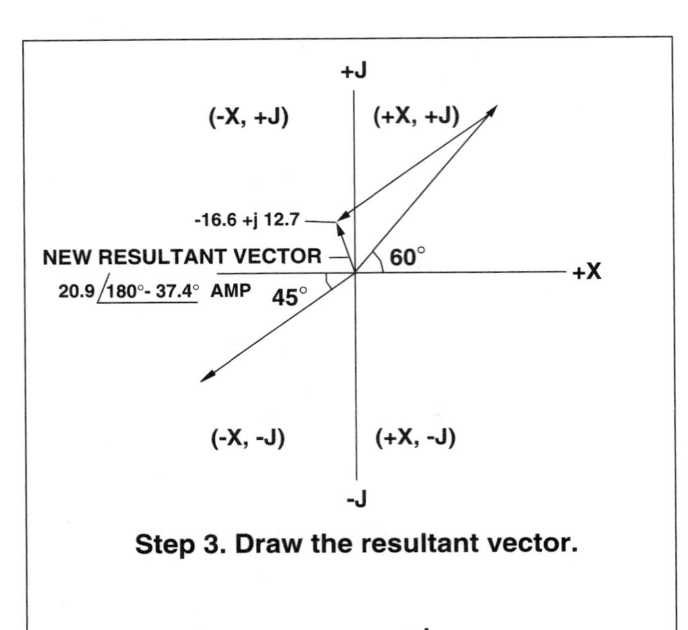

Step 3. Draw the resultant vector.

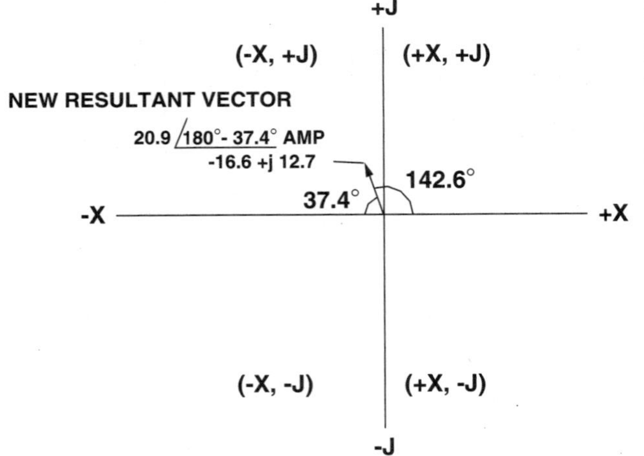

Step 4. Determine the angles from the signs of x and j.

Figure 3-1 *(Continued)*

It is necessary to change rectangular coordinates back into polar coordinate form; this conversion is done in accordance with this formula:

$$S \angle + \phi° = \sqrt{x^2 + y^2} \angle (\arctan y/x)$$

$$S \angle + \phi° = \sqrt{(16.6)^2 + (12.7)^2} \angle (\arctan 12.7/16.6)$$

$$S \angle + \phi° = \sqrt{436.85} \angle (\arctan 0.765)$$

$$S \angle + \phi° = (20.9) \angle (37.4°)$$

Note that 37.4° is the angle the resulting vector makes with the x axis, and $(-16.6, +j12.7)$ places the arrow tip of the vector in the second quadrant; thus the true angle from 0 is $(180° - 37.4°) = 142.6°$. Therefore, the resulting vector is $20.9 \angle (142.6°)$.

Another example of changing from rectangular form to polar form is shown in Fig. 3-2. This figure also graphically illustrates where each of the four quadrants are located and in which quadrants the x and jy values are $(+)$ or $(-)$.

CHANGE RECTANGULAR Z TO POLAR FORM

Z = 7.5 -j 756.94

$Z = \sqrt{(7.5)^2 + (756.94)^2}$ $\angle$ arctan (756.94/7.5)

$Z = \sqrt{573014}$ $\angle$ arctan (100.92)

$Z = 756.97 \angle 89.43°$ ohms

quadrant 2 -x + j y	quadrant 1 x + j y
-x - j y quadrant 3	x - j y quadrant 4

Figure 3-2 Use this methodology to solve for polar form of vector given the vector rectangular form.

Multiplying or Dividing Vectors

Vectors are most easily multiplied in polar form $(100 \angle 0°)$. When a polar vector $(100 \angle 0°)$ is multiplied by a scalar number (a number without an angle, such as 5), the result maintains the same angle. Thus:

$$(100 \angle 0°)\,(5) = 500 \angle 0°$$

When a vector in polar form is multiplied by another vector in polar form, the scalar numbers are multiplied together and the angles are added in this way:

$$(100 \angle 10°)\,(10 \angle 20°) = (1000 \angle 30°)$$

See Fig. 3-3 for a further illustration of this calculation.

When a vector in polar form is divided by another vector in polar form, the scalar numbers are divided and the angle of the denominator is subtracted from the angle of the numerator in this way:

$$(100 \angle 40°) \div (5 \angle 30°) = 20 \angle 10°)$$

See Fig. 3-4 for a further illustration of this calculation.

Problem: (0.576 -j 0.17) x (0.576 -j 0.17) = V; Solve for V.

1. Change vectors from rectangular to polar form before multiplying.

$$0.576\ -j\,0.17 = \sqrt{(0.576)^2 + (0.17)^2} \qquad \angle \text{ARCTAN}\,(-0.17/0.576)$$

$$\sqrt{0.3607} \qquad \angle \text{ARCTAN}\,(-0.2951)$$

$$0.6 \angle -16.4°$$

2. Then, multiply scalars together and add angles.

$$V = (0.6 \angle -16.4°) \quad \times \quad (0.6 \angle -16.4°)$$

$$V = (0.6 \times 0.6) \angle(-16.4° + (-16.4°)) \text{ units}$$

$$V = \underline{\underline{0.36 \angle -32.8°}} \text{ units}$$

Figure 3-3 To multiply vectors, first change them to rectangular form and then multiply the scalars together and sum angles.

Problem: (0.576 -j 0.17) / (0.576 -j 0.17) = V; Solve for V.

1. Change vectors from rectangular to polar form before dividing.

$$0.576\ -j\,0.17 = \sqrt{(0.576)^2 + (0.17)^2}\ \angle \text{ARCTAN}\,(-0.17/0.576)$$

$$= \sqrt{0.3607}\ \angle \text{ARCTAN}\,(-0.2951)$$

$$= 0.6\ \angle -16.4^\circ$$

2. Then, divide the first scalars by the second scalar and subtract the denominator angle from the numerator angle.

$$V = (0.6\ \angle -16.4^\circ)\ /\ (0.6\ \angle -16.4^\circ)$$

$$V = (0.6 / 0.6)\ \angle (-16.4^\circ - (-16.4^\circ))$$

$$V = \underline{\underline{1.0\ \angle 0^\circ}}\ \text{units}$$

Figure 3-4 To divide vectors, first change them to polar form and then divide the scalars and sum angles.

With these vector calculation tools, calculations of impedances and complex voltage and current values are possible.

Solving for Current and Power Factor in an ac Circuit Containing Only Inductive Reactance

Figure 3-5 shows the proper methodology to use in solving for current and power factor in an ac circuit that contains only one branch, an inductive reactance. Note that the impedance of the inductive reactance is in the $+j$ direction (since current in an inductance lags the voltage by 90 electrical degrees). The number of degrees is with reference to the $+x$ axis, which is at 0 electrical degrees, and positive degree counting from there is in a counterclockwise direction. For example, a vector from zero directly upward to $+j$ would be at an angle of $+90^\circ$, whereas a vector from zero directly downward to $-j$ would be at an angle of $+270^\circ$, or -90°.

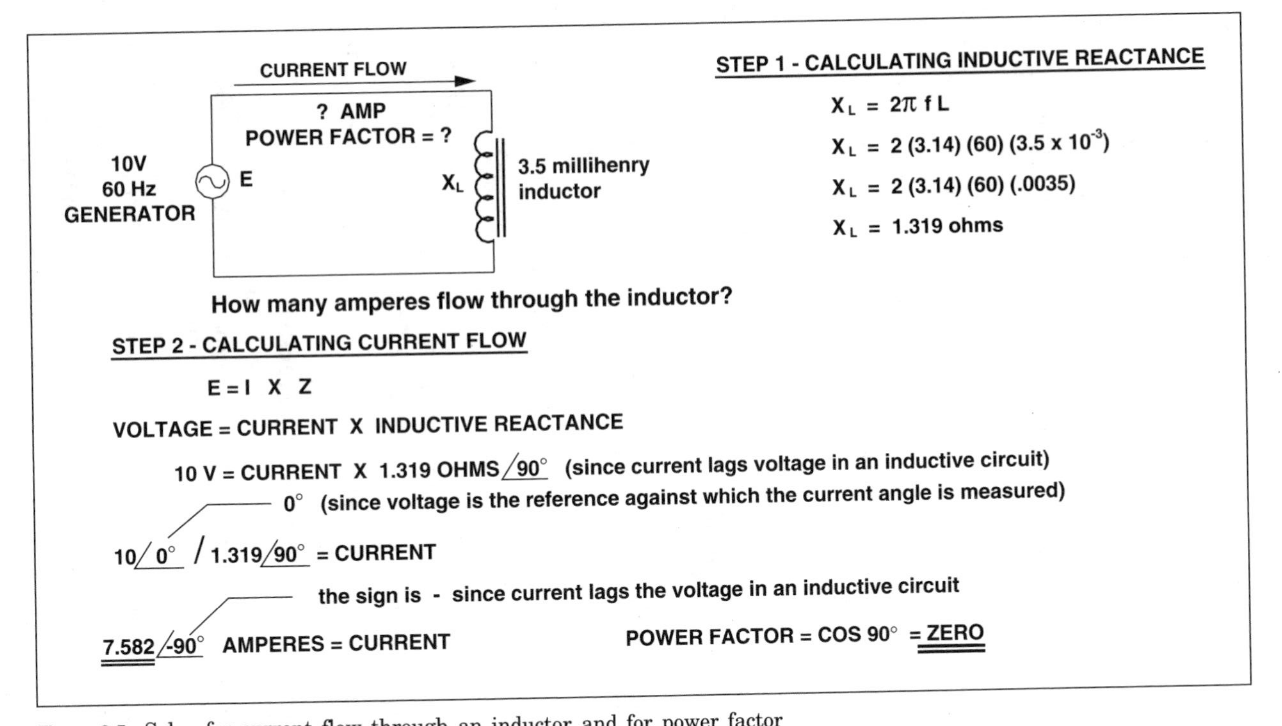

Figure 3-5 Solve for current flow through an inductor and for power factor voltage source and inductance values.

Solving for Current and Power Factor in an ac Circuit Containing Both Inductive Reactance and Resistance in Series with One Another

Figure 3-6 shows the proper methodology to use in solving for current and power factor in an ac circuit that contains only one branch having both an inductive reactance and a resistance. Note that the impedance of the inductive reactance is in the $+j$ direction (since current in an inductance lags the voltage by 90 electrical degrees), whereas the resistance of the resistor is at 0 electrical degrees, or exactly in phase with the voltage. Therefore, the impedance of the sum of the resistance plus the inductive reactance is the vector sum of the resistance at 0 electrical degrees plus the inductive reactance at +90 electrical degrees.

Solving for Current and Power Factor in an ac Circuit Containing Two Parallel Branches that Both Have Inductive Reactance and Resistance in Series with One Another

Figure 3-7 shows the proper methodology to use in solving for current and power factor in an ac circuit that contains two (or more) branches each having both an inductive reactance and a resistance. Note that the impedance of the inductive reactance is in the $+j$ direction (90 electrical degrees, which shows that the current lags the voltage by 90 electrical degrees in an inductive circuit), whereas the resistance of the resistor is at 0 electrical degrees, or exactly in phase with the voltage. Therefore, the impedance of the sum of the resistance plus the inductive reactance is the vector sum of the resistance at 0 electrical degrees plus the inductive reactance at +90 electrical degrees. The impedance of the overall circuit is generally most easily determined by calculating the impedance of each individual branch, calculating the current flow through each individual branch, and summing the branch currents to obtain total current. Then use Ohm's law to divide the source voltage at 0 electrical degrees (because it is the reference

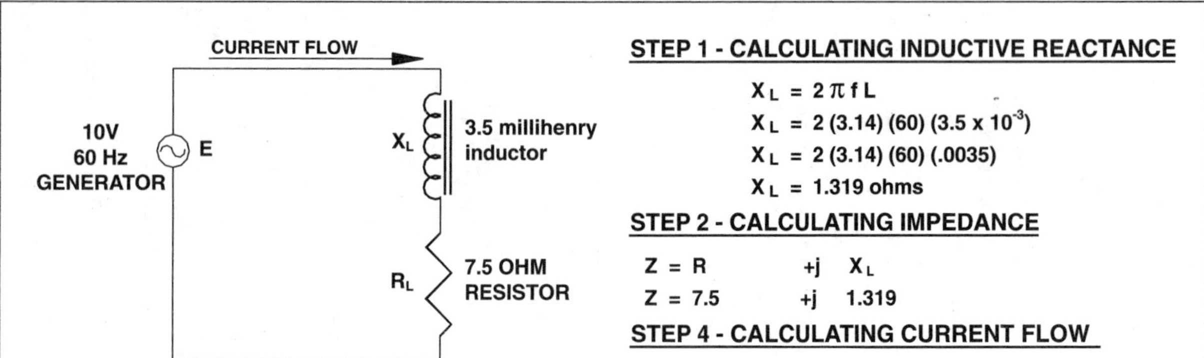

How many amperes flow through the inductor and at what power factor?

STEP 1 - CALCULATING INDUCTIVE REACTANCE

$X_L = 2 \pi f L$

$X_L = 2 (3.14) (60) (3.5 \times 10^{-3})$

$X_L = 2 (3.14) (60) (.0035)$

$X_L = 1.319$ ohms

STEP 2 - CALCULATING IMPEDANCE

$Z = R \quad +j \quad X_L$

$Z = 7.5 \quad +j \quad 1.319$

STEP 3 - CHANGE Z TO POLAR FORM

$Z = 7.5 \quad +j \quad 1.319$

$Z = \sqrt{(7.5)^2 + (1.319)^2} \quad \angle \arctan (1.319/7.5)$

$Z = \sqrt{57.99} \quad \angle \arctan (0.1759)$

$Z = 7.615 \angle 9.97^\circ$ ohms

STEP 4 - CALCULATING CURRENT FLOW

VOLTAGE = CURRENT X IMPEDANCE

$10\text{ V} \angle 0^\circ$ = CURRENT X $7.615 \angle 9.97^\circ$ OHMS

$10 \angle 0^\circ / 7.615 \angle 9.97^\circ$ = CURRENT

$1.313 \angle 0.0^\circ - 9.97^\circ$ AMPERES = CURRENT

$1.313 \angle -9.97^\circ$ AMPERES = CURRENT

STEP 5 - CALCULATING POWER FACTOR

POWER FACTOR = - COS 9.77° = .9849 LAGGING

Figure 3-6 Solve for current and power factor in a series circuit containing an inductor with resistance.

angle) by the combined total current flow, and the result of the division will be the impedance of the circuit as seen from the voltage-source terminals.

Solving for Current and Power Factor in an ac Circuit Containing Parallel Branches, One of Which Has Inductive Reactance and Resistance in Series with One Another and the Other of Which Has a Capacitive Reactance

Figure 3-8 shows the proper methodology to use in solving for current and power factor in an ac circuit that contains two (or more) branches each having both an inductive reactance and a resistance and one or more parallel branches that contain parallel capacitive reactances. Note that the impedance of the inductive reactance is in the $+j$ direction (so that the current will be at -90 electrical degrees, which is 90 electrical degrees lagging the voltage) and the impedance of the capacitive reactance is in the $-j$ direction (so that the current will be at $+90$ electrical degrees, which is 90 electrical degrees leading the voltage), whereas the resistance of the resistor is at 0 electrical degrees, or exactly in phase with the voltage.

While it is possible to "model" electrical power systems using these symbol tools and vector methodologies, a much more common and simpler method of solving for power factor in large electrical power systems is shown later in this chapter in the sections dealing with power factor correction. Prior to studying that, however, it is necessary to introduce standard ac electrical power system voltages and transformer connections that produce the selected voltages.

Electrical Power in Common ac Circuits

Power in ac circuits was treated in Chap. 1 from the perspective of theoretical circuits. In everyday electrical engineering work, however, power is treated in a much simpler manner that is explained in this section.

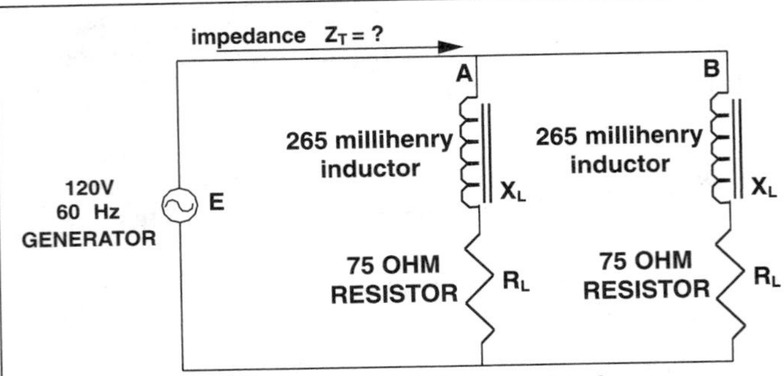

This solution is most simply done by solving for the current in each branch and summing them.

STEP 1 - CALCULATING INDUCTIVE REACTANCE

$$X_L = 2\pi f L$$
$$X_L = 2\,(3.14)\,(60)\,(265 \times 10^{-3})$$
$$X_L = 2\,(3.14)\,(60)\,(0.265)$$
$$X_L = 100 \text{ ohms}$$

STEP 2 - CALCULATING IMPEDANCE OF BRANCH A

$$Z = R \quad +j \quad X_L$$
$$Z = 75 \quad +j \quad 100$$

STEP 3 - CHANGE Z OF BRANCH A TO POLAR FORM

$$Z = \sqrt{(75)^2 + (100)^2} \quad \angle \arctan(100/75)$$
$$Z = \sqrt{15625} \quad \angle \arctan(1.33)$$
$$Z = 125 \angle 53.1° \text{ ohms}$$

Figure 3-7 Solve for current and power factor in a parallel circuit containing inductance and resistance.

Kilowatts do "real" work

By definition, a measured amount of power flowing for a certain amount of time can do a definite amount of real work, such as lifting an elevator. In the electrical industry, this real work is measured in kilowatthours, so the kilowatthour is the

STEP 4 - CALCULATING CURRENT FLOW IN BRANCH A

VOLTAGE = CURRENT X IMPEDANCE

$120\angle 0^\circ$ = CURRENT X $125\angle 53.1^\circ$ OHMS

$120\angle 0^\circ / 125\angle 53.1^\circ$ = CURRENT

$0.96\angle 0.0^\circ - 53.1^\circ$ AMPERES = CURRENT

$0.96\angle -53.1^\circ$ AMPERES = CURRENT

STEP 5 - SUM THE CURRENTS FROM BOTH BRANCHES

BRANCH A

$0.96\angle -53.1^\circ$

= .96 COS 53.1 -j .96 SIN 53.1

= [(0.96)(0.6) -j (0.96)(0.7997) = 0.576 -j 0.768

BRANCH B (by observation, same as branch A current)

$0.96\angle -53.1^\circ$

= .96 COS 53.1 -j .96 SIN 53.1

= [(0.96)(0.6) -j (0.96)(0.7997) = 0.576 -j 0.768

= 1.152 -j 1.536

$01.152 - j\,1.536 = \sqrt{(1.152)^2 + (1.1536)^2}$ $\angle$ARCTAN (1.536/1.152)

$= \sqrt{3.686}$ $\angle$ARCTAN (1.333)

$= 1.92\angle 53.13^\circ$ TOTAL AMPERES

STEP 9 - SOLVE FOR OVERALL IMPEDANCE

E = I X Z

$120\angle 0^\circ = 1.9200\angle 53.13^\circ$ X Z

$120\angle 0^\circ / 1.9200\angle 53.13^\circ$ = Z

$62.5\angle 0^\circ - 53.13^\circ$ = Z

$62.50\angle 53.13^\circ$ OHMS = Z_T

basic electrical utility meter billing unit. As its name implies, an electrical load of 1 kilowatt (kW) that is operating for 1 hour (h) consumes 1 kilowatthour (kWh). All electrical heating elements, all incandescent lighting, and the part of rotating motors that causes the actual rotation of the shaft are exam-

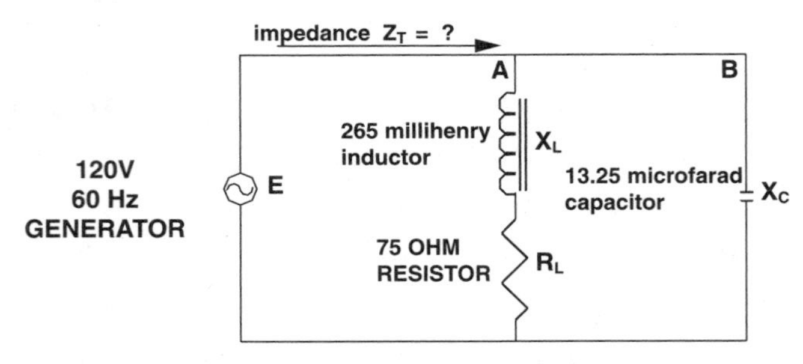

This solution is most simply done by solving for the current in each branch and summing them.

STEP 1 - CALCULATING CAPACITIVE REACTANCE

$X_C = 1/\ 2\pi\ f\ C$

$X_C = 1/\ 2\ (\ 3.14)\ (\ 60)\ (\ 13.25 \times 10^{-6})$

$X_C = 1/\ 2\ (\ 3.14)\ (\ 60)\ (\ .00001325\ )$

$X_C = 200$ ohms

STEP 2 - CALCULATING INDUCTIVE REACTANCE

$X_L = 2\pi\ f\ L$

$X_L = 2\ (3.14)\ (60)\ (265 \times 10^{-3})$

$X_L = 2\ (3.14)\ (60)\ (0.265)$

$X_L = 100$ ohms

STEP 3 - CALCULATING IMPEDANCE OF BRANCH A

$Z = R \quad +j \quad X_L$

$Z = 75 \quad +j \quad 100$

STEP 4 - CHANGE Z OF BRANCH A TO POLAR FORM

$Z = \sqrt{(75)^2 + (100)^2} \quad \angle \arctan(100/75)$

$Z = \sqrt{15625} \quad \angle \arctan(1.33)$

$Z = 125 \angle 53.1^\circ$ ohms

STEP 5 - CHANGE Z OF BRANCH B TO POLAR FORM

$Z = 200 \angle -90$

Figure 3-8 Solve for current and power factor in a parallel circuit containing inductance, resistance, and capacitance.

STEP 6 - CALCULATING CURRENT FLOW IN BRANCH A

VOLTAGE = CURRENT X IMPEDANCE

$120 \angle 0^\circ$ = CURRENT X $125 \angle 53.1^\circ$ OHMS

$120 \angle 0^\circ / 125 \angle 53.1^\circ$ = CURRENT

$0.96 \angle 0.0^\circ - 53.1^\circ$ AMPERES = CURRENT

$0.96 \angle -53.1^\circ$ AMPERES = CURRENT

STEP 7 - CALCULATING CURRENT FLOW IN BRANCH B

VOLTAGE = CURRENT X IMPEDANCE

$120 \angle 0^\circ$ = CURRENT X $200 \angle -90^\circ$ OHMS

$120 \angle 0^\circ / 200 \angle -90^\circ$ = CURRENT

$0.6 \angle 0.0^\circ - (-90)^\circ$ AMPERES = CURRENT

$0.6 \angle 90^\circ$ AMPERES = CURRENT

STEP 8 - SUM THE CURRENTS FROM BOTH BRANCHES

BRANCH A

$0.96 \angle -53.1^\circ$

= .96 COS 53.1 -j .96 SIN 53.1

= [(0.96)(0.6) -j (0.96)(0.7996) = 0.576 -j 0.77

BRANCH B

$0.6 \angle 90^\circ$

= .6 COS 90 +j 0.6SIN 90

= [(.6)(0) +j (0.6)(1)] = 0 +j 0.6

= 0.576 -j 0.17

$0.576 - j\,0.17 = \sqrt{(0.576)^2 + (0.17)^2} \angle \text{ARCTAN } (0.17/0.576)$

$\sqrt{0.3306} \angle \text{ARCTAN } (0.2951)$

$0.6 \angle 16.4^\circ$ TOTAL AMPERES

STEP 9 - SOLVE FOR POWER FACTOR.

POWER FACTOR = COS 16.4°

POWER FACTOR = .9593

STEP 10 - SOLVE FOR OVERALL IMPEDANCE

E = I X Z

$120 \angle 0^\circ = 0.6 \angle 16.4^\circ$ X Z

$120 \angle 0^\circ / 0.6 \angle 16.4^\circ$ = Z

$200 \angle 0^\circ - 16.4^\circ$ = Z

$200 \angle 16.4^\circ$ OHMS = Z_T

ples of kilowatt loads. The power factor of kilowatt loads is 1.0, or 100 percent.

There are types of ac electrical loads that have magnetic and/or capacitive components, and these ac loads exhibit leading or lagging power factors. The measurement of nonzero power factor loads is done using voltamperes (VA) instead of watts. The relationship of the voltampere to the watt is as follows: One watt equals one voltampere at 100 percent power factor, or at a power factor of 1.0. Recalling from Chap. 1 the definition of power factor, where the power factor is the cosine of the electrical angle (θ) between the voltage and the current, the calculation of wattage becomes

$$\text{Watts} = \text{voltamperes} \times \text{power factor}$$

$$= \text{voltamperes} \times \cos\theta$$

Figure 3-9 shows a common calculation for the kilowatts, reactive kilovoltamperes (kVAR), and power factor for a typical 480-V three-phase motor. It also shows in vector form the relationship of these three to one another. Notice that the quantity of amperes flowing to the motor can be minimized by having the power factor of the motor at 1.0, or 100 percent.

In the vector right triangle, one can observe that changing the motor power factor to 100 percent would require minimizing the reactive voltamperes. The following is a discussion of the methodology to achieve this desired result.

Leading and lagging voltamperes and reactive power

Simultaneous with electric current flow in a conductor is the existence of a magnetic field that surrounds the conductor. The magnetic field begins at the center of the conductor and extends outward to infinity. Current cannot flow until this magnetic field exists, so current *lags* the application of voltage by 90 electrical degrees, and it takes energy to build up the magnetic field. While the magnetic field exists, energy is stored within the magnetic field. On interruption of current flow (such as opening the switch in a circuit), the magnetic

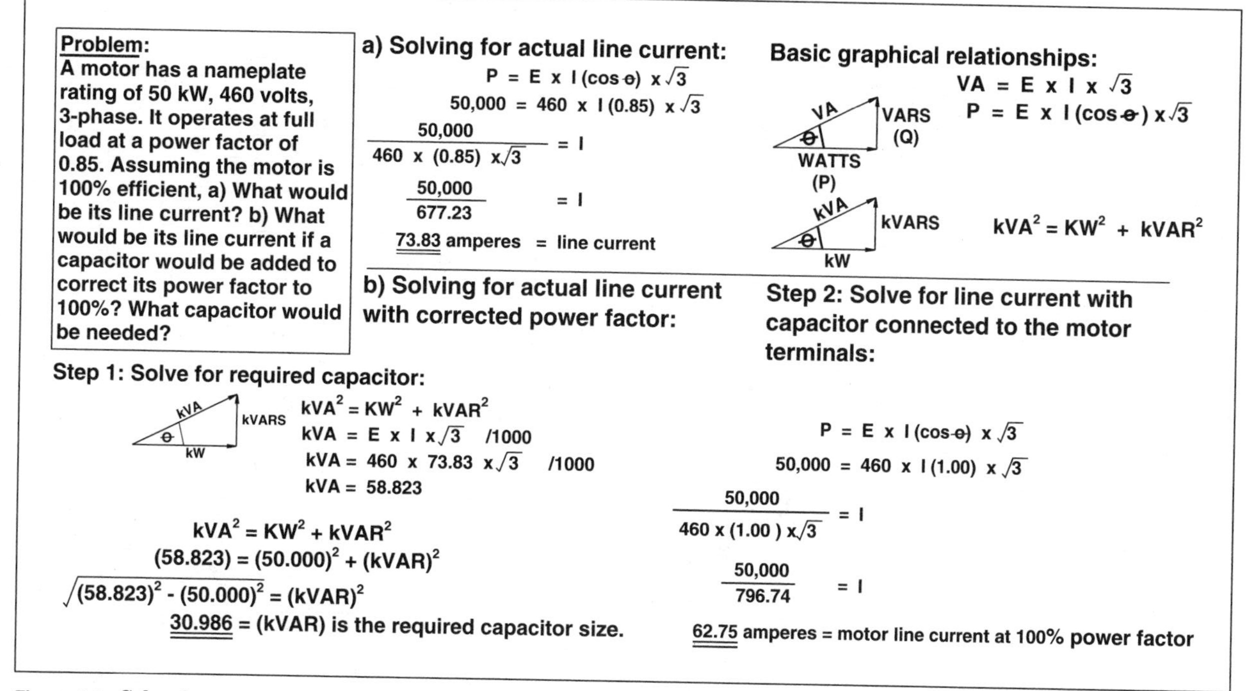

Figure 3-9 Solve for motor current and power factor given voltage and motor characteristics.

field collapses, *returning the energy back to the electrical power system.* The net result of this is that real power consumption (kilowatts) in a pure inductor (that has no resistance) would be zero.

The exact opposite happens in a circuit to which a capacitor is connected. When current flows into a capacitor, the current flow *leads* the voltage by 90 electrical degrees, which is the exact opposite of that for an inductance. The net result in an actual circuit is that energy is traded from the magnetic field to the capacitor and back again during every cycle of the voltage wave. If the capacitor and inductor can be sized to "hold" the same amount of energy, then the current flow in the circuit to them would be minimized because it must then only provide the real power energy (watts), and the net power factor of the load would be 100 percent. Electric current flows from the inductance to the capacitance and back again during every cycle, so the closer the capacitor can be placed to the inductor, the less will be the heat losses (I^2R) in the power supply conductor, and the smaller that conductor needs to be. Often this ideal situation is implemented simply by connecting the capacitors to the terminals of motors at the location of the motor junction box, as shown in Fig. 3-10 for one motor, but when doing this, one must remember that the relative kilovoltampere limitation shown on the motor nameplate must not be exceeded. Generally, the limitation is approximately 15 percent of the kilovoltampere value of the motor, or an individual motor full-load power factor of 0.95. Exceeding this limitation presents damage risks to the motor from shaft-twisting overtorque during starting and from coil overvoltage.

The next best solution is to connect the capacitors to the motor control center or panelboard bus, as shown in Figs. 3-11 through 3-13. These figures show a system of 10 motors having an overall power factor of 0.80 prior to the addition of power factor correction and with an overall power factor of 1.0 after power factor correction of the combined motor load. Note that in Fig. 3-13 the steps are shown to determine the feeder ampere rating by sizing the capacitor for unity power factor. This procedure is different from that followed in Fig. 3-12, where the factor of 1.25 [i.e., the multiplying factor from the bottom of *National Electrical Code* (NEC) Table 430-151 for an individual motor operating at 80 percent power factor] was not used because power factor was corrected at each motor (i.e., the branch-circuit current to each motor is as if the motor is a 100 percent power factor load; therefore, the note requiring this 1.25 factor at the bottom of NEC Table 430-151 was not applicable). In summary, power factor can be corrected at each motor or at the motor control center. There are benefits to correcting power factor at load centers, including being able to correct beyond 95 percent without causing motor problems and including a lowered cost per reactive kilovoltampere when larger capacitor units are installed.

Power Factor Correction to Normal Limits

If an existing electrical system has become increasingly loaded over time and its conductors are operating at their maximum operating limit, often it is possible to permit additional load simply by installing capacitors so that the out-of-phase (lagging) current does not have to come from the utility power source but instead can come from capacitors that are connected near the load. Figure 3-13 shows how this application is made.

A review of the power triangle shown in Fig. 3-14 shows that it takes much more capacitance to improve from a power factor of 0.95 to 1.0 than it takes to improve from a power factor of 0.85 to 0.90. Since capacitors cost money, the amount of capacitance that should be added to improve the power factor of the system generally is dictated by consideration of the variables in the utility bill or by current-limitation considerations in the supply conductor. Frequently, it is desirable to change the power factor from some lagging value to an improved value, but not quite to 1.0. Figure 3-15 shows a quick method of calculating the amount of capacitance required to change from an existing measured power factor to an improved one.

PROBLEM:
SOLVE FOR BRANCH CIRCUIT WIRE AMPERES WITH, AND WITHOUT, POWER FACTOR CORRECTION CAPACITORS AT THE MOTOR.

WITHOUT POWER FACTOR CORRECTION CAPACITORS AT THE MOTOR.

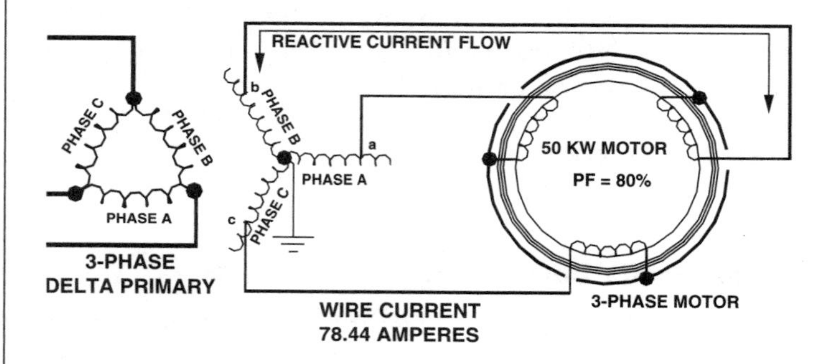

WITHOUT CAPACITORS, WIRES MUST CARRY REACTIVE CURRENT

$$P = E \times I \times \sqrt{3} \times \text{POWER FACTOR}$$

$$50000 = 460 \times I \times \sqrt{3} \times 0.80$$

$$\frac{50000}{(460)(0.80)(\sqrt{3})} = I$$

$$78.44 \text{ AMPERES} = I$$

SOLUTION: BRANCH CIRCUIT AMPERE FLOW IS REDUCED FROM 78.44 AMPERES TO 62.75 AMPERES WHEN POWER FACTOR IS CORRECTED TO 100% AT THE MOTOR. NOTE: MOTOR NAMEPLATE DATA MAY PROHIBIT CORRECTING ABOVE 95%.

Figure 3-10 Solve for reduced line current by placing power factor correction capacitors at the motor terminal box.

WITH POWER FACTOR CORRECTION CAPACITORS AT THE MOTOR.

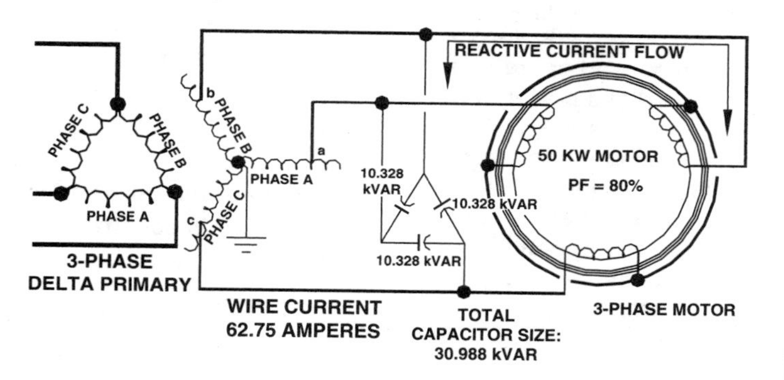

CAPACITORS DELIVER REACTIVE CURRENT SO THAT IT NO LONGER MUST COME FROM UTILITY

$$P = E \times I \times \sqrt{3} \times \text{POWER FACTOR}$$

$$50000 = 460 \times I \times \sqrt{3} \times 1.00$$

$$\frac{50000}{(460)(1.00)(\sqrt{3})} = I$$

$$62.75 \text{ AMPERES} = I$$

Problem:
Each of ten motors has a nameplate rating of 40 horsepower, 460 volts, 3-phase, and operates at full load at a power factor of 0.80. Assuming the National Electrical Code must be followed, find the required feeder ampacity for the power panelboard, PP, that serves the group of ten motors.

Step 1:
Solve for motor current for each motor by reference to NEC Table 430-150:

Step 2:
Solve for motor current for ten motors, in accordance with NEC 430-24.

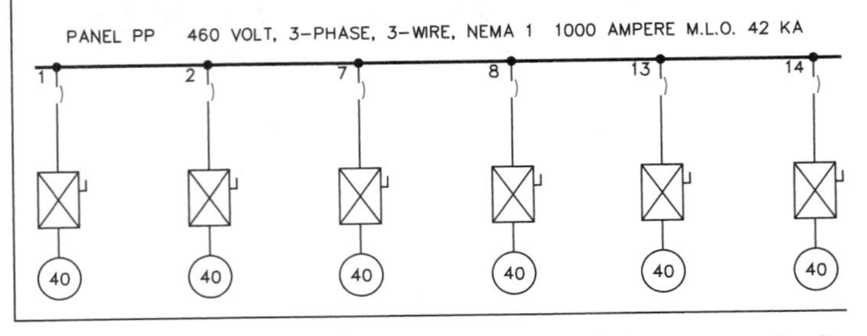

Figure 3-11 Solve for feeder current given motor horsepower and voltage and quantity of motors.

Real Power (Kilowatts), Apparent Power (Kilovoltamperes), Demand, and the Electrical Utility Bill

Although there are very many ways to calculate an electrical utility bill, for large electrical power systems, almost all billing methods include provisions for kilowatt billing and for peak-demand (kilovoltampere) billing. These are related to one another through the power factor, since at unity power factor peak kilowatts equal peak kilovoltamperes.

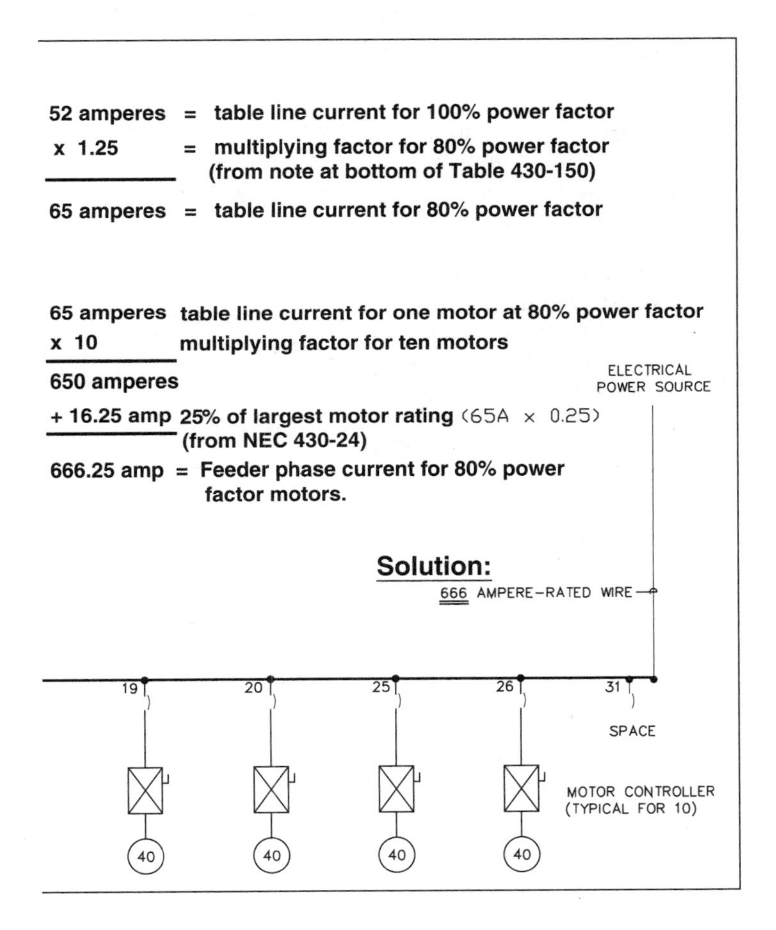

The peak demand represents a billable item because the utility must have enough capability in its system to carry the current required by the kilovoltampere load regardless of its kilowatt rating. Further, if the peak electrical demand exists for only, say, 1 hour every year, it still must be provided for by making the utility system large enough to meet this requirement. The required ampere and kilovoltampere sizes of electrical utility systems and the peak kilovoltampere demand of electrical power systems at customer

Problem:

Each of ten motors has a nameplate rating of 40 horsepower, 460 volts, 3-phase, and operates at full load at a power factor of 0.80, corrected to 100% locally at each motor with a capacitor. Assuming the National Electrical Code must be followed, find the required feeder ampacity for the power panelboard, PP, that serves the group of ten motors.

Step 1:
Solve for motor current for each motor by reference to NEC Table 430-150:

Step 2:
Solve for motor current for ten motors, in accordance with NEC 430-24.

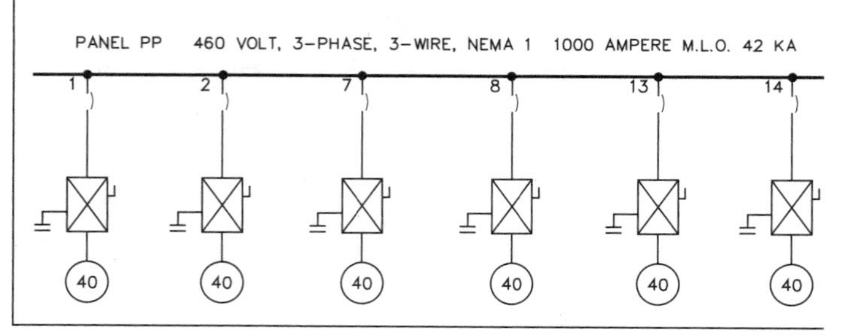

Figure 3-12 Solve for reduced feeder current by placing power-factor-correction capacitors at the motor control center.

premises can be minimized by keeping power factor values within the customer electrical power systems as near to 100 percent as possible. Although a detailed discussion of this is beyond the scope of this book, the peak kilovoltampere demand measured by the utility electric meters can be lessened by load shedding (turning off certain loads for short times) and by the addition of on-site generation.

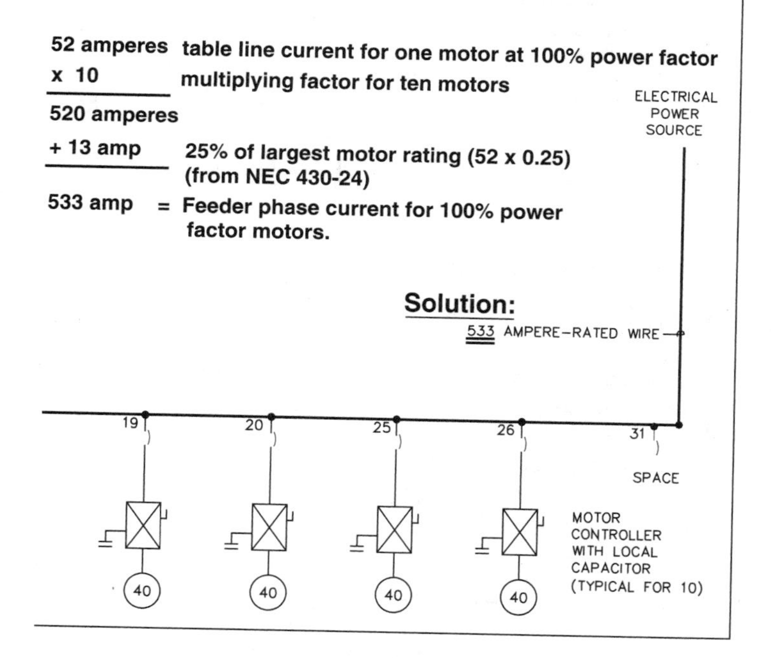

Power Factor Correction System Design in an Electrical Power System Containing No Harmonics

Installing power factor correction can be done by either adding capacitors or installing rotating synchronous condensers, but the most common method of correcting power factor is through the addition of capacitors. In electrical systems that contain only linear loads (those which do not

Problem:
Each of ten motors has a nameplate rating of 40 horsepower, 460 volts, 3-phase, and operates at full load at a power factor of 0.80. Assuming the National Electrical Code must be followed, find the required feeder ampacity for the power panelboard, PP, that serves the group of ten motors.

Step 1:
Solve for motor current for each motor by reference to NEC Table 430-150:

52 amperes
x 1.25
65 amperes

Step 2:
Solve for motor current for ten motors, in accordance with NEC 430-24.

65 amperes
x 10
650 amperes
+ 16.25 amp
666.25 amp

PANEL PP 460 VOLT, 3-PHASE, 3-WIRE,
NEMA 1 1000 AMPERE M.L.O. 42 KA

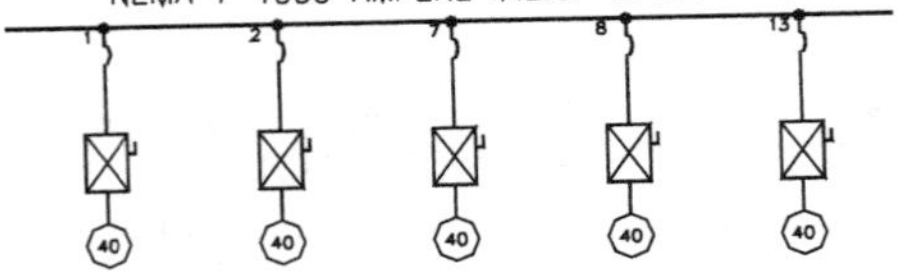

1 2 7 8 13

40 40 40 40 40

Step 3: Solve for required capacitor:

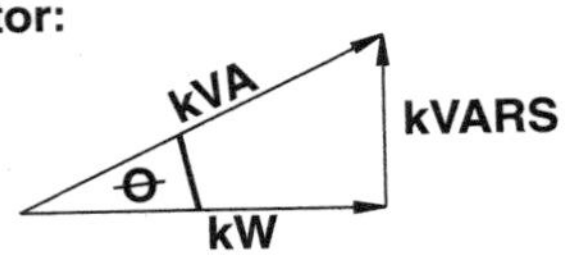

kVA kVARS θ kW

Find kVA:
kVA = E x I x $\sqrt{3}$
kVA = 460 x 666.25 x $\sqrt{3}$ /1000
kVA = 530.8 kVA

Find kW:
kW = E x I (cos θ) x $\sqrt{3}$
kW = (460) x (666.25) x (0.80) x $\sqrt{3}$ /1000
kW = 424.7 kW

Step 4: Solve for line current with capacitor connected to the Panel.

$$P = E \times I (\cos \theta) \times \sqrt{3}$$

$$424{,}700 = 460 \times I (1.00) \times \sqrt{3}$$

$$\frac{424{,}700}{460 \times (1.00) \times \sqrt{3}} = I$$

$$\frac{424{,}700}{796.74} = I$$

533 amperes = line current at 100% power factor

Figure 3-13 Solve for feeder current at the motor control center with power factor correction capacitors at the motor control center.

= table line current for 100% power factor
= multiplying factor for 80% power factor
(from note at bottom of Table 430-151)
= table line current for 80% power factor

table line current for one motor at 80% power factor
multiplying factor for ten motors

25% of largest motor rating (65 x 0.25)
(from NEC 430-24)
= Feeder phase current for 80% power factor motors.

Solution:

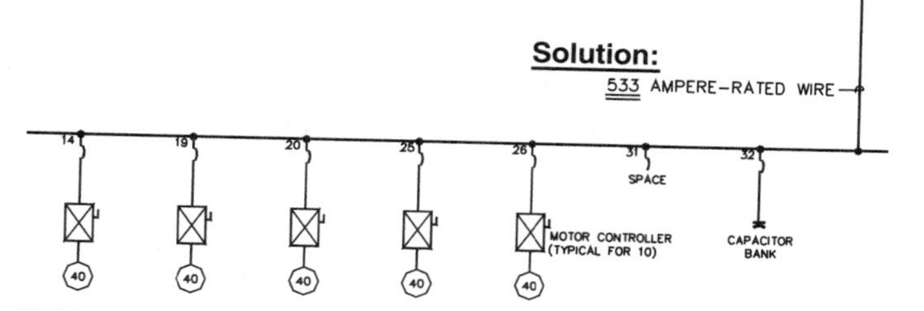

Find kVAR capacitor rating:

$$kVA^2 = KW^2 + kVAR^2$$

$$(532.2)^2 = (425.8)^2 + (kVAR)^2$$

$$\sqrt{(530.8)^2 - (424.7)^2} = (kVAR)$$

$\underline{\underline{318.4}}$ = (kVAR) is the required capacitor size to make kVA equal to kW, which is what 100% power factor means.

Problem:
Solve for the capacitor size needed to correct a true power load of 500 kW from 95% to 100%.

Solve for required capacitor:

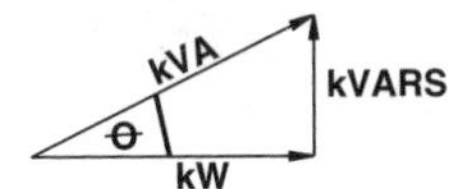

Find kW:

kW = 500 kW

Find kVA:

kW = kVA x p.f.

500 = kVA x (0.95)

$\frac{500}{0.95}$ = kVA

526.3 = kVA

Find kVAR value for system:

$$kVA^2 = KW^2 + kVAR^2$$

$$(526.3)^2 = (500)^2 + (kVAR)^2$$

$$\sqrt{(526.3)^2 - (500)^2} = (kVAR)$$

$$\underline{\underline{164.3}} = kVAR$$

164.3 kVAR is required to correct 500 kW at 95% p.f. to 100% p.f. (where kVA = kW).

Problem: Solve for the capacitor size needed to correct the power factor of a true load of 500 kW from 85% to 90%.

Step 1
Solve for reactive power at 85% power factor.

Find kW:

kW = 500 kW

Find kVA:

kW = kVA x p.f.

500 = kVA x (0.85)

$\frac{500}{0.85}$ = kVA

588.2 = kVA

Find kVAR value for system:

$$kVA^2 = KW^2 + kVAR^2$$

$$(588.2)^2 = (500)^2 + (kVAR)^2$$

$$\sqrt{(588.2)^2 - (500)^2} = (kVAR)$$

$$\underline{\underline{309.8}} = kVAR$$

309.8 kVAR is the reactive power of this system of 500 kW at 85% p.f.

Step 2
Solve for reactive power at 90% power factor.

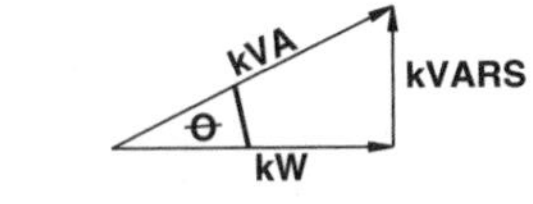

Find kW:

kW = 500 kW

Find kVA:

kW = kVA x p.f.

500 = kVA x (0.90)

$\frac{500}{0.90}$ = kVA

555.5= kVA

Find kVAR value for system:

$$kVA^2 = KW^2 + kVAR^2$$

$$(555.5)^2 = (500)^2 + (kVAR)^2$$

$$\sqrt{(555.5)^2 - (500)^2} = (kVAR)$$

$$\underline{\underline{242}} = kVAR$$

242 kVAR is the reactive power of this system of 500 kVA at 90% p.f.

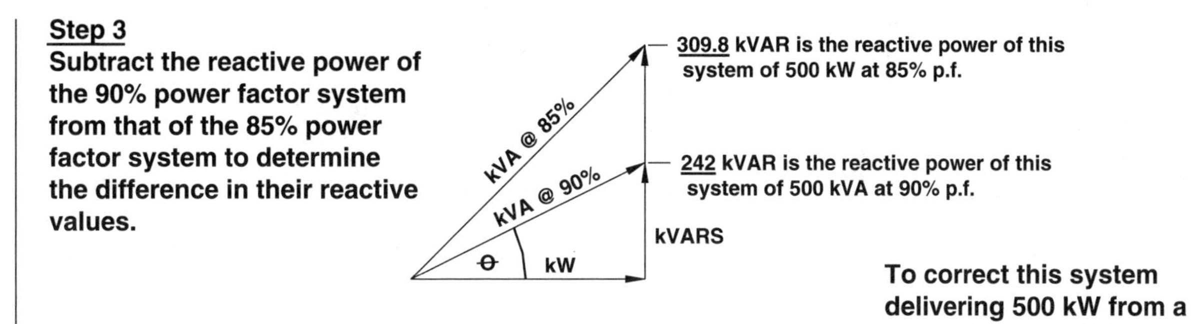

Figure 3-14 Solve for capacitor value to correct power factor to a predetermined value given loads and initial power factor.

Problem: Solve for the capacitor size needed to correct the power factor of a true load of 500 kW from 85% to 90%.

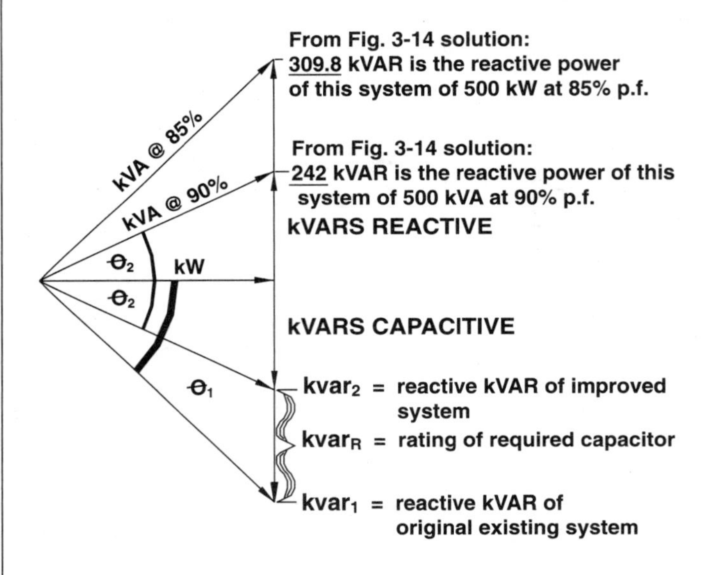

where:

$kvar_R$ = rating of required capacitor

$kvar_1$ = reactive kVAR of original existing system

$kvar_2$ = reactive kVAR of improved system

θ_1 = original (existing power factor angle) in degrees (ex. arccos (0.75) = 41.40°)

θ_2 = new improved power factor angle in degrees (ex. arccos (0.90) = 25.81°)

kW = the true load of the existing system

Figure 3-15 Use one-step formula to solve for capacitor value to correct power factor to a predetermined value.

$\arccos 0.85 = 31.78^\circ = \theta_1$

$\arccos 0.90 = 25.84^\circ = \theta_2$

$kvar_R = kW \times (\tan \theta_1 - \tan \theta_2)$

$kvar_R = 500 \times (\tan 31.78^\circ - \tan 25.84^\circ)$

$kvar_R = 500 \times (0.6195 - 0.4843)$

$kvar_R$ = 67.6 kVAR of capacitance is required

Solution:

To correct this system delivering 500 kW from a power factor of 85% to a power factor of 90% requires 67.6 kVAR of capacitance.

VOLTAGE	230	460	2400	4.2 – 13.8 kV
	7.5	15	15	90
	15	30	45	180
	30	60	90	900
	45	90	135	1800
	60	120	180	2700
	90	180	600	3600
	180	360	900	4500
	270	540	1200	5400
	450	900	1800	
	630	1260	2700	
			3600	

Figure 3-16 Use this listing of commonly available capacitor sizes and common voltages for standard capacitor bank designs.

create harmonic currents in the system), capacitors can be connected directly to the electrical power system. Capacitors can be purchased at almost any voltage and at almost any reactive kilovoltampere rating, although each manufacturer offers standard values as normal supply items. See Fig. 3-16 for a listing of commonly available sizes of capacitors in standard voltage ratings. It is of value to note, however, that almost any size capacitor is available as a special-order item.

In applying capacitors to electrical power systems, note that the installation of three 50-kVAR capacitor "cans" in either a wye or delta configuration appears as a 150-kVAR capacitor "bank" because capacitor reactive kilovoltampere values simply add algebraically to one another as long as the rated voltage for each capacitor is applied to that capacitor in the connection scheme used.

Capacitance value varies with the square of the voltage, since a little higher voltage physically causes the electrons to pack much more tightly against the dielectric within the capacitors. Figure 3-17 shows how to calculate the resulting value of a capacitor that is connected to an elec-

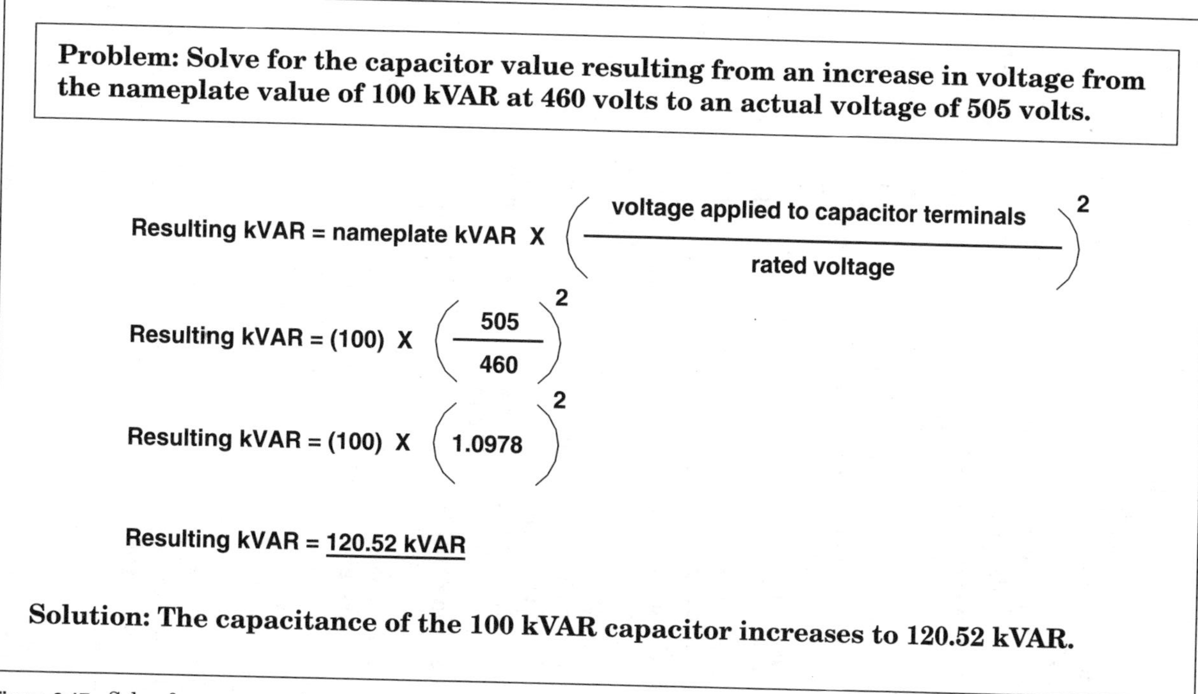

Figure 3-17 Solve for new capacitor value given capacitor voltage nameplate rating and system voltage.

trical power system having a voltage that is different from the voltage rating of the capacitor. Figure 3-18 shows additional useful formulas for use with capacitors, when needed.

Capacitors draw large amounts of inrush current during initial energization, and the electrical system protection devices must permit this current to flow while still protecting the capacitor system from short circuit. While the NEC requires fuses and circuit breakers to most electrical devices and requires that conductors be sized in accordance with the long-time amperage rating of the appliance or device, an exception is made in the case of capacitors. The NEC only requires that overcurrent protection for capacitor banks and the conductors to them be "as low as practicable." History in the electrical industry has shown that "slow blow" fuses of the dual-element type will work satisfactorily in these applications when they are sized at approximately 200 percent of the full-load long-time amperage rating of the capacitor. The NEC requires that the conductors to the capacitor bank be sized at 135 percent of the full-load long-time amperage rating of the capacitor. Figure 3-19 shows how to calculate the size of conductors, switches, and fuses to a 480-V capacitor bank.

The loads within some electrical installations are deenergized regularly. For example, in a shopping center, the lighting and air-conditioning systems are generally deenergized after hours and reenergized the following morning. A close review of the charge-discharge characteristics of capacitors would reveal that when capacitors are the predominant loads connected to electrical power systems, the voltage of the system begins to rise. This is simply due to the charged plates of the capacitor discharging during the next half-cycle, acting as voltage generators similar to rotating generators, and this voltage rise begins noticeably when the power factor of the system is *leading* 10 percent or more (i.e., the power factor is 90 percent leading, and the voltage rise increases as the power factor changes to 80 percent leading, and so on). To prevent excessive voltage rise in electrical systems to which capacitors are connected, a power

$$X_C = \frac{1}{2\pi fC} = \frac{(1000)\,(kV)^2}{kVAR}$$

For parallel capacitors, kVAR values add:

$$kVAR_{Total} = kVAR_1 + kVAR_2 + \ldots kVAR_n$$

where:

X_C is capacitive reactance in ohms
f is the frequency in Hertz per second
C is capacitance in farads
kVAR is capacitance in kilovars

Figure 3-18 These are some capacitor calculation formulas that are often useful.

factor controller normally is used to switch capacitors "on" when needed for power factor correction and "off" when not needed. Capacitors are arranged and connected in steps that permit none, some, or all of the capacitors to be energized at any one time.

When energized initially, capacitors appear to be short circuits because of their large values of inrush current. Similarly, if a capacitor is charged and then a short circuit is connected to its line terminals, almost infinite instantaneous current would flow from the capacitor through the short circuit. Capacitor banks must be specially designed when they are controlled with power factor controllers that provide steps within the bank. A normal capacitor has fuses at its line terminals, and the other capacitors in a capacitor bank also have fuses at each of their line terminals. If one capacitor in a bank is energized and operating and then is switched to be in parallel with a capacitor that has been deenergized, all the electrons from the charged capacitor will flow into the discharged capacitor or short-time resonance can occur, either of which can cause melting and clearing of the fuses at each capacitor. To prevent this from happening, inductive reactance (coils) must be connected between steps of a multistep capacitor bank, as shown in Fig. 3-20.

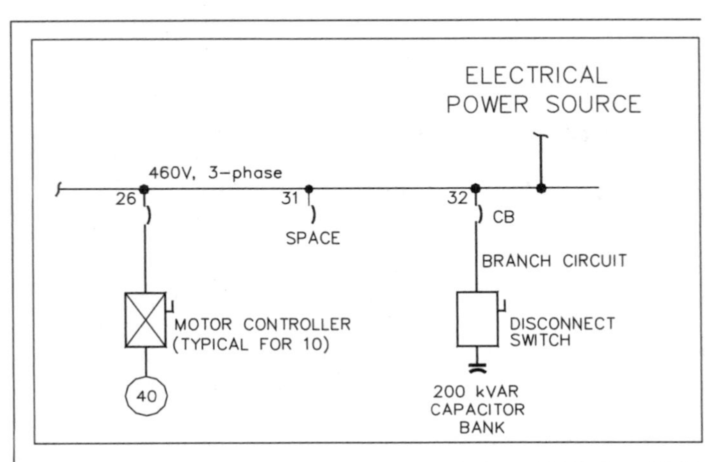

STEP 3: CALCULATE BRANCH CIRCUIT WIRE AMPERE RATING

251 AMPERES
X 1.35 NEC SEC. 460–8a

339 AMPERES

THE WIRE MUST BE RATED FOR AT LEAST 339 AMPERES.

STEP 4: SELECT AMPERE RATING OF DISCONNECT SWITCH.

251 AMPERES
X 1.35 NEC SEC. 460–8c.

339 AMPERES

THE SWITCH MUST BE RATED FOR AT LEAST 339 AMPERES. STANDARD SWITCH SIZES ARE 30, 60, 100, 200, & 400 AMPERE. SELECT THE NEXT SIZE LARGER THAN 339 AMPERE: 400 AMPERE

Figure 3-19 Solve for the ratings of conductors, switches, and fuses given capacitor size and voltage.

Power Factor Correction System Design in an Electrical Power System Containing Harmonics

Every electrical power system has a natural parallel resonance frequency. At this frequency, energy is traded back and forth from the capacitance in the system to the inductance in the system. Although all electrical power system parts have some capacitance, particularly when shielded

STEP 1: CALCULATE CAPACITOR CURRENT

$$KVAR = E_{KV} \times I \times \sqrt{3}$$

$$200 = (0.460) \times I \times \sqrt{3}$$

$$\frac{200}{(0.460) \times \sqrt{3}} = I$$

251 AMPERES = I

STEP 2: CALCULATE THE CIRCUIT BREAKER LONG–TIME TRIP RATING.

251 AMPERES
X 1.75 TO X 2 (VALUES FROM HISTORICAL INDUSTRIAL DATA)
502 AMPERES. STANDARD RATINGS FROM NEC 240–6 ARE 300, 350, 400, 450, 500, AND 600. SELECT 500/3P THERMAL MAG. BREAKER

STEP 5: SELECT SUITABLE BRANCH CIRCUIT WIRES.

FROM TABLE 310–16 IN A 40 C ENVIRONMENT, THREE COPPER 500 KCMIL THWN WIRES WOULD SUFFICE. SINCE THEY ARE RATED TO CARRY 380 AMPERES.

STEP 6: SELECT RIGID STEEL CONDUIT OF SUITABLE SIZE.

FROM TABLES 4 AND 5 OF CHAPTER 9 OF THE NEC: CROSS SECTIONAL AREA FOR 3 – 500 KCMIL THWN: 3 X 1.0082 SQ. IN. AT 40% FILL, THE 3.0246 SQ. IN. FIT INTO A 3" CONDUIT.

medium-voltage cable is involved, most of it is contained within the capacitors. Similarly, although all electrical power systems parts have some inductance, most of it is contained within the transformer and motor coils.

The natural resonance frequency of a system having no capacitors is normally quite high (frequently greater than 50 times the fundamental frequency) due to the low value of capacitance in the system. When installing power factor

Problem: Insert minimum value of inductor to prevent capacitor step switching surges and natural frequency resonance between steps in a capacitor bank.

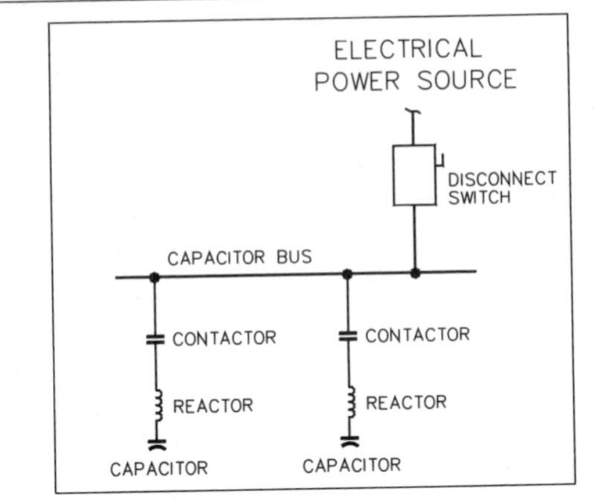

Select the reactor value from this table of typical inductance values between steps in a capacitor bank.

RATED VOLTAGE	INDUCTANCE IN MICROHENRIES
0 – 15 kV	30
15.5 kV – 38 kV	40
38.5 kV – 48 kV	50
48.5 kV – 72.5 kV	60
73 kV– 120 kV	70

Note: These are not detuning reactors. Parallel resonance scans must be made after the capacitor bank is designed to make certain that the capacitors with any combination of steps energized are not parallel-resonant with the rest of the electrical power system, with any of its load switching permutations.

Figure 3-20 Solve for minimum inductance value between capacitor "steps" to prevent fuse blowing due to inrush current.

correction capacitors, the natural parallel resonance frequency of the electrical power system is reduced. At 100 percent power factor, the natural parallel resonance frequency of an electrical power system can be expected to be approximately at the fifth harmonic [i.e., if the frequency of the system is 60 hertz (Hz), then the fifth harmonic is 5 times 60, or 300 Hz].

The natural resonance frequency of a system is of little importance as long as only 60-Hz voltages and currents are present in the electrical power system, since the system capacitance and inductance will not begin to trade power back and forth, or oscillate, unless the oscillation is initiated by a harmonic source elsewhere in the electrical power system. Engineers can either field measure the harmonic contents of an electrical power system or can observe or computer model the system to determine whether harmonic currents will exist there and what the frequencies of those harmonic currents will be. Some guidelines to forecasting the harmonic currents in an electrical power system are shown in Fig. 3-21. This figure shows the harmonic currents that are generated by certain nonlinear loads, from which the harmonic currents flow back to the electrical power source. In flowing back into the source impedance of the electrical power source, the harmonic currents create ($I_{\text{harmonic}} \times Z_{\text{source}}$) harmonic voltages that "ride" on top of the fundamental 60-Hz sinusoidal waveform, creating "voltage distortion."

Calculating the Parallel Harmonic Resonance of an Electrical Power System Containing Capacitors

When the addition of capacitors is calculated to create a parallel harmonic resonance at a frequency that exists or is forecast to exist in an electrical system, then reactors can be added in series with the capacitors to "detune" the capacitor bank from the specific parallel resonance frequency to another frequency that does not exist in the system.

HARMONIC CURRENTS GENERATED BY LOADS

PULSE #n	CHARACTERISTIC HARMONICS (n ± 1)
2	3, 5, 7, 9, 11, 13, 15, 17, 19...
6	5, 7, 9, 11, 13, 15, 17, 19...
12	11, 13, 15, 17, 19...
18	17, 19...

Typical Harmonic Spectrum for 6-Pulse Converter Current

Harmonic	Theoretical Magnitude	Actual Magnitude
5	20.00%	17.50%
7	14.28%	11.10%
11	9.00%	4.50%
13	0.07%	2.90%
17	0.06%	1.50%
19	0.05%	1.00%
23	0.04%	0.90%
25	0.04%	0.80%

Notes:
1) The magnitude is approximately 1/harmonic number.
2) The triplen harmonics of 9 and 15 are normally cancelled in delta transformer coils.

Figure 3-21 Solve for anticipated harmonic currents given type of nonlinear load.

While the most accurate method of determining the natural parallel resonance frequency of an electrical power system is with a computer program within which the system is modeled, it is possible to calculate the approximate parallel resonance frequency of the system with the formula shown in Fig. 3-22. However, the exact behavior of the electrical power system after adding detuned capacitors of specific series resonant values is most accurately determined with a definite-purpose computer program.

Resulting Values of Adding Harmonic Currents or Voltages

In dc systems, voltages add by simple algebra, as shown in Fig. 1-7. In ac systems that have no harmonic contents

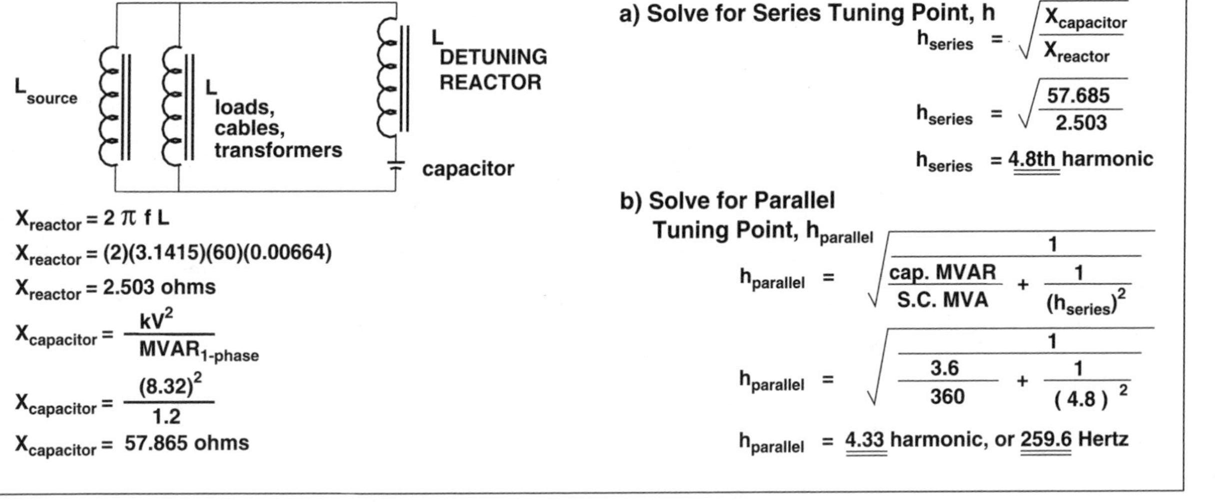

Figure 3-22 Solve for series filter or capacitor bank detuning frequency and parallel system resonance.

(these are known as *linear*), ac voltages of the same frequency and phase also add by simple algebra. However, in ac systems containing harmonics, the individual voltages or currents of each frequency must be added vectorially. The scalar sum of these vector additions can be determined by the square root of the sum of the squares of these vector values, as demonstrated in Fig. 3-23.

Acceptable Levels of Harmonic Current and Voltage

The most commonly accepted document regarding permissible values of harmonic currents and harmonic voltages is Institute of Electrical and Electronics Engineers (IEEE) Standard 519-1992 (sometimes listed as IEEE 519-1993 because it was completed in 1992 and issued in 1993). This document provides acceptable requirements for total harmonic current distortion (THCD) at the point of common coupling (Pcc), where the utility interfaces with the plant or building system, and these are shown in its Chapter 10 and summarized in its Table 10.3, and replicated here in Fig. 3-24. In Chapter 11, IEEE 519 also provides acceptable requirements for total harmonic voltage distortion (THVD) at the Pcc, but when the customer meets the current distortion requirements of Table 10.3, then it is up to the electrical utility to provide sufficiently low source impedance to meet the requirements of Chapter 11 THVD.

In electrical power systems containing harmonic currents and harmonic voltages, there are two different power factors. *Displacement power factor* is defined as watts divided by the product of the fundamental voltage and the fundamental current. That is:

$$\text{Displacement power factor} = \frac{\text{watts}}{\text{60-Hz voltage} \times \text{60-Hz current}}$$

The second type of power factor in electrical power systems containing harmonics is the *true power factor*. When harmonic currents are present, the true power factor is

Problem: A conductor is carrying 34 amperes of 60 hertz current, 15 amperes of 3rd harmonic current, 21 amperes of 5th, 17 amperes of 7th, and 10 amperes of 11th harmonic current. What is the total current in the conductor?

$$I_{total} = \sqrt{\sum_{h=1}^{n} (I_{harmonic})^2}$$

$$I_{total} = \sqrt{(34)^2 + (15)^2 + (21)^2 + (17)^2 + 10)^2}$$

$$I_{total} = \sqrt{2211}$$

$$I_{total} = \underline{\underline{47.02}} \text{ amperes}$$

Figure 3-23 Solve for total current given harmonic currents.

defined as the wattage divided by the product of the rms voltage multiplied by the rms current. That is:

$$\text{True power factor} = \frac{\text{watts}}{\text{true rms voltage} \times \text{true rms current}}$$

The installation of harmonic current producers such as variable-speed drives causes the power factor value to be lowered, whereas the removal of harmonic currents and power factor improvement are both accomplished by the installation of harmonic filters containing capacitors.

The Harmonic Current-Flow Model

Harmonic current can be thought of as originating at nonlinear loads, such as variable-speed drives, and flowing toward the power source, as shown in Fig 3-25. This is exactly the opposite of 60-Hz power flow that originates at the power source and flows to the loads. If harmonic current

CURRENT DISTORTION

IEEE 519-1993 TABLE 10.3 CURRENT DEMAND* DISTORTION LIMITS

I_{sc}/I_L AT Pcc	MAXIMUM HARMONIC CURRENT IN PERCENT OF I_L					
	INDIVIDUAL HARMONIC ORDER					
	<11	11–15	19–21	23–33	>33	TDD
<20	4.0%	2.0%	1.5%	0.6%	0.3%	5.0%
20–50	7.0%	3.5%	2.5%	1.0%	0.5%	8.0%
50–100	10.0%	4.5%	4.0%	1.5%	0.7%	12.0%
100–1000	12.0%	5.5%	5.0%	2.0%	1.0%	15.0%
>1000	15.0%	7.0%	6.0%	2.5%	1.4%	20.0%

* TOTAL DEMAND DISTORTION IS HARMONIC CURRENT IN % OF MAXIMUM DEMAND LOAD CURRENT FOR 15–30 MINUTE DEMAND.

I_L IS THE AVERAGE CURRENT OF THE MAXIMUM DEMAND FOR THE PRECEEDING 12 MONTHS.

LARGER CUSTOMERS (I_L) HAVE MORE STRINGENT LIMITS BECAUSE THEY REPRESENT A LARGER PORTION OF THE SYSTEM LOAD.

THE TABLE IS FOR 6-PULSE RECTIFIERS. FOR 12-PULSE, 18-PULSE, ETC, INCREASE CHARACTERISTIC HARMONICS BY:
THE VALUE OF THE SQUARE ROOT OF q/6, WHERE q = 12, 18, ETC. THUS FOR 12-PULSE, INCREASE BY 1.414.

Figure 3-24 Replication of Table 10.3 of IEEE 519 showing maximum allowable current distortion values.

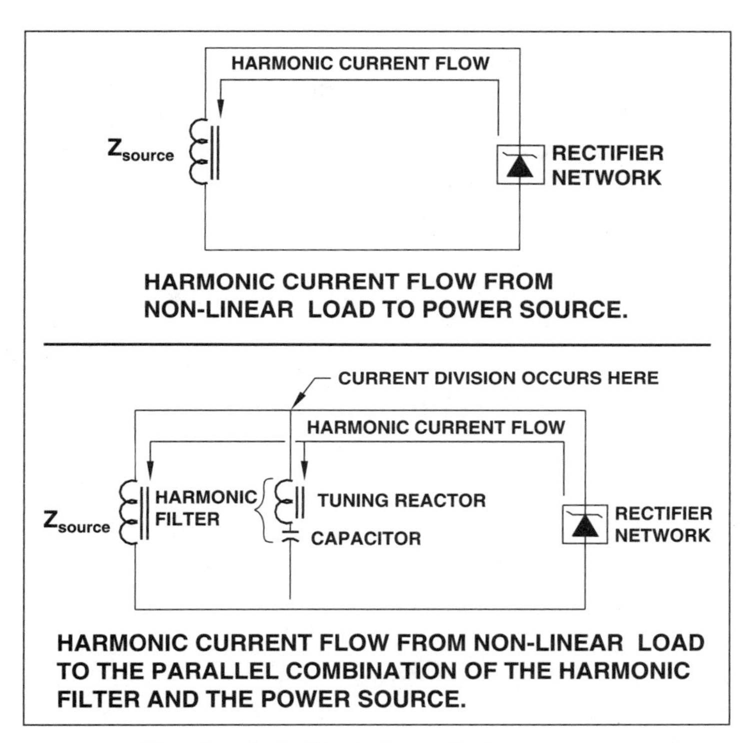

Figure 3-25 Place harmonic filter between the harmonic current source and the power source to form a harmonic current divider.

distortion at the Pcc is too great, then the current-divider principle can be implemented to divert some of the harmonic current into harmonic filters, also shown in Fig. 3-25. By connecting series combinations of inductance and capacitance whose series resonant frequency impedance approaches zero at a certain frequency, most of the harmonic current of that frequency can be redirected into the capacitor instead of having to flow to the power source. When the "now diverted" harmonic current does not flow into the impedance of the power source, it does not create harmonic voltage distortion. The proper method of computation of THCD is shown in Fig. 3-26, and THVD is calculated in the same way.

Problem: A conductor is carrying 34 amperes of 60 hertz current, 1.5 amperes of 3rd harmonic current, 2.1 amperes of 5th, 1.7 amperes of 7th, and 1.0 ampere of 11th harmonic current. What is the total harmonic current distortion (THCD) in the conductor?

$$\text{THCD} = \frac{\sqrt{\sum_{h=2}^{n} (I_{\text{harmonic}})^2}}{I_1}$$

$$\text{THCD} = \frac{\sqrt{(1.5)^2 + (2.1)^2 + (1.7)^2 + 1.0)^2}}{34}$$

$$\text{THCD} = \frac{\sqrt{10.55}}{34}$$

$$\text{THCD} = \underline{\underline{0.0955}}, \text{ or } \underline{\underline{9.55}}\ \%$$

Figure 3-26 Solve for total harmonic current distortion given harmonic currents.

Effects of Harmonic Current on Transformers

Harmonic current most seriously affects transformer operation by causing extra heating from skin effect in coil conductors, extra eddy currents in core laminations, and excessive hysteresis (molecules rubbing against one another, similar to microwave oven operation). After engineers have either measured or computer-forecast the current value of each frequency of current that a transformer will conduct, they can specify a "k-rated transformer" that is specially constructed to handle these harmonic currents and their effects. A harmonic current content of k-4 is one that would cause heating equal to that which would have been caused by 1.140 times the load current had it all been fundamental (60-Hz) current. The correct manner of determining the required k-rating of a transformer is most easily demonstrated by example, as shown in Fig. 3-27.

Problem: If the measured (or computer-forecast) current is as follows, what K-rating of transformer would be required to carry this load?

True rms amperes = 73.3 amperes

h_1 = 52.45 amperes **h_7 = 9.44 amperes** **h_{13} = 4.77 amperes**

h_3 = 42.27 amperes **h_9 = 3.72 amperes**

h_5 = 24.97 amperes **h_{11} = 5.51 amperes**

Solving for K-rating:

$$K = \sum_{h=1}^{\infty} \left[I_{h_{PU}} \right]^2 h^2$$

h	I_{RMS}	I_{PU}	I^2_{PU}	h^2	$I^2_{PU}h^2$
1	52.45	0.715	0.511	1	0.511
3	42.27	0.577	0.333	9	3.0
5	24.97	0.341	0.116	25	2.9
7	9.44	0.129	0.016	49	0.784
9	3.72	0.051	0.003	81	0.243
11	5.51	0.075	0.006	121	0.726
13	4.77	0.065	0.004	169	0.676
			1.000		8.84

Solution:
The transformer must have a K-rating of 8.84 or greater.

Figure 3-27 Solve for the transfomer k-rating given known values of harmonic currents.

Effects of Harmonic Voltage on Motors

Of all loads, harmonic voltage most seriously affects electrical motors. The reason for this is that every other odd frequency tries to make the motor rotor change in rotational direction, and the rotor is at "locked rotor" condition at every frequency above the fundamental frequency. The question about how much THVD a motor can withstand without deleterious effects contains too many variables to calculate effectively, but research into historical documentation shows that operating a fully loaded motor from a voltage containing 10 percent THVD is equivalent in terms of extra motor heating to operating the motor at 115 percent load. Thus motors having a 1.15 service factor can only be loaded safely to 100 percent load when operating from a voltage source containing 10 percent THVD.

Harmonic Current Flow through Transformers

While most harmonic currents travel through transformers from the harmonic current–creating loads to the electrical power supply, some are captured within the transformer. Balanced triplen harmonic currents of the third, ninth, and fifteenth harmonics are captured within the delta winding of a transformer, where they simply circulate and heat the delta winding. The only triplen harmonics that travel through a delta-wye, wye-delta, or wye-delta-wye transformer are unbalanced triplen harmonic currents. Therefore, a good way of eliminating a large portion of the harmonic currents is simply to insert a transformer with a delta winding into the power system to the load. Another good way of "canceling" fifth and seventh harmonic currents from several loads is to connect some of them to delta-delta and some to delta-wye transformers, causing a 30° phase shift and a vector addition to almost zero of fifth and seventh harmonic currents. This is exactly the methodology used when installing a 12-pulse variable-speed drive (VSD) instead of a 6-pulse drive, for the 12-pulse drive requires another transformer winding that is 30° phase-shifted from the first transformer's secondary winding.

Harmonic Filters

Tuning of a series resonant filter is done as shown in Fig. 3-28; this figure also shows the actual three-phase values and connections in the filter. The theory behind the series-tuned filter is simply to short circuit one particular harmonic current so that the harmonic current will flow into the filter instead of back to the power source. If more than one harmonic current exceeds IEEE 519-1992, Table 10.3 values, then more than one filter normally is used. Frequently, the simultaneous application of a fifth, seventh, eleventh, and thirteenth harmonic filter is made, and if there are some higher-frequency components too, such as the seventeenth harmonic, then the thirteenth harmonic filter is fitted with a shunt resistor around the detuning reactor, thus forming a "high pass" filter for all higher-frequency currents. The reactor in each filter is designed and constructed with a Q value, and each series filter has a Q value such that the greater the resistance and capacitance, the lower is the Q value. As shown in Fig. 3-29, the higher the Q, the steeper are the skirts of the resonant curve, and the smaller is the range of harmonics that will flow through the filter. A Q of 12 is broadband in tuning, whereas a Q of 50 to 150 is narrow in tuning. A Q of 12 to 20 would be used when the fifth harmonic filter is intended to also conduct some seventh harmonic current. Generally, the harmonic filter is tuned to just below the harmonic current that it is intended to conduct. For example, the fifth harmonic filter generally is tuned to approximately the 4.7th–4.8th harmonic. The exact tuning, however, ultimately depends on

1. The resulting THCD at the Pcc with that specific filter in the system.
2. The parallel resonance scan that is computer-modeled of the system with the filter connected to the system. The wrong values of capacitance here can cause voltage rises

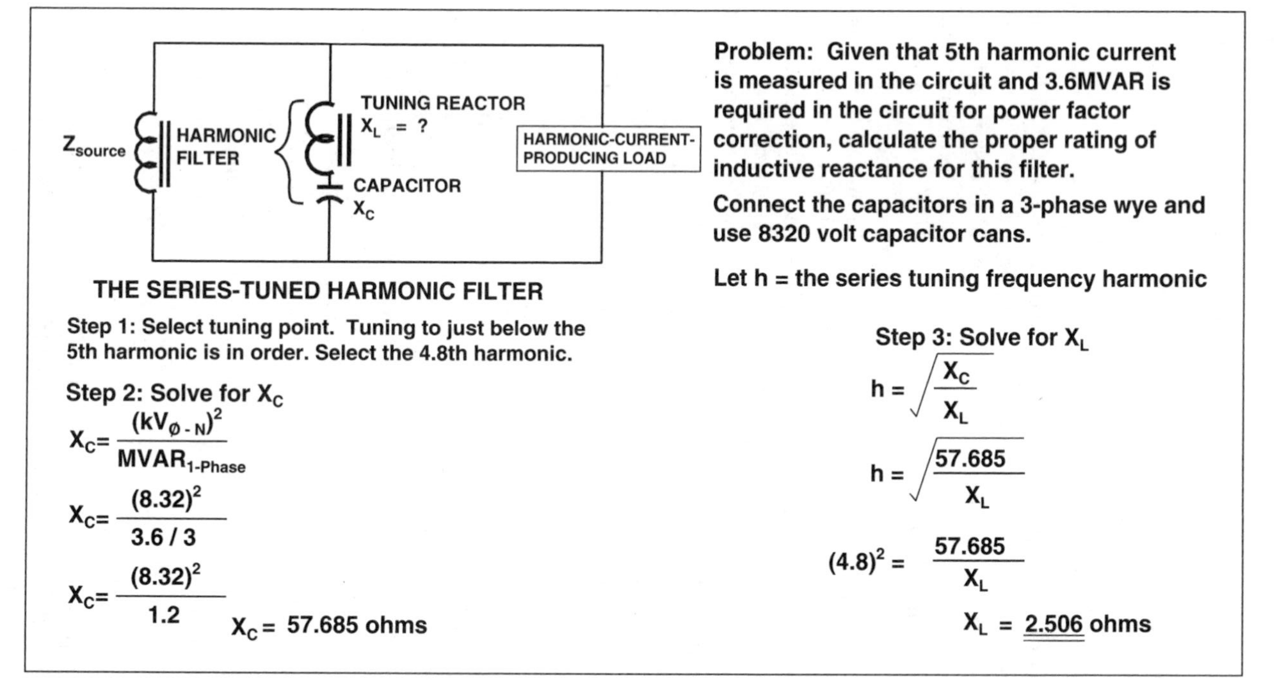

Figure 3-28 Solve for harmonic filter reactor value given a frequency and capacitor size.

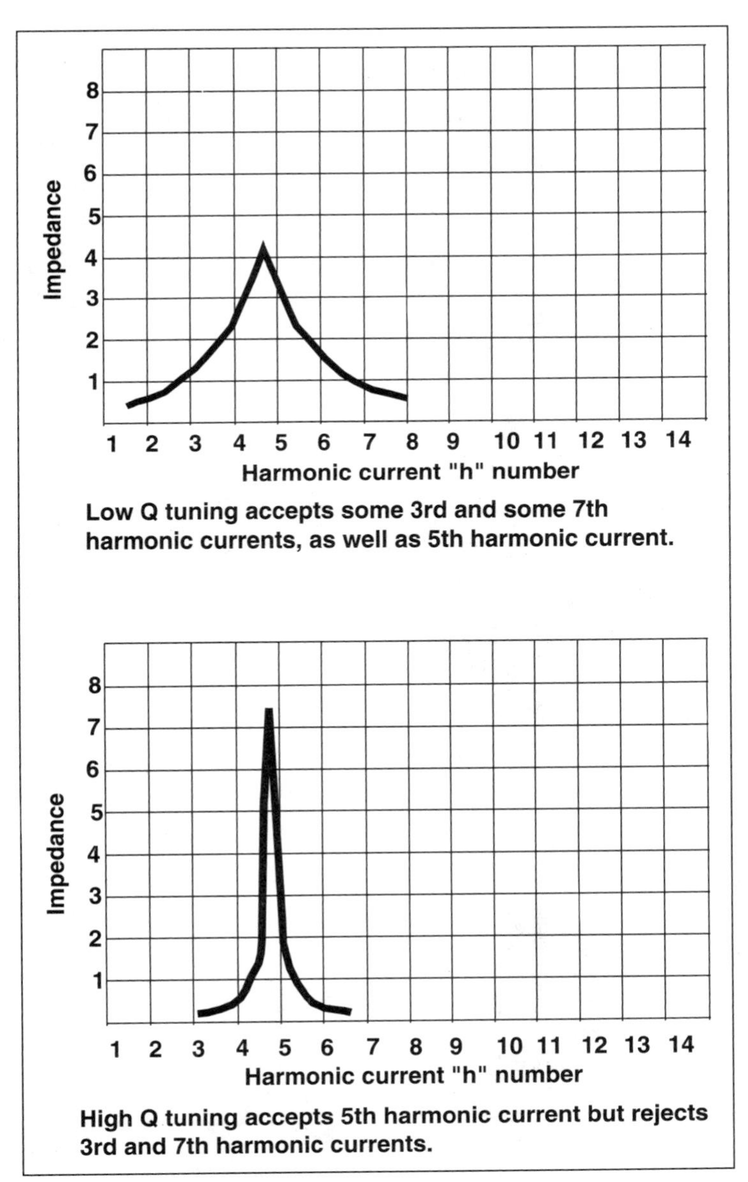

Figure 3-29 The Q of a filter defines how many harmonic current frequencies are filtered.

SYMPTOM	CAUSE	REMEDY
TOTAL HARMONIC CURRENT DISTORTION EXCEEDS IEEE 519 LIMITS	1. NONLINEAR LOADS ARE INTRODUCING 5TH AND 7TH HARMONIC CURRENTS INTO THE ELECTRICAL POWER SYSTEM	1. CHANGE 6-PULSE UNITS TO 12-PULSE UNITS. 2. PUT APPROXIMATELY ONE HALF OF THE NON-LINEAR LOADS ON PHASE-SHIFTING TRANSFORMERS. 3. ADD 5TH OR 5TH AND 7TH HARMONIC CURRENT FILTERS.
TOTAL HARMONIC VOLTAGE DISTORTION EXCEEDS IEEE 519 LIMITS	1. TOTAL HARMONIC CURRENT DISTORTION EXCEEDS IEEE 519' LIMITS	1. REDUCE HARMONIC CURRENT CONTENT.
	2. IMPEDANCE OF POWER SOURCE IS TOO HIGH.	2. REDUCE THE SOURCE IMPEDANCE BY ADDING POWER GENERATION OR INCREASING INCOMING TRANSFORMER CAPACITY.
HARMONIC CURRENT DISTORTION IS MEASURED ONLY AT LIGHT LOADS	1.THE HARMONIC SOURCES ARE "ON" ONLY DURING TIME PERIODS OF LIGHT LOADS, SUCH AS AT NIGHT.	1. MOVE CAPACITORS OR CHANGE THEIR VALUES BY CAPACITOR SWITCHING.

	2. THE SYSTEM VOLTAGE IS INCREASED AT LIGHT LOAD, WITH THE RESULTING INCREASED HARMONIC CURRENT GENERATION BY TRANSFORMERS ON THE SYSTEM.	2. REDUCE THE SYSTEM VOLTAGE AT LIGHT LOAD BY USING VOLTAGE REGULATORS OR BY DE-ENERGIZING CAPACITORS.
CAPACITORS ARE FAILING NEAR NON-LINEAR LOADS	THE ELECTRICAL POWER SYSTEM IS PARALLEL RESONANT AT A LOW FREQUENCY (3RD, 5TH, 7TH) CAUSING THE PEAK VOLTAGE TO EXCEED CAPACITOR INSULATION CAPABILITIES.	DETUNE THE SYSTEM BY CHANGING CAPACITOR SIZE OR ADDING REACTORS, OR CONVERT THE CAPACITORS TO A HARMONIC FILTER.
CAPACITOR FUSES ARE BLOWING AND HIGH HARMONIC CURRENT EXISTS	1. THE SYSTEM IS RESONANT.	1. DETUNE THE SYSTEM BY ADDING REACTORS IN SERIES WITH THE CAPACITORS.
	2. THE CAPACITOR IS BEING USED AS A FILTER AND THE CAPACITY OF THE HARMONIC SOURCE IS TOO GREAT.	2. INCREASE THE CAPACITOR CAPACITY.
MISOPERATION OF ELECTRONIC CONTROLS DUE TO VOLTAGE DISTORTION	THE VOLTAGE IS DISTORTED IN SUCH A WAY THAT THE ZERO CROSSINGS ARE ABNORMAL, AND MULTIPLE ZERO CROSSINGS OCCUR	DECREASE THE IMPEDANCE OF THE POWER SUPPLY. APPLY A HARMONIC FILTER TO CAPTURE THE HARMONIC CURRENT BEFORE IT CAN CAUSE HARMONIC VOLTAGE DISTORTION.

Figure 3-30 Solve for the cause of given problems that are related to harmonics.

due to parallel resonance. In these cases, additional capacitance or changed inductor values are required to eliminate the unwanted parallel resonance.

As with any series resonant ac circuit, the sum of the effective voltages across each component in the circuit is greater than the source voltage. Accordingly, the voltage impressed across the capacitor is greater than the source voltage. Keeping in mind that capacitance increases as the voltage increase squared, as shown in Fig. 3-17, capacitance increases from nameplate value when capacitors are connected into a filter configuration. All professional computer modeling software accommodates this voltage rise by increasing the capacitive value of the capacitors accordingly.

Harmonics Symptoms, Causes, and Remedies

A summary of symptoms, causes, and remedies relating to harmonic currents and harmonic voltages is given in Fig. 3-30.

NOTES

NOTES

Chapter

4

Conductors

Conductors, Conductor Resistance, Conductor and Cable Impedance, and Voltage Drop

Conductors are the lifeline of every electrical system, and an electrical system only operates as well as the conductors that connect the power source to the loads. In this chapter, calculations are provided to assist in the selection of conductors.

Calculating the One-Way Resistance of a Wire

To calculate either the voltage drop or the heat losses in a conductor, one must first determine the resistance of the conductor. This section provides a method for determining conductor resistance, considering its shape, its length, the material of which it is made, and the temperature at which its resistance is to be determined.

To begin with, Fig. 4-1 shows a table containing physical characteristics and direct-current (dc) resistance at 75°C of American Wire Gauge (AWG) and circular mil conductors, and Fig. 4-2 is a table containing cross-references in wire sizes from AWG to square millimeters, the wire size convention used in International Electrotechnical Commission

SIZE (AWG OR KCMIL)	AREA (CIRCULAR MILS)	STRANDS	DIAMETER (INCHES)	AREA (SQ. IN.)	DC RESISTANCE AT 75 DEG. C COPPER (uncoated) (OHMS/1000 FT.)	DC RESISTANCE AT 75 DEG. C ALUMINUM (OHMS/1000 FT.)
18	1620	1	0.040	0.001	7.77	12.8
18	1620	7	0.046	0.002	7.95	13.1
16	2580	1	0.051	0.002	4.89	8.05
16	2580	7	0.058	0.003	4.99	8.21
14	4110	1	0.064	0.003	3.07	5.06
14	4110	7	0.073	0.004	3.14	5.17
12	6530	1	0.081	0.005	1.93	3.18
12	6530	7	0.092	0.006	1.98	3.25
10	10380	1	0.102	0.008	1.21	2.00
10	10380	7	0.116	0.011	1.24	2.04

8	16510	1	0.128	0.013	0.764	1.26
8	16510	7	0.146	0.017	0.778	1.28
6	26240	7	0.184	0.027	0.491	0.808
4	41740	7	0.232	0.042	0.308	0.508
3	52620	7	0.26	0.053	0.245	0.403
2	66360	7	0.292	0.067	0.194	0.319
1	83690	19	0.332	0.087	0.154	0.253
0	105600	19	0.373	0.109	0.122	0.201
00	133100	19	0.418	0.138	0.0967	0.159
000	167800	19	0.470	0.173	0.0766	0.126
0000	211600	19	0.528	0.219	0.0608	0.100
250	250000	37	0.575	0.260	0.0515	0.0847
300	300000	37	0.630	0.312	0.0429	0.0707
350	350000	37	0.681	0.364	0.0367	0.0605
500	500000	37	0.813	0.519	0.0258	0.0424
750	750000	61	0.998	0.782	0.0171	0.0282
1000	1000000	61	1.152	1.042	0.0129	0.0212

Figure 4-1 Mechanical and electrical characteristics of copper and aluminum wires.

AWG	kcmil	Actual Eq. Metric sq. mm.	Approximate Equivalent sq. mm. (use this column for trade sizes of wire)
18	1.62		
16	2.58		
14	4.11		
12	6.5	3.3	--------
--------	7.7	--------	4
10	10.4	5.3	--------
--------	11.5	--------	6
8	16.5	8.4	--------
--------	19.4	--------	10
6	26.2	13	--------
--------	30.8	--------	16
4	41.7	21	--------
--------	48.9	--------	25
2	66.4	34	--------
--------	67.7	--------	35
1	83.7	42	--------
--------	91.6	--------	50
0	106	53	--------
--------	132	--------	70
00	133	67	--------
000	168	85	--------
--------	184	--------	95
0000	212	107	--------
--------	232	-------	120
--------	250	127	--------
--------	285	--------	150
--------	300	152	--------
--------	350	177	--------
--------	357	--------	185
--------	400	203	--------
--------	469	--------	240
--------	500	253	--------
--------	589	--------	300
--------	600	304	--------
--------	700	355	--------
--------	753	--------	400
--------	800	405	--------
--------	950	--------	500
--------	1000	507	--------

Figure 4-2 Equivalent AWG and square millimeter wires.

(IEC) countries and within Australia. The table shows wire sizes that can be calculated using the methodology shown in Fig. 4-3.

Some conductors, such as rectangular or square tubular bus bars, are not round, so their characteristics are not shown in the table. For all nonstandard shapes, it is necessary to know how to convert from AWG and circular mil wire sizes to square millimeter sizes. The proper methods of converting from square inches to square mils, from square inches to circular mils, or from square inches to square millimeters are shown in Fig. 4-4. For reference, the physical and electrical characteristics of common copper and aluminum bus bars are shown, respectively, in Figs. 4-5 and 4-6.

Sometimes the conductor with which one is dealing is not copper or aluminum, and sometimes its size or shape is very unusual. In such cases, actual calculation of the resistance of the conductor must be done. Begin by considering the specific resistance of the conductor material, which is usually given in terms of resistivity, using the symbol ρ (Greek lower-case rho) for ohm-meters. Figure 4-7 shows the resistivity of some of the more common electrical conductor materials, such as silver, copper, aluminum, tungsten, nickel, and iron. Figure 4-8 shows how to calculate the resistance of a conductor that is made of a noncopper material.

Temperature also has an effect on the electrical resistance of conductors. When their operating temperature will be different from 20°C (on which the table in Fig. 4-7 is based), a further calculation is required to determine the resistance at the operating temperature. This calculation is shown in Fig. 4-9, and this figure also contains values for the variables required for each conductor material in this calculation.

There is another factor that affects the apparent resistance of a conductor and is most notable in the resistance of a round conductor such as a wire. When electric current flows through a wire, lines of flux form beginning at the center of the wire and extending out to infinity, but most of the lines of magnetic flux are concentrated at the center of

Problem: What is the equivalent square millimeter wire size to a #1 AWG wire?

Incorrect method due to the larger wire O.D. that results from stranding:

Step #1: Determine wire O.D. in inches.

From the Wire Characeristic table of Fig. 4-1,
O.D. of #1 AWG wire is 0.332 inches.

Step #2: Determine wire O.D. in mils.

mils = 0.332 X 1000
mils = 332

Step #3: Determine wire area in circular mils.

c.m. = $(\text{mils})^2$
c.m. = $(332)^2$
c.m. = 110224 circular mils

Step #4: Change circular mils to square millimeters.

sq. mm. = (c. m.) X .00050671
sq. mm. = (110224) X .00050671
sq. mm. = 55.85 sq. mm.
This is incorrect!

Correct method:

Step #1: Determine wire area in circular mils.

From the Wire Characeristic table of Fig. 4-1,
c.m. = 83690 circular mils

Step #2: Change circular mils to square millimeters.

sq. mm. = (c. m.) X .00050671
sq. mm. = (83690) X .00050671
sq. mm. = 42.40 sq. mm.
This is correct!

Figure 4-3 Solve for the equivalent square millimeter wire size given AWG size.

Problem: Solve for the equivalent area in square inches, in square mils, in circular mils, and in square millimeters of a 1/4" X 4" copper bus bar.

Step #1: Determine conductor area in square inches.

Area = length X width
Area = 4 in. X 0.25 in.
Area = 1 sq. in.

Step #2: Determine conductor area in square mils.

The bus bar measures 250 thousandths by 4000 thousandths, or 250 mils by 4000 mils.

sq. mils = (length in mils) X (width in mils)
sq. mils = (250) X (4000)
sq. mils = 1000000 square mils

Step #3: Determine conductor area in circular mils.

$$\text{c.m.} = \frac{\text{square mils}}{0.7854}$$

$$\text{c.m.} = \frac{1000000}{0.7854}$$

c.m. = 1273236 circular mils

Step #4: Change circular mils to square millimeters.

sq. mm. = (c. m.) X .00050671
sq. mm. = (1273236) X .00050671
sq. mm. = 645.16 sq. mm.

Figure 4-4 Solve for square inches, square mils, circular mils, and square millimeters given bus bar dimensions.

the wire. As the frequency of the alternating current increases, the amount of flux increases even more. With the center of the wire essentially "occupied" with concentrated lines of magnetic flux, electron flow is impeded there to the extent that most of the electron flow in wires carrying large amounts of current is along the surface of the wire. This type of electron flow is known as *skin effect,* and its inclusion into the resistance value of a wire is said to change the

CHARACTERISTICS OF COMMON RECTANGULAR BUS BARS
98% CONDUCTIVITY COPPER BUS BARS

SIZE	AMPACITY	AREA SQ. IN.	AREA CIRC. MILS	WEIGHT LB./FT.	D.C. RESISTANCE MICRO-OHMS PER FT.
1/8" X 1"	247	0.125	159200	0.485	66.01
1/8" X 2"	447	0.25	318300	0.97	33.04
1/8" X 3"	696	0.375	477500	1.45	22.02
1/8" X 4"	900	0.5	636600	1.94	16.52
1/4" X 1"	366	0.25	318300	0.97	33.04
1/4" X 2"	647	0.5	636600	1.94	16.52
1/4" X 3"	973	0.75	955000	2.91	11.01
1/4" X 4"	1220	1	1273000	3.88	8.25
1/4" X 6"	1660	1.5	1910000	5.81	5.51
	@30 DEG C AMBIENT, WITH A 30 DEG C				

Figure 4-5 Characteristics of copper bus bars.

CHARACTERISTICS OF COMMON RECTANGULAR BUS BARS
61 % CONDUCTIVITY ALUMINUM BUS BARS

SIZE	AMPACITY	AREA SQ. IN.	AREA CIRC. MILS	WEIGHT LB./FT.	D.C. RESISTANCE MICRO-OHMS PER FT.
1/4" X 1"	310	0.25	318300	0.294	57.16
1/4" X 2"	550	0.5	636600	0.588	25.58
1/4" X 3"	775	0.75	955000	0.882	19.05
1/4" X 4"	990	1	1273000	1.176	14.29
1/4" X 6"	1400	1.5	1910000	1.764	9.527
	@40 DEG C AMBIENT, WITH A 30 DEG C RISE				

Figure 4-6 Characteristics of aluminum bus bars.

resistance value into the alternating-current (ac) resistance value of the wire. The ac resistance of the wire is increased when the wire is enclosed within a raceway that itself concentrates magnetic flux, such as rigid steel conduit. Note that the ac resistance value of a wire still does not include the inductive reactance or the capacitive reactance components of the wire impedance. For convenience, Fig. 4-10 is

Bus bars must be braced for short-circuit current.

CONDUCTOR MATERIAL	RESISTIVITY (OHM-METERS @ 20°C) $\times 10^{-8}$
Silver	1.64
Copper	1.72
Aluminum	2.83
Tungsten	5.5
Nickel	7.8
Iron	12.0

Figure 4-7 Resistivity of common electrical conductors.

a table that shows ac wire resistance and impedance in a 60-Hz system operating at 75°C.

Calculating the Impedance of a Cable

The impedance of a cable or a set of conductors is the vector sum of

$$R + jX_L - jX_C$$

where R is the conductor resistance calculated in the last section, and X_L is the inductive reactance of the cable. The

Problem: Determine the resistance at 20 Deg. C of 656 ft of 4 sq. mm. aluminum wire.

Step #1: Convert feet to meters:

$$\text{meters} = \frac{\text{feet}}{3.28}$$

$$\text{meters} = \frac{656}{3.28}$$

$$\text{meters} = 200 \text{ meters}$$

Step #2: Convert sq. mm. to sq. meters:

$$\frac{1 \text{ sq. meter}}{x \text{ sq. meter}} = \frac{1000000 \text{ sq. mm.}}{4 \text{ sq. mm.}}$$

$$\frac{(1 \text{ sq. meter})(4 \text{ sq. mm.})}{1000000 \text{ sq. mm.}} = x \text{ sq. meter}$$

$$0.000004 = x \text{ sq. meter}$$

$$4.0 \times 10^{-6} = x \text{ sq. meter}$$

Step #3: Calculate resistance

$$\text{Resistance} = \frac{\rho \text{ X length in meters}}{\text{Area in square meters}}$$

$$\text{Resistance} = \frac{(2.83 \times 10^{-8}) \text{ X } (200)}{4.0 \times 10^{-6}}$$

$$\text{Resistance} = 142 \times 10^{-2}$$

$$\text{Resistance} = \underline{\underline{1.42}} \text{ ohms}$$

Figure 4-8 Solve for the resistance of a conductor given resistivity, cross-sectional area, and length.

Problem: Determine the resistance at 100 Deg. C of a copper wire that measures 12 ohms at 20 Deg. C.

Let R20 be the conductor resistance at 20 Deg. C.
Let R100 be the conductor resistance at 100 Deg. C.
Let T20 be the Temperator of 20 Deg. C.
Let T100 be the temperature at 100 Deg. C.

VALUES OF x FOR COMMON CONDUCTORS	
CONDUCTOR MATERIAL	x (Deg. C)
Copper	234.5
Aluminum	236
Silver	243

$$\frac{R100}{R20} = \frac{x + T100}{x + T20}$$

$$\frac{R100}{12} = \frac{(234.5) + 100}{(234.5) + 20}$$

$$\frac{R100}{12} = \frac{334.5}{254.5}$$

$$\frac{(12) \times (334.5)}{254.5} = R100$$

$$\underline{\underline{15.772}} = R100$$

The wire increases in resistance from 12 ohms to 15.772 ohms when the temperature rises from 20 Deg. C to 100 Deg. C.

Figure 4-9 Solve for the resistance of a conductor at a temperature other than 20°C.

AC RESISTANCE AND IMPEDANCE OF COPPER AND ALUMINUM WIRES

(3 WIRES IN A CONDUIT)

(TABLE UNITS ARE IN OHMS PER 1000 FEET)

WIRE SIZE (AWG)	WIRE SIZE (KCMIL)	INDUCTIVE REACTANCE			AC RESISTANCE - COPPER WIRE		
		PVC CONDUIT	ALUMINUM CONDUIT	STEEL CONDUIT	PVC CONDUIT	ALUMINUM CONDUIT	STEEL CONDUIT
14	4.11	0.058	0.058	0.073	3.1	3.1	3.1
12	6.53	0.054	0.054	0.068	2	2	2
10	10.38	0.05	0.05	0.063	1.2	1.2	1.2
8	16.51	0.052	0.052	0.065	0.78	0.78	0.78
6	26.24	0.051	0.051	0.064	0.49	0.49	0.49
4	41.74	0.048	0.048	0.06	0.31	0.31	0.31
3	52.62	0.047	0.047	0.059	0.25	0.025	0.25
2	66.36	0.045	0.045	0.057	0.19	0.2	0.2
1	83.69	0.046	0.046	0.057	0.15	0.16	0.16
0	105.6	0.044	0.044	0.055	0.12	0.13	0.12
00	133.1	0.043	0.043	0.054	0.1	0.1	0.1
000	167.8	0.042	0.042	0.052	0.077	0.082	0.079
0000	211.6	0.041	0.041	0.051	0.062	0.067	0.063
N.A.	250	0.041	0.041	0.052	0.052	0.057	0.054
N.A.	300	0.041	0.041	0.051	0.044	0.049	0.045
N.A.	350	0.04	0.04	0.05	0.038	0.043	0.039
N.A.	500	0.039	0.039	0.048	0.027	0.032	0.029

NOTE 1 IMPEDANCE IS R(COS POWER FACTOR ANGLE) + X SIN (POWER FACTOR ANGLE); THIS TABLE ASSUMES THE POWER FACTOR IS 85%, LAGGING.

NOTE 2 WIRE SIZES LARGER THAN #8 ARE STRANDED

Figure 4-10 ac resistance and impedance values of 600-V copper and aluminum wire in conduit.

inductive reactance of a cable is directly related to the magnetic flux coupling of one wire in the cable with the other wires in the cable. If the wires are very close together, the fluxes from each conductor add to zero in three-dimensional space, and the inductive reactance of the cable is very small. If, however, the wires are spaced far apart, the lines of flux do not cancel so readily, and the inductive reactance of the

AC RESISTANCE-ALUMINUM WIRE			IMPEDANCE-COPPER WIRE			IMPEDANCE-ALUMINUM WIRE		
PVC CONDUIT	ALUMINUM CONDUIT	STEEL CONDUIT	PVC CONDUIT (NOTE 1)	ALUMINUM CONDUIT (NOTE 1)	STEEL CONDUIT (NOTE 1)	PVC CONDUIT (NOTE 1)	ALUMINUM CONDUIT (NOTE 1)	STEEL CONDUIT (NOTE 1)
N.A.	N.A.	N.A.	2.67	2.67	2.67	N.A.	N.A.	N.A.
3.2	3.2	3.2	1.7	1.7	1.7	2.8	2.8	2.8
2	2	2	1.1	1.1	1.1	1.8	1.8	1.8
1.3	1.3	1.3	0.69	0.69	0.7	1.1	1.1	1.1
0.81	0.81	0.81	0.44	0.45	0.45	0.71	0.72	0.72
0.51	0.51	0.51	0.29	0.29	0.3	0.46	0.46	0.46
0.4	0.41	0.4	0.23	0.24	0.24	0.37	0.37	0.37
0.32	0.32	0.32	0.19	0.19	0.2	0.3	0.3	0.3
0.25	0.26	0.25	0.16	0.16	0.16	0.24	0.24	0.25
0.2	0.21	0.2	0.13	0.13	0.13	0.19	0.2	0.2
0.16	0.16	0.16	0.11	0.11	0.11	0.16	0.16	0.16
0.13	0.13	0.13	0.088	0.092	0.094	0.13	0.13	0.14
0.1	0.11	0.1	0.074	0.078	0.08	0.11	0.11	0.11
0.085	0.09	0.086	0.066	0.07	0.073	0.094	0.098	0.1
0.071	0.76	0.072	0.059	0.063	0.065	0.082	0.086	0.088
0.061	0.66	0.063	0.053	0.058	0.06	0.073	0.077	0.08
0.043	0.048	0.045	0.043	0.048	0.05	0.057	0.061	0.064
0.029	0.034	0.031	0.036	0.04	0.043	0.045	0.049	0.052

circuit is increased. Figure 4-11 can be used to approximate the inductive reactance of a cable. Values from this figure are directly applicable to nonarmored cable, cable with non-magnetic armor, and cable in a nonmagnetic raceway. For cable with magnetic armor or cable drawn into a magnetic raceway, correction factors found in the same figure must be applied.

Bus duct is used to carry large values of current.

Service conductors to commercial building in the form of bus duct.

An individual can calculate the impedance of a cable, but this work has already been done by cable manufacturers for many specific types of cable. For convenience, typical cable impedance values are shown in Fig. 4-12 for 600-volt (V) cable, for 5-kV cable, and for 15-kV cable.

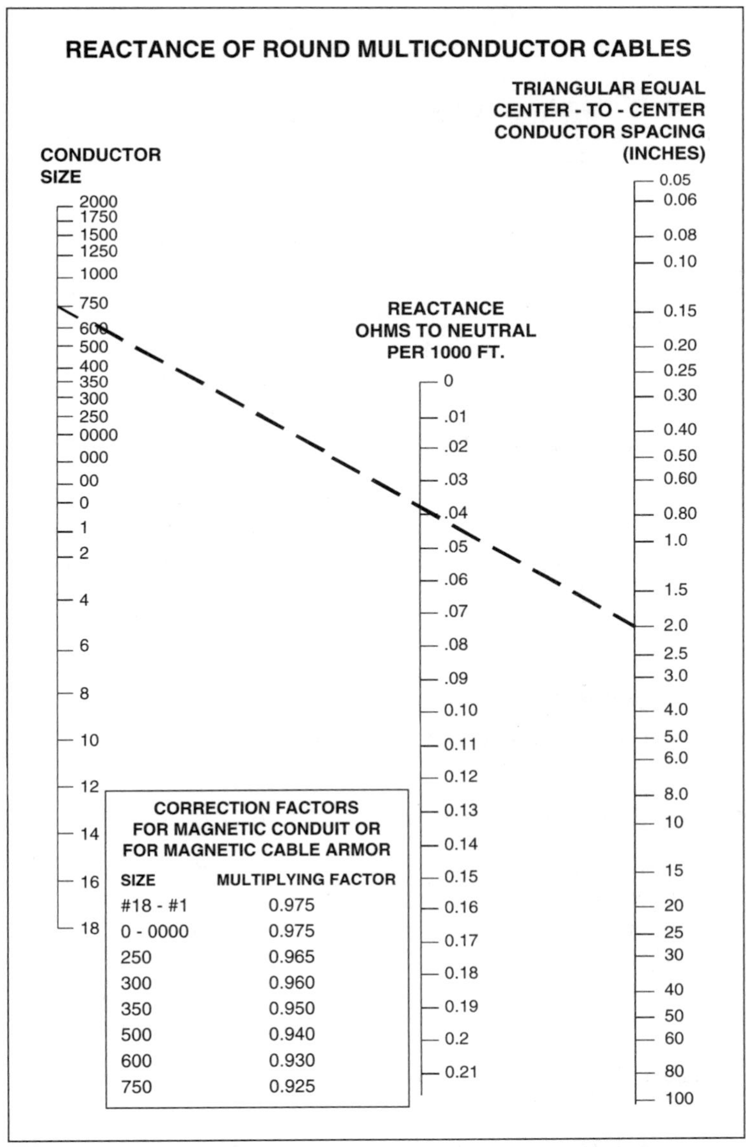

SIZE	MULTIPLYING FACTOR
#18 - #1	0.975
0 - 0000	0.975
250	0.965
300	0.960
350	0.950
500	0.940
600	0.930
750	0.925

Figure 4-11 Solve for inductive reactance of a cable given conductor size and dimensions.

IMPEDANCE OF COPPER CABLE

(3/CONDUCTOR CABLE)

TABLE UNITS ARE IN OHMS TO NEUTRAL PER 1000 FEET

WIRE SIZE (AWG)	WIRE SIZE (KCMIL)	600 VOLT	5kV	15kV
14	4.11	0.0369	N.A.	N.A.
12	6.53	0.0347	N.A.	N.A.
10	10.38	0.0321	N.A.	N.A.
8	16.51	0.0334	0.0537	N.A.
6	26.24	0.034	0.0492	N.A.
4	41.74	0.0321	0.0457	N.A.
3	52.62	0.0307	0.0441	N.A.
2	66.36	0.0298	0.0424	0.0518
1	83.69	0.0309	0.0413	0.0297
0	105.6	0.0305	0.0392	0.0497
00	133.1	0.0294	0.039	0.0462
000	167.8	0.0287	0.0379	0.0445
0000	211.6	0.0281	0.0367	0.0427
N.A.	250	0.0287	0.0358	0.0441
N.A.	300	0.028	0.0349	0.0399
N.A.	350	0.0276	0.034	0.0394
N.A.	500	0.0268	0.0322	0.0368
N.A.	750	0.0265	0.0306	0.0367

Note: All conductors are stranded

Figure 4-12 Impedances of 600-V, 5-kV, and 15-kV copper cable.

Calculating Voltage Drop in a Cable

There are several valuable individual voltage-drop calculations that commonly encountered when working with electrical systems. It is important that the calculations be done correctly and that the engineer can check the solution to a current problem against the answer to an example problem. For these reasons, the following figures are provided as examples:

Figure 4-13: Calculate voltage drop in a single-phase dc circuit.

Figure 4-14: Calculate the approximate voltage drop in a single-phase ac circuit at unity power factor in a plastic conduit.

Problem: A solid #14 copper two-wire cable that is 750 feet long supplies a 125 ohm load resistor for heating purposes at 75 deg. C. Find the voltage drop in the cable; and find the resulting voltage across the load resistor.

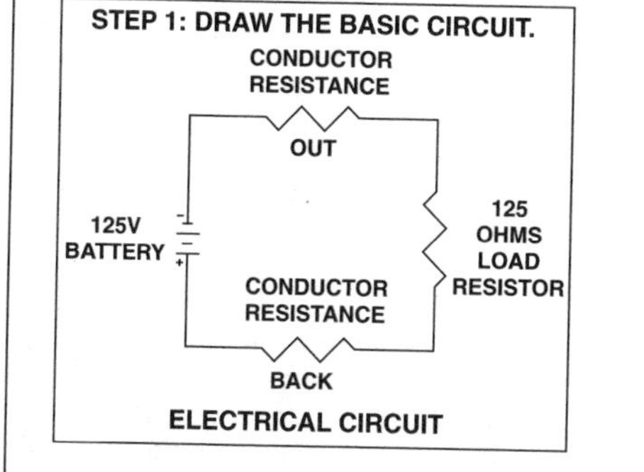

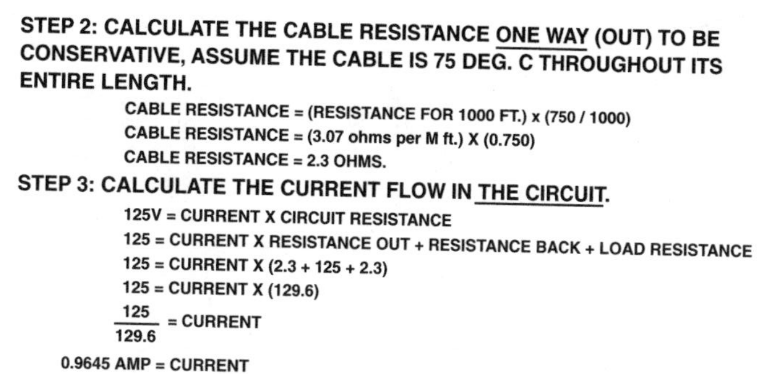

STEP 2: CALCULATE THE CABLE RESISTANCE ONE WAY (OUT) TO BE CONSERVATIVE, ASSUME THE CABLE IS 75 DEG. C THROUGHOUT ITS ENTIRE LENGTH.

CABLE RESISTANCE = (RESISTANCE FOR 1000 FT.) x (750 / 1000)
CABLE RESISTANCE = (3.07 ohms per M ft.) X (0.750)
CABLE RESISTANCE = 2.3 OHMS.

STEP 3: CALCULATE THE CURRENT FLOW IN THE CIRCUIT.

125V = CURRENT X CIRCUIT RESISTANCE
125 = CURRENT X RESISTANCE OUT + RESISTANCE BACK + LOAD RESISTANCE
125 = CURRENT X (2.3 + 125 + 2.3)
125 = CURRENT X (129.6)

$$\frac{125}{129.6} = \text{CURRENT}$$

0.9645 AMP = CURRENT

STEP 4: CALCULATE THE VOLTAGE DROP IN THE CABLE ONE WAY (OUT).

VOLTAGE DROP IN CABLE = CURRENT X CABLE RESISTANCE
VOLTAGE DROP IN CABLE = (0.9645) X (2.3)
VOLTAGE DROP IN CABLE = 2.218 VOLTS ONE WAY

STEP 5: CALCULATE THE VOLTAGE DROP IN THE CABLE BOTH OUT AND BACK.

VOLTAGE DROP IN CABLE = (VOLTAGE DROP 1 WAY) x (2)
VOLTAGE DROP IN CABLE = (2.218) X (2)
VOLTAGE DROP IN CABLE = 4.436 VOLTS

STEP 4: CALCULATE THE VOLTAGE DROPPED ACROSS THE LOAD.

VOLTAGE DROP ACROSS LOAD = (SOURCE VOLTAGE) - (CABLE VOLTAGE DROP)
VOLTAGE DROP ACROSS LOAD = (125) - (4.436)
VOLTAGE DROP ACROSS LOAD = 120.56 VOLTS

Figure 4-13 Solve for voltage drop in a dc circuit given wire size, temperature, and load characteristics.

Problem:
From a 115VAC circuit breaker, a solid #12 copper two-wire cable in plastic conduit that is 750 feet long supplies a motor load that requires 1.3 kW. Find the approximate voltage drop in the cable; and find the resulting voltage supplied to the motor load (ignoring constant kVA motor action).

STEP 1: DRAW THE BASIC CIRCUIT.

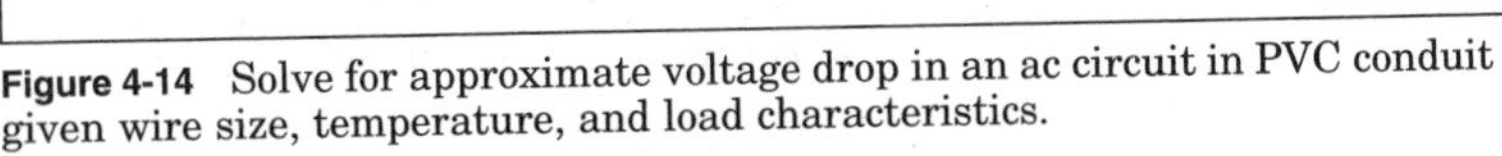

STEP 2: CALCULATE THE CABLE RESISTANCE ONE WAY (OUT)
ASSUME THE CABLE WILL OPERATE AT 75 DEG. C THROUGHOUT ITS ENTIRE LENGTH.
CABLE IMPEDANCE = (IMPEDANCE FOR 1000 FT.) x (750 / 1000)
CABLE IMPEDANCE= (1.7 OHMS PER M FT.) X (0.750)
CABLE IMPEDANCE = 1.275 OHMS.

STEP 3: CALCULATE THE CURRENT FLOW IN THE CIRCUIT.
POWER = VOLTAGE X CURRENT
THE MOTOR DRAWS 1.3 kW, THEREFORE, FOR A CLOSE APPROXIMATION OF LINE CURRENT: 1300 = (115 VOLTS) x (CURRENT) (ignoring power factor)

$$\frac{1300}{115} = \text{CURRENT}$$

11.3 AMP = CURRENT

STEP 4: CALCULATE THE VOLTAGE DROP IN THE CABLE ONE WAY (OUT).
VOLTAGE DROP IN CABLE = CURRENT X CABLE IMPEDANCE
VOLTAGE DROP IN CABLE = (11.3) X (1.275)
VOLTAGE DROP IN CABLE = 14.41 VOLTS ONE WAY

STEP 5: CALCULATE THE VOLTAGE DROP IN THE CABLE BOTH OUT AND BACK.
VOLTAGE DROP IN CABLE = (VOLTAGE DROP 1 WAY) x (2)
VOLTAGE DROP IN CABLE = (14.41) X (2)
VOLTAGE DROP IN CABLE = 28.82 VOLTS

STEP 6: CALCULATE THE VOLTAGE DROPPED ACROSS THE LOAD.
VOLTAGE DROP ACROSS LOAD = (SOURCE VOLTAGE) - (CABLE VOLTAGE DROP)
VOLTAGE DROP ACROSS LOAD = (115) - (28.82)
VOLTAGE DROP ACROSS LOAD = 86.18 VOLTS

Figure 4-14 Solve for approximate voltage drop in an ac circuit in PVC conduit given wire size, temperature, and load characteristics.

Figure 4-15: Calculate a more accurate voltage drop in a single-phase ac circuit at unity power factor in a nonmagnetic conduit.

Figure 4-16: Calculate voltage drop in a single-phase ac circuit at unity power factor in a steel conduit to an induction furnace load. Note that the furnace still tries to draw 21 kilowatts (kW); thus the line current must increase to offset the cable voltage drop.

Figure 4-17: Calculate voltage drop in a single-phase ac circuit at unity power factor in an aluminum conduit.

Figure 4-18: Calculate voltage drop in a three-phase ac circuit at less than unity power factor using unarmored type TC cable.

Figure 4-19: Calculate voltage drop in a three-phase ac circuit at less than unity power factor using armored type MC cable.

Figure 4-20: Calculate voltage regulation in an ac circuit.

Calculating dc Resistance in a Bus Bar

Calculating the resistance of a bus bar is similar to calculating the resistance of a wire, except that the cross-sectional area is of a different shape and is larger. This calculation is shown in Fig. 4-21. (Refer to Fig. 4-5 for data.)

Calculating Heat Loss in a Conductor

Heat loss is always calculated simply as I^2R, so the solution to a heat-loss problem in a circuit is to calculate the dc resistance (even if it is in an ac circuit) of each wire, solve for the current flow through the wire (even if the current flow is out of phase with the voltage), calculate the I^2R heat loss in each wire individually, and then simply add the heat losses in the wires. This methodology is shown in Fig. 4-22.

Wires and Cables

Most of the electrical systems in the world use wires and cables to transport electrical energy, so it is important to

(*Text continues on p. 152.*)

Problem: From a 115VAC circuit breaker, a solid #12 copper two-wire cable that is 560 feet long supplies a motor load that requires 1.3 kW. Find the actual voltage drop in the cable; and find the resulting actual voltage supplied to the motor load (considering constant kVA motor action).

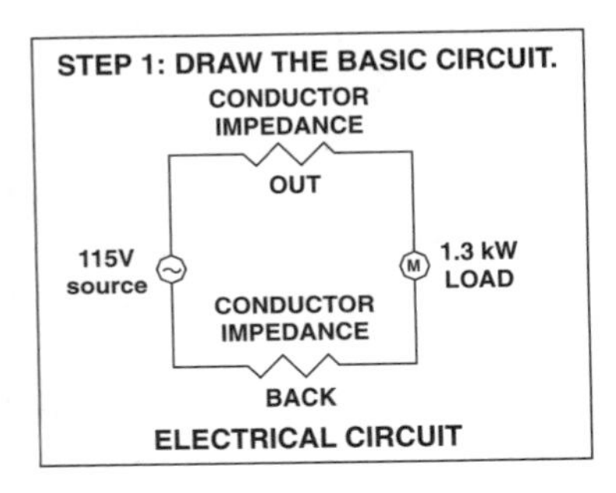

STEP 2: CALCULATE THE CABLE RESISTANCE ONE WAY (OUT)
ASSUME THE CABLE WILL OPERATE AT 75 DEG. C THROUGHOUT ITS ENTIRE LENGTH.
CABLE IMPEDANCE = (IMPEDANCE FOR 1000 FT.) x (560 / 1000)
CABLE IMPEDANCE= (1.7 OHMS PER M FT.) X (0.750)
CABLE IMPEDANCE = 1.275 OHMS.

STEP 3: CALCULATE THE APROXIMATE INITIAL CURRENT FLOW IN THE CIRCUIT.
POWER = VOLTAGE X CURRENT
THE MOTOR DRAWS 1.3 kW, THEREFORE, FOR A CLOSE APPROXIMATION OF LINE CURRENT: 1300 = (115 VOLTS) x (CURRENT)

$$\frac{1300}{115} = \text{CURRENT}$$

11.3 AMP = CURRENT

STEP 4: CALCULATE THE VOLTAGE DROP IN THE CABLE ONE WAY (OUT).
VOLTAGE DROP IN CABLE = CURRENT X CABLE IMPEDANCE
VOLTAGE DROP IN CABLE = (11.3) X (1.275)
VOLTAGE DROP IN CABLE = 14.41 VOLTS ONE WAY

STEP 5: CALCULATE THE VOLTAGE DROP IN THE CABLE BOTH OUT AND BACK.
VOLTAGE DROP IN CABLE = (VOLTAGE DROP 1 WAY) x (2)
VOLTAGE DROP IN CABLE = (14.41) X (2)
VOLTAGE DROP IN CABLE = 28.82 VOLTS

STEP 6: CALCULATE THE VOLTAGE DROPPED ACROSS THE LOAD.
VOLTAGE DROP ACROSS LOAD = (SOURCE VOLTAGE) – (CABLE VOLTAGE DROP)
VOLTAGE DROP ACROSS LOAD = (115) - (28.82)
VOLTAGE DROP ACROSS LOAD = 86.18 VOLTS

Figure 4-15 Solve for accurate voltage drop in ac circuit to motor in nonmagnetic conduit.

STEP 7: CALCULATE THE ACTUAL LINE CURRENT DRAWN BY THE MOTOR AT ITS LOW TERMINAL VOLTAGE.

POWER = VOLTAGE X CURRENT

THE MOTOR DRAWS 1.3 kW, THEREFORE, FOR A CLOSER VALUE OF ACTUAL LINE CURRENT: 1300 = (86.18 VOLTS) x (CURRENT)

$$\frac{1300}{86.18} = \text{CURRENT}$$

15.08 AMP = CURRENT

STEP 8: CALCULATE THE VOLTAGE DROP IN THE CABLE <u>ONE WAY</u> (OUT).

VOLTAGE DROP IN CABLE = CURRENT X CABLE IMPEDANCE

VOLTAGE DROP IN CABLE = (15.08) X (1.275)

VOLTAGE DROP IN CABLE = 19.23 VOLTS ONE WAY

STEP 9: CALCULATE THE VOLTAGE DROP IN THE CABLE BOTH OUT AND BACK.

VOLTAGE DROP IN CABLE = (VOLTAGE DROP 1 WAY) x (2)

VOLTAGE DROP IN CABLE = (19.231) X (2)

VOLTAGE DROP IN CABLE = <u>38.46 VOLTS</u>

STEP 10: CALCULATE THE VOLTAGE DROPPED ACROSS THE LOAD.

VOLTAGE DROP ACROSS LOAD = (SOURCE VOLTAGE) – (CABLE VOLTAGE DROP)

VOLTAGE DROP ACROSS LOAD = (115) - (38.46)

VOLTAGE DROP ACROSS LOAD = <u>76.53 VOLTS</u>

STEP 11: CALCULATE THE ACTUAL LINE CURRENT DRAWN BY THE MOTOR AT ITS LOW TERMINAL VOLTAGE.

POWER = VOLTAGE X CURRENT

THE MOTOR DRAWS 1.3 kW, THEREFORE, FOR A CLOSER VALUE OF ACTUAL LINE CURRENT: 1300 = (76.53 VOLTS) x (CURRENT)

$$\frac{1300}{76.53} = \text{CURRENT}$$

16.98 AMP = CURRENT

STEP 12: CALCULATE THE VOLTAGE DROP IN THE CABLE <u>ONE WAY</u> (OUT).

VOLTAGE DROP IN CABLE = CURRENT X CABLE IMPEDANCE

VOLTAGE DROP IN CABLE = (16.98) X (1.275)

VOLTAGE DROP IN CABLE = 21.66 VOLTS ONE WAY

STEP 13: CALCULATE THE VOLTAGE DROP IN THE CABLE BOTH OUT AND BACK.

VOLTAGE DROP IN CABLE = (VOLTAGE DROP 1 WAY) x (2)

VOLTAGE DROP IN CABLE = (21.66) X (2)

VOLTAGE DROP IN CABLE = <u>43.32 VOLTS</u>

STEP 14: CALCULATE THE VOLTAGE DROPPED ACROSS THE LOAD.

VOLTAGE DROP ACROSS LOAD = (SOURCE VOLTAGE) - (CABLE VOLTAGE DROP)

VOLTAGE DROP ACROSS LOAD = (115) - (43.32)

VOLTAGE DROP ACROSS LOAD = <u>71.68 VOLTS</u>

COMMENT ON FINAL SOLUTION: THREE ITERATIONS WILL NORMALLY SUFFICE FOR THIS TYPE OF CALCULATION. THE RESULTS SHOW THAT A LARGER WIRE SIZE SHOULD BE SELECTED AND THE CALCULATION DONE AGAIN.

Problem:
From a 480VAC circuit breaker, a solid #8 copper two-wire cable that is 230 feet long supplies an induction furnace load that requires 21 kW. Find the approximate voltage drop in the cable; and find the resulting approximate voltage supplied to the induction furnace (ignoring current harmonics).

STEP 1: DRAW THE BASIC CIRCUIT.

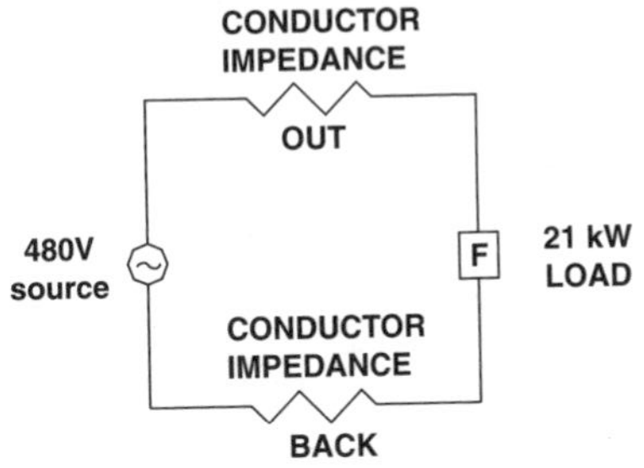

STEP 2: CALCULATE THE CABLE RESISTANCE ONE WAY (OUT)

ASSUME THE CABLE WILL OPERATE AT 75 DEG. C THROUGHOUT ITS ENTIRE LENGTH.
CABLE IMPEDANCE = (IMPEDANCE FOR 1000 FT.) x (230 / 1000)
CABLE IMPEDANCE= (0.7 OHMS PER M FT.) X (0.230)
CABLE IMPEDANCE = 0.161 OHMS.

STEP 3: CALCULATE THE CURRENT FLOW IN THE CIRCUIT.

POWER = VOLTAGE X CURRENT
THE FURNACE DRAWS 21kW, THEREFORE, LINE CURRENT IS:
21000 = (480 VOLTS) x (CURRENT)

$$\frac{21000}{480} = \text{CURRENT}$$

43.75 AMP = CURRENT

STEP 4: CALCULATE THE VOLTAGE DROP IN THE CABLE ONE WAY (OUT).

VOLTAGE DROP IN CABLE = CURRENT X CABLE IMPEDANCE
VOLTAGE DROP IN CABLE = (43.75) X (0.161)
VOLTAGE DROP IN CABLE = 7.043 VOLTS ONE WAY

STEP 5: CALCULATE THE VOLTAGE DROP IN THE CABLE BOTH OUT AND BACK.

VOLTAGE DROP IN CABLE = (VOLTAGE DROP 1 WAY) x (2)
VOLTAGE DROP IN CABLE = (7.043) X (2)
VOLTAGE DROP IN CABLE = 14.08 VOLTS

STEP 6: CALCULATE THE VOLTAGE DROPPED ACROSS THE LOAD.

VOLTAGE DROP ACROSS LOAD = (SOURCE VOLTAGE) - (CABLE VOLTAGE DROP)
VOLTAGE DROP ACROSS LOAD = (480) - (14.08)
VOLTAGE DROP ACROSS LOAD = 465.91 VOLTS

Figure 4-16 Solve for approximate voltage drop in an ac circuit in magnetic conduit given wire size, temperature, and load characteristics.

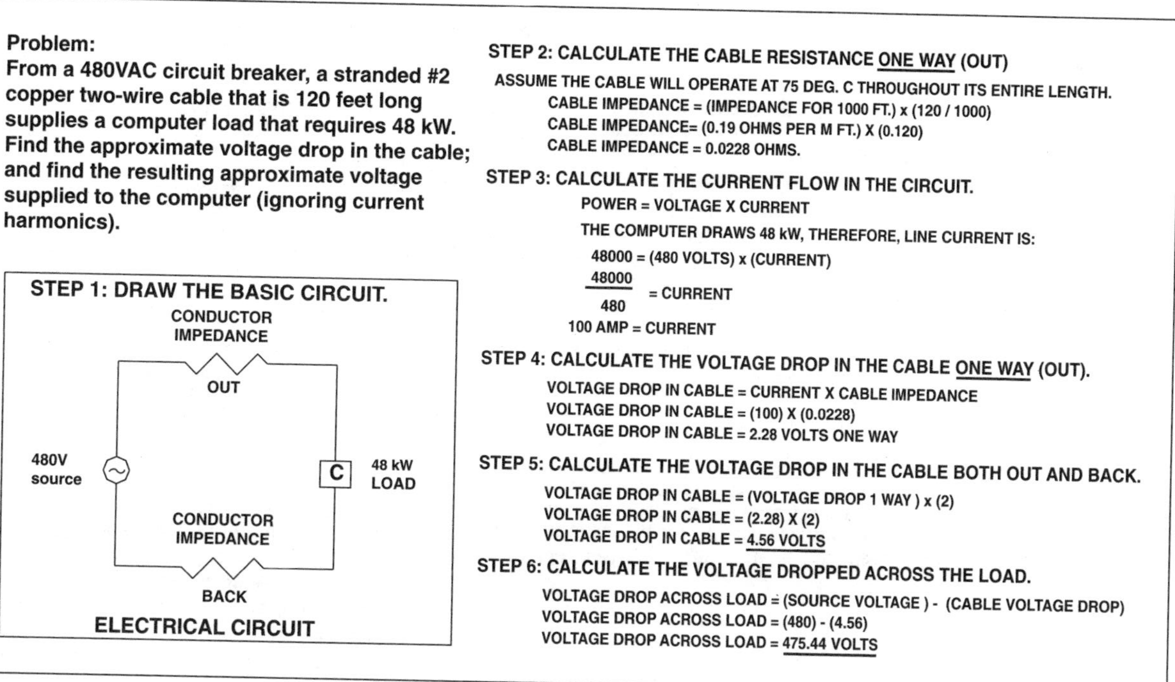

Figure 4-17 Solve for voltge drop in a single-phase ac circuit at unity PF in an aluminum conduit.

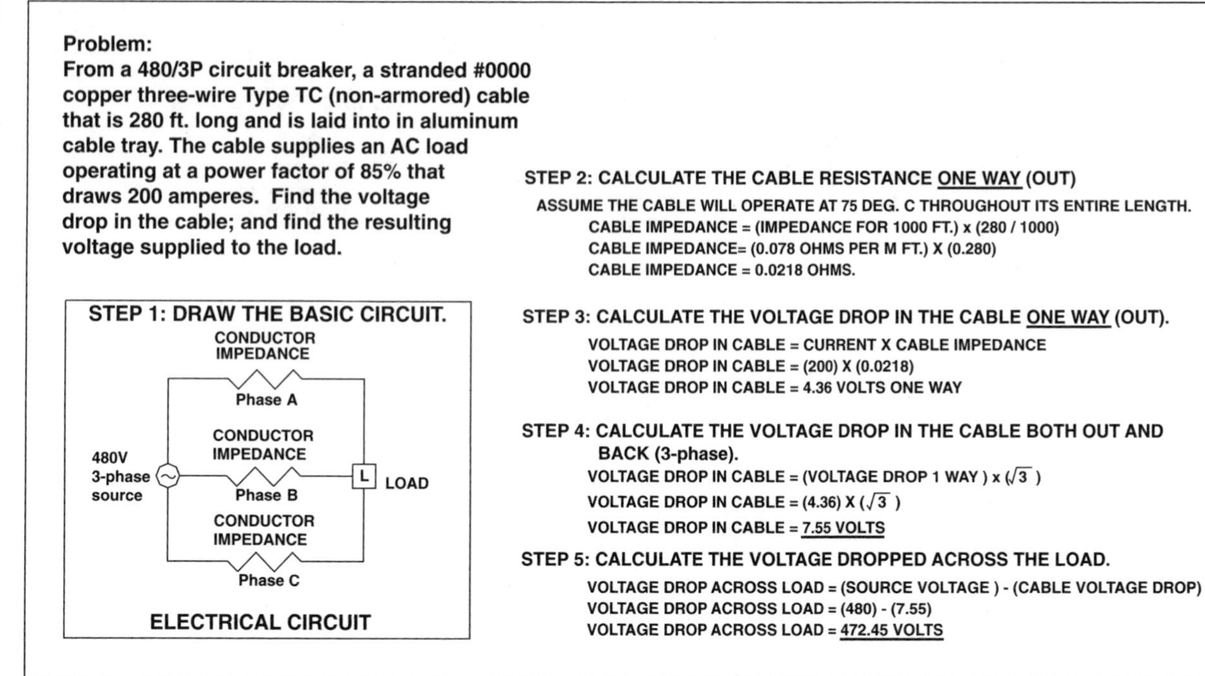

Figure 4-18 Solve for voltage drop in a three-phase ac circuit in nonarmored cable in cable tray.

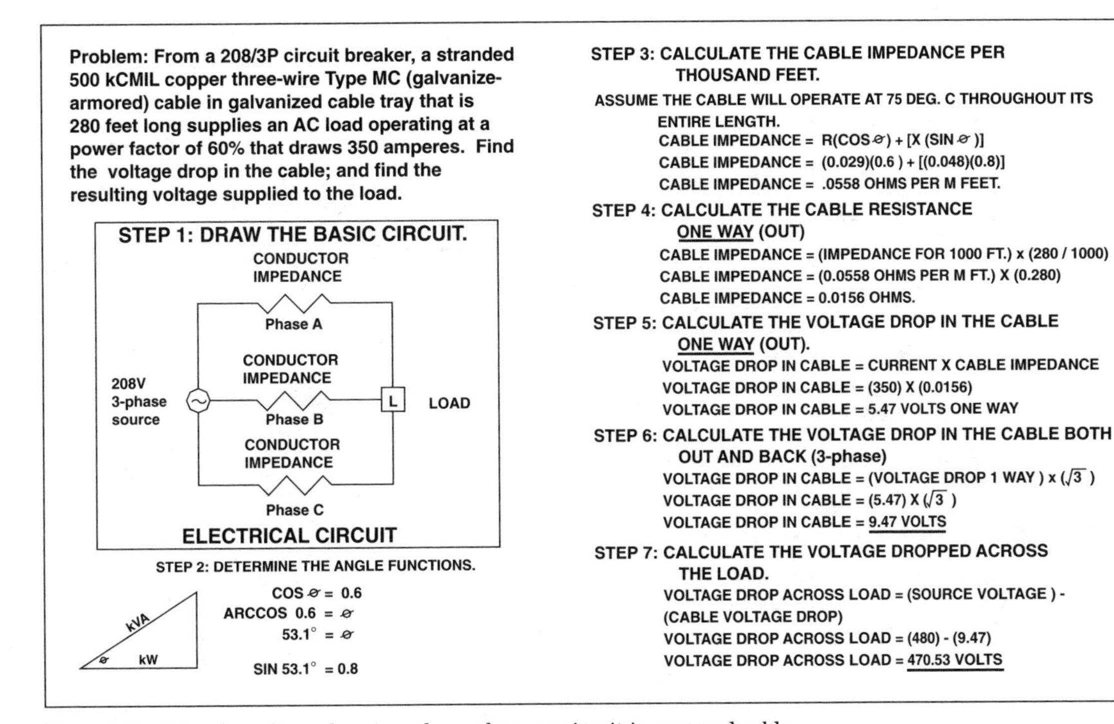

Figure 4-19 Solve for voltage drop in a three-phase ac circuit in armored cable in cable tray.

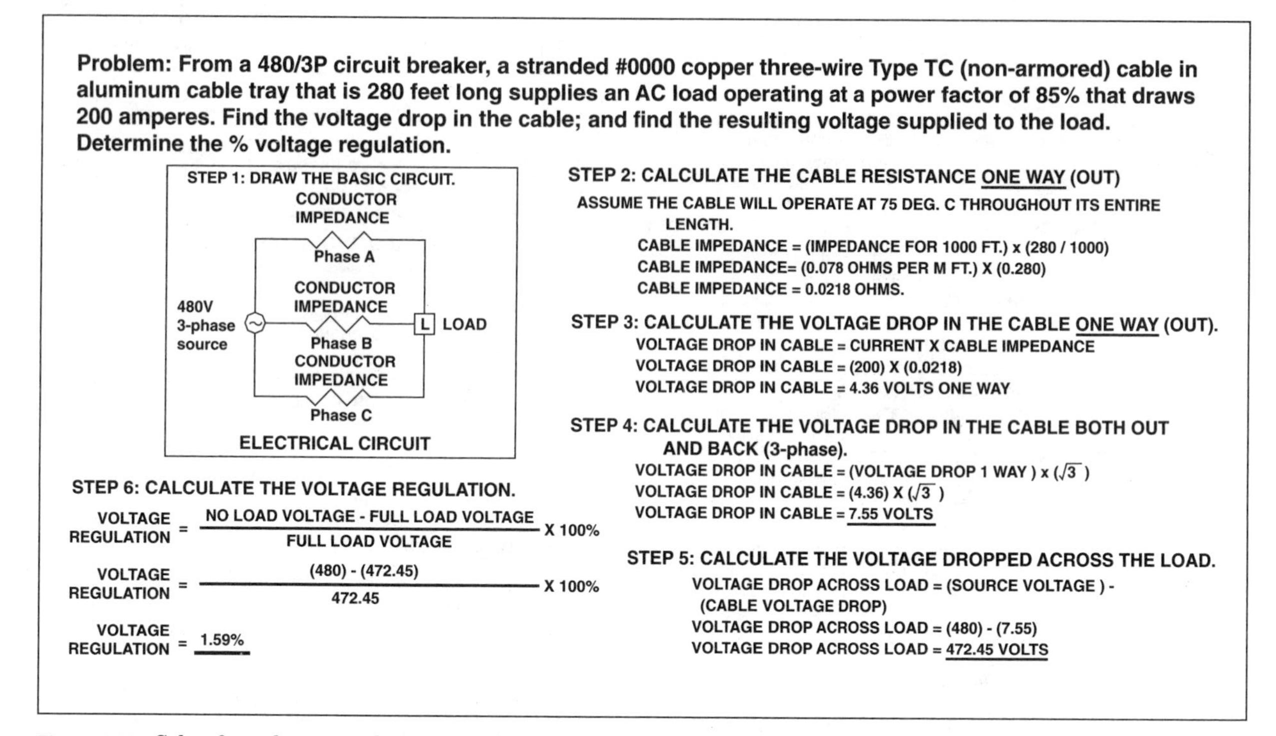

Figure 4-20 Solve for voltage regulation in a three-phase ac circuit in non-magnetic cable and cable tray.

Cabling in a cable tray is a normal wiring method in industrial facilities.

Medium-voltage cables leaving cable tray and entering switchgear building.

Problem: A 1/4" X 4" copper bus bar is to be installed in a DC circuit. It will be 200 feet long. Calculate the DC resistance of the bus bar.

STEP 1: CALCULATE THE BUS BAR RESISTANCE ONE WAY:

ASSUME THE BUS BAR WILL OPERATE AT 30 DEG. C (AMBIENT TEMPERATURE) THROUGHOUT ITS ENTIRE LENGTH.

RESISTANCE = (RESISTANCE FOR 1 FT.) x (200)
RESISTANCE = (0.00000825 OHMS PER FT.) X (200 FT)
BUS BAR RESISTANCE = 0.00165 OHMS.

Figure 4-21 Solve for the resistance of a copper bar given its dimensions and temperature.

Problem: A 1/4" X 4" copper bus bar is to be installed in a DC circuit. It will be 200 feet long. Calculate the DC resistance of the bus bar and its kW losses when it carries 1000 amperes.

STEP 1: CALCULATE THE BUS BAR RESISTANCE ONE WAY:

ASSUME THE BUS BAR WILL OPERATE AT 30 DEG. C (AMBIENT TEMPERATURE) THROUGHOUT ITS ENTIRE LENGTH.

RESISTANCE = (RESISTANCE FOR 1 FT.) x (200)
RESISTANCE = (0.00000825 OHMS PER FT.) X (200 FT)
BUS BAR RESISTANCE = 0.00165 OHMS.

STEP 2: CALCULATE THE HEAT LOSSES:

POWER (HEAT) LOSSES = $(\text{CURRENT})^2$ X RESISTANCE
POWER (HEAT) LOSSES = $(1000)^2$ X (0.00165)
POWER (HEAT) LOSSES = 1.65 kW

Figure 4-22 Solve for the heat losses in a copper bar given its dimensions and ampere load.

understand how wires and cables work, what types of insulation are suitable for what applications and environments, and how many amperes can be transported safely through each size of cable.

Figure 4-23 is a replication of selected parts of Tables 310-16 and 310-17 of the *National Electrical Code*. The proper way to read this table is best illustrated through an example: At the top of the table are several columns, each containing a Celsius temperature rating. On the left side of the table are wire sizes, graduated in AWG from small to large. Within the table are ampacity values. The entire table is based on the conductor operating within an ambient temperature of 30°C (which is 86°F). Selecting the row for No. 8 AWG and referring to the 75°C column, the table is read properly in this way: When a No. 8 conductor conducts 50 amperes (A) within an ambient temperature of 30°C and there are no more than three current-carrying conductors in the cable or raceway, then the wire will increase in temperature over a long time period from 30 to 75°C. (See Fig. 4-24 for more examples.)

Several things affect wire ampacity and temperature rating, and each of these must be considered in making wire sizing calculations:

- Insulation temperature capability
- Ambient temperature
- Quantity of wires that carry current in a conduit or cable
- Location where the heat *cannot* escape from the cable very rapidly
- Location where the heat *can* escape from the cable very rapidly
- Duration of current flow (Heat cannot build up over a very short time period with moderate amounts of overcurrent, but the repeated starting of motors, for example, can cause wires to heat excessively.)
- Wire terminal maximum temperature rating

Insulation temperature capability

Some insulations simply begin to melt when they increase in temperature to what would be considered to be "warm," or 65°C, whereas other insulations such as silicon rubber appear to be almost impervious to temperature. In Fig. 4-9, the increase in conductor resistance with temperature was calculated, but the main component of conductor impedance

AMPACITIES OF COPPER CABLE RATED 0 - 2 kV
TABLE UNITS ARE IN AMPERES
(3/CONDUCTOR CABLE OR 3 CURRENT-CARRYING CONDUCTORS ***IN A RACEWAY***)

AMBIENT TEMPERATURE OF 30 DEG. C (86 DEG F)

	TEMPERATURE RATING OF INSULATION		
WIRE SIZE (AWG) (Kcmil)	**60 Deg C**	**75 Deg C**	**90 Deg C**
18	N.A.	N.A.	4
16	N.A.	N.A.	18
14	20	20	25
12	25	25	30
10	30	35	40
8	40	50	55
6	55	65	75
4	70	85	95
3	85	100	110
2	95	115	130
1	110	130	150
0	125	150	170
00	145	175	195
000	165	200	225
0000	195	230	260
250	215	255	390
300	240	285	320
350	260	310	350
500	320	380	430
750	400	475	535
INSULATION TYPES	TW, UF	THW, RH, THWN, XHHW	SIS, FEP, MI, RHH, RHW-2, THHN, THHW, THW-2, THWN-2, USE-2, XHH, XHHW-2

Figure 4-23 Solve for wire ampacity given wire size, voltage rating, insulation temperature rating, and ambient temperture.

SINGLE CONDUCTOR IN **FREE AIR**

AMBIENT TEMPERATURE OF 30 DEG. C (86 DEG F)

TEMPERATURE RATING OF INSULATION		
60 Deg C	**75 Deg C**	**90 Deg C**
N.A.	N.A.	18
N.A.	N.A.	24
25	30	35
30	35	40
40	50	55
60	70	80
80	95	105
105	125	140
120	145	165
140	170	190
165	195	220
195	230	260
225	265	300
260	310	350
300	360	405
450	405	455
375	455	505
420	505	570
515	620	700
655	785	885
TW, UF	THW, RH, THWN, XHHW	SIS, FEP, MI, RHH, RHW-2, THHN, THHW, THW-2, THWN-2, USE-2, XHH, XHHW-2

Cables of different systems are installed into different cable trays.

was shown to be due to magnetic effects instead of resistance. In this section, the actual maximum temperature of the insulation is considered, since the insulation system must remain intact to keep the conductor from forming a short circuit. To do this, consideration must be made of two different forms of current flow through the conductor:

1. Short-circuit current flow, such as during a fault condition
2. Long-time-duration overload current

Figures 4-24 through 4-26 are step-by-step illustrations of how to correctly use NEC Tables 310-16 and 310-17.

There are many types of insulation, ranging from thermoplastic to packed mineral. Insulation limitations deal with temperature, moisture, and chemical attack. Each type of insulation exhibits its own unique capability to withstand heat from short-circuit currents. If the capability of the insulation is exceeded, then the insulation is damaged, and it simply allows the phase conductor to form a short circuit to ground. Since conductor temperature is a function of current squared times time, the amount of time and current can be

(a) To determine the AWG size of wire insulated to operate at a maximum of 60°C for a load of 36 A in free air, see NEC Table 310-17 (parts of which are replicated in Fig. 4-23), noting that this is the table for "Allowable Ampacities of Single-Insulated Conductors Rated 0 through 2000 Volts in Free Air, Based on Ambient Air Temperature of 30°C." Start at the top of the column with the heading **60°C** and proceed downward, row by row (where each row represents one AWG wire size, beginning with #18 AWG), until an ampere value that is greater than or equal to 36 is encountered. The first ampere value in the **60°**C column that is greater than or equal to the required 36 A is 40 A. Follow the row from the 40 A number toward the left to the first column in the table, and determine the answer: **#10 AWG is the proper wire size**. The table is read in this way: When conducting 40 A, #10 AWG wire in free air will increase in temperature to 60°C in an ambient of 30°C.

(b) To determine the AWG size of wire insulated to operate at a maximum of 60°C for a load of 36 A in conduit, see NEC Table 310-16 (parts of which are replicated in Fig. 4-23), noting that this is the table for "Allowable Ampacities of Single-Insulated Conductors Rated 0 through 2000 Volts in Raceway, Cable, or Earth, Based on Ambient Air Temperature of 30°C." Start at the top of the column with the heading **60°C** and proceed downward, row by row (where each row represents one AWG wire size, beginning with #18 AWG), until an ampere value that is greater than or equal to 36 is encountered. The first ampere value in the **60°C** column that is greater than or equal to the required 36 A is 40 A. Follow the row from the 40 A number toward the left to the first column in the table, and determine the answer: **#8 AWG is the proper wire size.** The table is read in this way: When conducting 40 A, #8 AWG wire in conduit will increase in temperature to 60°C in an ambient of 30°C.

Figure 4-24 Solve for required 60°C wire size in free air and in conduit given ampere load and temperatures.

used to determine the thermal damage limits of a conductor. Accordingly, once the available short-circuit current has been determined from a short-circuit calculation, consideration must be given to the amount of time the short-circuit current can flow. This duration is determined by the contact parting time of the circuit breaker that is to interrupt the short-circuit current. Circuit breakers exhibit from 1.5-cycle to as long as 6-cycle contact-parting/current-interrupting times, where 6 cycles requires 0.1 second at a system frequency of 60 Hz (cycles per second). Armed with the short-circuit current

To determine the AWG size of wire insulated to operate at a maximum of 75°C for a load of 86 A in conduit, see NEC Table 310-16 (parts of which are replicated in Fig. 4-23), noting that this is the table for "Allowable Ampacities of Single-Insulated Conductors Rated 0 through 2000 Volts in Raceway, Cable, or Earth, Based on Ambient Air Temperature of 30°C." Start at the top of the column with the heading **75°C** and proceed downward, row by row (where each row represents one AWG wire size, beginning with #18 AWG), until an ampere value that is greater than or equal to 86 is encountered. The first ampere value in the **75°C** column that is greater than or equal to the required 86 A is 100 A. Follow the row from the 100 A number toward the left to the first column in the table, and determine the answer: **#3 AWG is the proper wire size.** The table is read in this way: When conducting 100 A, #3 AWG wire in conduit will increase in temperature to 75°C in an ambient of 30°C.

Figure 4-25 Solve for required 75°C wire size in conduit given ampere load and temperatures.

To determine the ampacity of a 350 kCMIL wire insulated to operate at a maximum of 90°C, see NEC Table 310-16 (parts of which are replicated in Fig. 4-23), noting that this is the table for "Allowable Ampacities of Single-Insulated Conductors Rated 0 through 2000 Volts in Raceway, Cable, or Earth, Based on Ambient Air Temperature of 30°C." Start in the left column at the row labeled 350 kCMIL and proceed to the right to the column with the heading **90°C**, and read the ampere value of that conductor as 350 amperes. The table is read in this way: When conducting 350 amperes, 350 kCMIL wire in conduit will increase in temperature to 90°C in an ambient of 30°C.

Figure 4-26 Solve for required 90°C wire size in conduit given ampere load and temperatures.

availability and the circuit breaker interrupting time, an engineer can refer to Fig. 4-27 to determine the smallest size of conductor that can be used in a given installation (based on short-circuit current). Simply enter Fig. 4-27 at the left side (along the *y* axis) of the figure at the value that corresponds to the available short-circuit current; then follow the horizontal line to the right until intersecting the circuit breaker contact-parting/current-interrupting time. From this point, simply follow the vertical line down to read the minimum

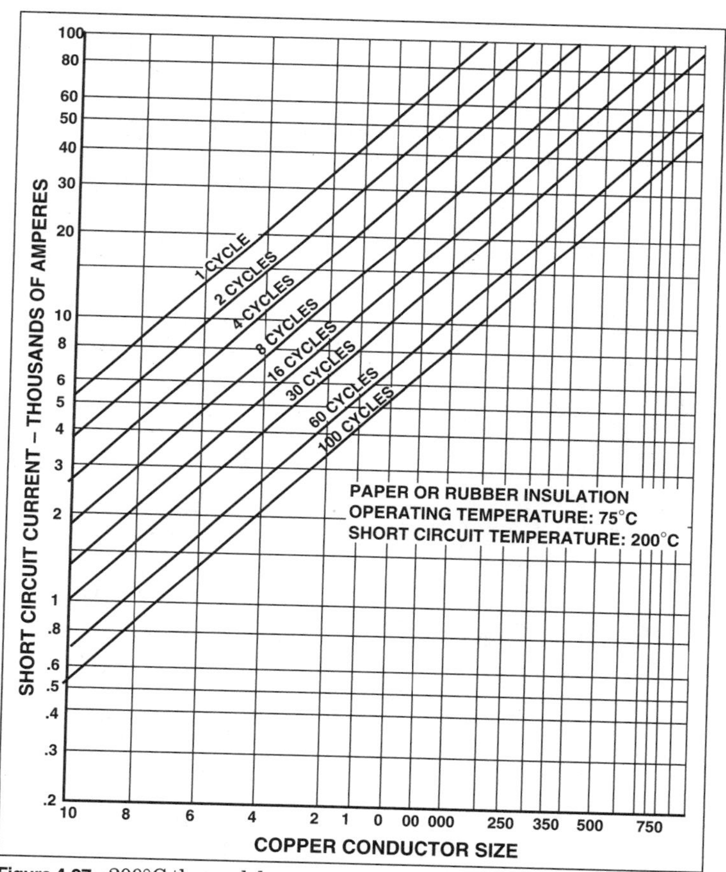

Figure 4-27 200°C thermal damage curve set for paper or rubber insulation.

conductor size on the x axis of the graph. Note that Fig. 4-27 is for paper or rubber insulation, and these insulations support conductor temperatures of up to 200°C without damage. For thermoplastic insulation systems, Fig. 4-28 must be used instead. Observation of a few calculations of this type will show that generally 600-V cables are considered to be expendable from the point of view of this calculation. However, this calculation is the basis for the minimum

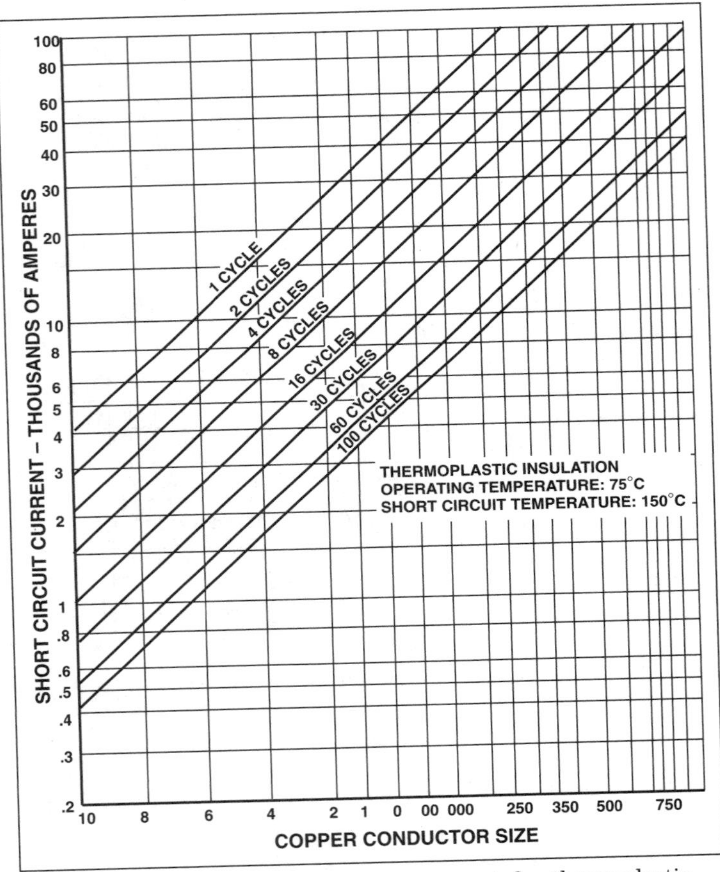

Figure 4-28 200°C thermal damage curve set for thermoplastic insulation.

conductor size for given voltages, as required by Table 310-5 of the *National Electrical Code.*

Determining Wire Size Given Insulation Type, Circuit Breaker Clearing Time, and Short-Circuit Current

An example of the use of Fig. 4-28 is in order here. The example problem has a 13.8-kV system feeder conductor down-

Cables in ladder cable tray.

stream of a 4-cycle circuit breaker that must interrupt 20 kA and asks that the minimum size of 15-kV cable with thermoplastic insulation be determined. In the solution, enter Fig. 4-28 along the y axis at 20 (thousands of amperes), and proceed along that line horizontally to the right until intercepting the line representing the 4-cycle breaker. Then proceed downward vertically to the x axis until reading between No. 1 AWG and No. 0 copper cable. Select the next larger cable, the No. 0 size, for the installation.

Selecting the Proper Insulation for an Environment

Before a conductor with a certain insulation can be selected for an application, the first things that must be determined are that no chemical in the intended environment will cause the insulation to degrade and whether or not the conductor will be placed in a dry, damp, or wet location. Figure 4-29 is a listing of the more common insulation types and environments that would prevent their proper operation, along with specific information as to whether each insulation type is usable in damp or wet locations. Note that the figure shows

TYPE LETTER	COMPOSITION	LIMITED APPLICATION
TW	THERMOPLASTIC	N.A.
UF	THERMOPLASTIC	N.A.
THHW	THERMOPLASTIC	N.A.
THW	THERMOPLASTIC	N.A.
THWN	THERMOPLASTIC & NYLON	N.A.
XHHW	THERMOSET	N.A.
USE	THERMOPLASTIC	N.A.
SA	SILICONE	N.A.
SIS	THERMOSET	N.A.
MI	COMPACTED MAGNESIUM OXIDE	HIGH TEMPERATURES
THHN	THERMOPLASTIC & NYLON	N.A.
THHW	THERMOPLASTIC	N.A.
THW-2	SPECIAL THERMOPLASTIC	N.A.
THWN-2	SPECIAL THERMOPLASTIC & NYLON	N.A.
USE-2	SPECIAL THERMOPLASTIC	N.A.
XHHW	THERMOSET	N.A.
XHHW-2	SPECIAL THERMOSET	N.A.
SF-1	SILICONE	HIGH TEMPERATURES
TF	THERMOPLASTIC	LIGHTING FIXTURES
TFFN	THERMOPLASTIC & NYLON	LIGHTING FIXTURES
C	THERMOPLASTIC	LAMP CORD
G	THERMOSET	PORTABLE POWER CABLE
S	THERMOSET	PORTABLE POWER CABLE
SJ	THERMOSET	PORTABLE POWER CABLE
SO	THERMOSET	PORTABLE POWER CABLE

Figure 4-29 Select proper insulation system given environment within which wire will be installed.

that the maximum operating temperature of many insulation types is reduced from 90 to 75°C when the insulation is submerged in water. Accordingly, when the conductor size is selected, a conductor that is rated for 90°C in dry locations must be treated as if it were only rated for 75°C.

Figure 4-30 is a summary of the physical characteristics of different materials from which cable jackets are commonly made.

DETERMINE AMPACITY FROM NEC TABLE	REMARKS
310-16 OR 310-17	
310-16 OR 310-17	UNDERGROUND FEEDER CABLE
310-16 OR 310-17	MOISTURE AND HEAT RESISTANT
310-16 OR 310-17	
310-16 OR 310-17	WITH A NYLON PROTECTIVE LAYER
310-16 OR 310-17	FLAME RETARDANT
310-16 OR 310-17	UNDERGROUND FEEDER CABLE
310-16 OR 310-17	FOR HIGH TEMPERATURE APPLICATIONS
310-16 OR 310-17	FLAME RETARDANT, FOR SWITCHBOARD WIRING
310-16 OR 310-17	WITH COPPER OR STAINLESS STEEL SHEATH, ALMOST FLAMEPROOF
310-16 OR 310-17	HEAT RESISTANT, WITH A NYLON PROTECTIVE LAYER
310-16 OR 310-17	
310-16 OR 310-17	RATED FOR 90 DEG C IN WET LOCATIONS
310-16 OR 310-17	RATED FOR 90 DEG C IN WET LOCATIONS
310-16 OR 310-17	RATED FOR 90 DEG C IN WET LOCATIONS
310-16 OR 310-17	MOISTURE RESISTANT
310-16 OR 310-17	RATED FOR 90 DEG C IN WET LOCATIONS
402-5	LIGHTING FIXTURE WIRE
402-5	LIMITED TO 60 DEG. C
402-5	WITH A NYLON PROTECTIVE LAYER, RATED AT 90 DEG C
400-5(a)	LAMP CORD, NOT FOR HARD USAGE LOCATIONS
400-5(a)	CORD FOR EXTRA HARD USAGE
400-5(a)	HARD SERVICE CORD
400-5(a)	HARD SERVICE CORD
400-5(a)	HARD SERVICE CORD FOR CONTACT WITH OIL

Ambient temperature

If the No. 8 conductor in the preceding example is in an ambient temperature that is very hot, say, 74°C, then it could carry only a very small current before it would reach its maximum operating temperature of 75°C. Figure 4-31 shows an example of how to consider the ambient temperature of the conductor surroundings in determining the allowable ampacity of the conductor simply by applying the respective factor to the conductor table ampacity.

	NEOPRENE RUBBER	HYPALON	CHLORINATED POLYETHYLENE	POLYVINYL CHLORIDE
TENSILE STRENGTH	2100-2200 PSI	2100-2200 PSI	1800-1900 PSI	1900-2000 PSI
TENSILE STRENGTH AFTER AGING	DECREASES 25%	INCREASES 10%	DECREASES 1/3	INCREASES 5%
MOISTURE RESISTANCE	POOR	POOR TO FAIR	GOOD	BEST
FLAME RESISTANCE	GOOD	GOOD	GOOD	POOR
RESISTANCE TO OIL	FAIR TO GOOD	FAIR TO GOOD	BEST	FAIR
IS FLEXIBLE IN COLD TEMPERATURES TO	-40 DEG. C	-40 DEG. C	-40 DEG. C	-10 DEG C.
DISTORTS WITH HIGH TEMPERATURE AT 120 DEG. C	NO	NO	YES	YES
RELATIVE PULLING TENSION	1	1	0.3	0.5
BEGINS TO DECOMPOSE AT	200 DEG. C	225 DEG. C	280 DEG C	160 DEG. C

Figure 4-30 Physical characteristics of cable jacket materials.

AMBIENT TEMPERATURE CORRECTION FACTORS (multiply the table ampacity by the following appropriate factor)				
AMBIENT TEMPERATURE	AMBIENT TEMPERATURE	MAXIMUM INSULATION TEMPERATURE RATING IN DEG. C		
DEG. C	DEG. F	60	75	90
21-25	70-77	1.08	1.05	1.04
26-30	78-86	1	1	1
31-35	87-95	0.91	0.94	0.96
36-40	96-104	0.82	0.88	0.91
41-45	105-113	0.71	0.82	0.87
46-50	114-122	0.58	0.75	0.82
51-55	123-131	0.41	0.67	0.76
56-60	132-140	N.A.	0.58	0.71
61-70	141-158	N.A.	0.33	0.58
71-80	159-176	N.A.	N.A.	0.41

Figure 4-31 Select ambient temperature correction factor for other than 30°C.

Quantity of wires that carry current in a conduit or cable

When a conductor carries current, it exudes heat equal to I^2R, where I is the current in amperes and R is the resistance of the conductor. According to the thermal flow laws of thermodynamics, the heat flows away from the conductor to cooler objects or gases, where the objects or gases absorb the heat energy and increase in temperature themselves. This heat flow can become a problem when there is more than one conductor in a cable or raceway because the heat from one conductor passes to the other conductor, heating it while it is heating itself from I^2R heat losses. The net effect is a much hotter conductor than would have been caused simply by the self-heat losses of one conductor operating alone. The ampacity table on the left side of Fig. 4-32 considers that three conductors are in a raceway or cable and that all three are carrying equal current of the values shown in the table. When more than three current-carrying conductors exist and are operational in the same cable or conduit, then the conductors in the cable or raceway reach their maximum operating temperature while carrying less than the current values found in the table. Figure 4-33 shows the table used to calculate the

AMPACITIES OF COPPER CABLE RATED 0 - 2 kV
(3/CONDUCTOR CABLE OR 3 CURRENT-CARRYING CONDUCTORS ***IN A RACEWAY***)
AMBIENT TEMPERATURE OF 30 DEG. C (86 DEG F)
(amperes)

	TEMPERATURE RATING OF INSULATION		
WIRE SIZE (AWG) (Kcmil)	**60 Deg C**	**75 Deg C**	**90 Deg C**
18	N.A.	N.A.	14
16	N.A.	N.A.	18
14	20	20	25
12	25	25	30
10	30	35	40
8	40	50	55
6	55	65	75
4	70	85	95
3	85	100	110
2	95	115	130
1	110	130	150
0	125	150	170
00	145	175	195
000	165	200	225
0000	195	230	260
250	215	255	390
300	240	285	320
350	260	310	350
500	320	380	430
750	400	475	535
INSULATION TYPES	TW, UF	THW, RH, THWN, XHHW	SIS, FEP, MI, RHH, RHW-2, THHN, THHW, THW-2, THWN-2, USE-2, XHH, XHHW-2

Figure 4-32 Wire ampacities given wire size, voltage rating, ambient temperature, and insulation temperature rating.

QUANTITY OF CURRENT-CARRYING CONDUCTORS	DERATED PERCENT OF ORIGINAL AMPERE CONDUCTOR AMPACITY
3	100%
4 - 6	80%
7 - 9	70%
10 - 20	50%
21 - 30	45%
31 - 40	40%
41 OR MORE	35%

Figure 4-33 Select correction factor for more than three current-carrying conductors in a raceway.

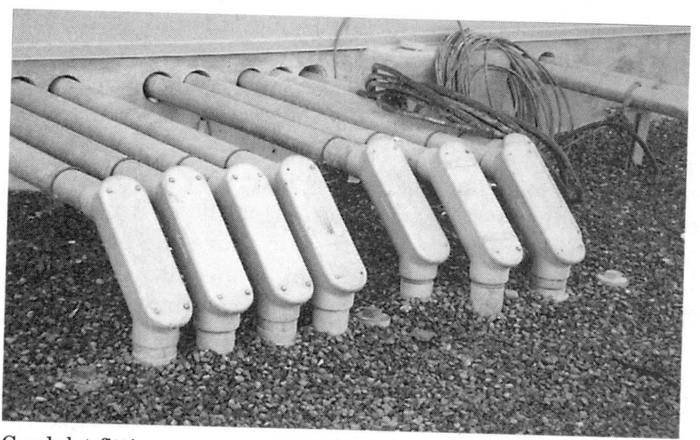

Condulet fittings are used to turn short-radius bends.

allowable ampacity where there are more than three current-carrying conductors in a raceway or cable.

Comparing the left side of the table in Fig. 4-23 (conductors in conduit) with the right side of the table (conductors in free air), it is apparent how locating a conductor where it can be cooled greatly increases its current-carrying capability.

Note that, at first glance, neutral conductors that carry only imbalance current would not be counted as current-carrying conductors because the current they would carry simply would be subtracted from the current of a phase conductor in the cable or raceway, and this would be true if the conductors all carried only fundamental current (60 Hz). If the load served by the conductors generates third harmonic current, such as does arc-type lighting, the third harmonic is present in the neutral conductor from each of the three phases. The net effect of this is that the neutral conductor heats at almost the same rate as do the phase conductors, and so the neutral conductor must be counted as a current-carrying conductor.

Location where the heat *cannot* escape from the cable very rapidly

In the thermal flow equation, British thermal units (Btus) of heat must flow from the operating conductor through thermal resistance. If the thermal resistance is very high, then the heat cannot flow away, and the operating conductor increases in temperature. Consider the case of a cable that is either buried in the earth or is pulled into a conduit that is buried in the earth. Some forms of earth carry heat very well, and others are essentially thermal insulators. Heat flows from buried cables through the earth and to the air above the earth. The actual computer modeling to determine the ultimate allowable line current for a cable or cables buried underground is beyond the scope of this book, but suffice it to state that the Neher-McGrath method [American Institute of Electrical Engineers (AIEE) Paper 57-660] of ampacity calculation should be referred to when a cable is buried deeper than 36 inches (in) or when more than one cable is installed near another.

Location where the heat *can* escape from the cable very rapidly

When a continuous supply of cool air is available to flow over a conductor, then thermal flow can occur continuously to maintain a cool conductor temperature, even when the conductor is carrying current. Thermal flow is improved so much when a single insulated conductor is located in "free air" that it takes a great deal more current to elevate the conductor to its maximum operating temperature.

Wire terminal maximum temperature rating

Although some insulation systems are capable of continuous operation at very hot temperatures, such as at 200°C, most wire terminals are not. In fact, most wire terminals for use below 600 V are only rated for operation at 75°C, and most wire terminals for use on medium-voltage cables are only rated for operation at 90°C. When a wire that is rated for use at 90°C is terminated on a wire terminal that is rated at 75°C, then the wire must be sized as if it can only operate at up to 75°C.

Figure 4-34 shows how to solve for a 60°C copper wire to carry 36 A in (*a*) free air and (*b*) in conduit. Figure 4-35 shows how to solve for the proper 75°C wire size to carry 86 A. And Fig. 4-36 shows how to determine the ampacity of 350 kCM at 90°C.

Where a raceway contains more than three current-carrying conductors, a final adjustment factor must be used to derate the current-carrying rating of the conductors in that raceway. This reduction in current is due to the mutual heating effect of the conductors and the tendency of the raceway to contain the heat from them. As is shown in Fig. 4-33, these final adjustment factors are as follows:

Quantity of current-carrying conductors	Reduced ampacity arrived at using all other considerations and factors
4–6	80%
7–9	70%
10–20	50%
21–30	45%
31–40	40%
41 and more	35%

For example, if a set of 15 current-carrying No. 12 TW wires in a 2.5-in conduit have already been derated to 0.82×25 A = 20.5 A for being in a room having an ambient temperature of greater than 96°F, to determine the final ampere-carrying capability of these wires, a further derating to 50 percent of the 20.5 A must be made. This leaves the No. 12 TW wires with an ampacity of only 10.25 A.

(a) To determine the AWG size of copper wire insulated to operate at a maximum of 60°C for a load of 58 A in free air, see NEC Table 310-17 (parts of which are replicated in Fig. 4-23), noting that this is the table for "Allowable Ampacities of Single-Insulated Conductors Rated 0 through 2000 Volts in Free Air, Based on Ambient Air Temperature of 30°C." Start at the top of the column with the heading **60°C** and proceed downward, row by row (where each row represents one AWG wire size, beginning with #18 AWG), until an ampere value that is greater than or equal to 58 is encountered. The first ampere value in the **60°C** column that is greater than or equal to the required 36 A is 60 A. Follow the row from the 60 A number toward the left to the first column in the table, and determine the answer: **#8 AWG is the proper wire size**. The table is read in this way: When conducting 60 A, #8 AWG wire in free air will increase in temperature to 60°C in an ambient of 30°C.

(b) To determine the AWG size of copper wire insulated to operate at a maximum of 60°C for a load of 58 A in conduit, see NEC Table 310-16 (parts of which are replicated in Fig. 4-37), noting that this is the table for "Allowable Ampacities of Single-Insulated Conductors Rated 0 through 2000 Volts in Raceway, Cable, or Earth, Based on Ambient Air Temperature of 30°C." Start at the top of the column with the heading **60°C** and proceed downward, row by row (where each row represents one AWG wire size, beginning with #18 AWG), until an ampere value that is greater than or equal to 58 is encountered. The first ampere value in the **60°C** column that is greater than or equal to the required 58 A is 70 A. Follow the row from the 70 A number toward the left to the first column in the table, and determine the answer: **#4 AWG is the proper wire size.** The table is read in this way: When conducting 70 A, #4 AWG wire in conduit will increase in temperature to 60°C in an ambient of 30°C.

Figure 4-34 Solve for required wire size in free air and in conduit given ampere load and temperatures.

Aluminum Conductors

Although most of the conductors installed today in the below 600 V class are copper, many of these conductors are also made of aluminum. Accordingly, it is of value to note the different electrical characteristic ampacities between copper and aluminum shown in Fig. 4-37.

Conductor and Cable Selection

Conductors should be selected to be impervious to the environment in which they will be located, and the same is true

To determine the AWG size of copper wire insulated to operate at a maximum of 75°C for a load of 86 A in conduit, see NEC Table 310-16 (parts of which are replicated in Fig. 4-23), noting that this is the table for "Allowable Ampacities of Single-Insulated Conductors Rated 0 through 2000 Volts in Raceway, Cable, or Earth, Based on Ambient Air Temperature of 30°C." Start at the top of the column with the heading **75°C** and proceed downward, row by row (where each row represents one AWG wire size, beginning with #18 AWG), until an ampere value that is greater than or equal to 86 is encountered. The first ampere value in the **75°C** column that is greater than or equal to the required 86 A is 100 A. Follow the row from the 100 A number toward the left to the first column in the table, and determine the answer: **#3 AWG is the proper wire size.** The table is read in this way: When conducting 100 A, #3 AWG wire in conduit will increase in temperature to 75°C in an ambient of 30°C.

Figure 4-35 Solve for required wire size in conduit given ampere load and 75°C insulation temperature.

To determine the ampacity of a 350 kCMIL copper wire insulated to operate at a maximum of 90°C, see NEC Table 310-16 (parts of which are replicated in Fig. 4-37), noting that this is the table for "Allowable Ampacities of Single-Insulated Conductors Rated 0 through 2000 Volts in Raceway, Cable, or Earth, Based on Ambient Air Temperature of 30°C." Start in the left column at the row labeled 350 kCMIL and proceed to the right to the column with the heading **90°C**, and read the ampere value of that conductor as 350 amperes. The table is read in this way: When conducting 350 amperes, 350 kCMIL wire in conduit will increase in temperature to 90°C in an ambient of 30°C.

Figure 4-36 Solve for wire ampacity in conduit of a given wire size with 90°C insulation temperature.

for the conductor insulation system. For example, copper that will be installed within a sulfur recovery unit industrial plant would be subject to attack from the sulfur in the process flow; therefore, this copper should be of the "tinned" type, where each strand of copper is coated with a combination of tin and lead. (Tin and lead are the primary metals in electrical solder.) Similarly, the most widely used insulation today, polyvinyl chloride (PVC), would not be usable at all for connections to an outdoor load in the Arctic Circle. This is so because when PVC gets cold, it becomes

AMPACITIES OF COPPER AND ALUMINUM CABLE RATED 0 - 2 kV

TABLE UNITS ARE IN AMPERES

WIRE SIZE (AWG) (Kcmil)	(3/CONDUCTOR CABLE OR 3 CURRENT-CARRYING CONDUCTORS ***IN A RACEWAY***) AMBIENT TEMPERATURE OF 30 DEG. C (86 DEG F) — TEMPERATURE RATING OF INSULATION						SINGLE CONDUCTOR IN **FREE AIR** AMBIENT TEMPERATURE OF 30 DEG. C (86 DEG F) — TEMPERATURE RATING OF INSULATION					
	60Deg C COPPER	60Deg C ALUMINUM	75 Deg C COPPER	75 Deg C ALUMINUM	90 Deg C COPPER	90 Deg C ALUMINUM	60 Deg C COPPER	60 Deg C ALUMINUM	75 Deg C COPPER	75 Deg C ALUMINUM	90 Deg C COPPER	90 Deg C ALUMINUM
18	N.A.	N.A.	N.A.	N.A.	14	N.A.	N.A.	N.A.	N.A.	N.A.	18	N.A.
16	N.A.	N.A.	N.A.	N.A.	18	N.A.	N.A.	N.A.	N.A.	N.A.	24	N.A.
14	20	N.A.	20	N.A.	25	N.A.	25	N.A.	30	N.A.	35	N.A.
12	25	20	25	20	30	25	30	25	35	30	40	35
10	30	25	35	30	40	35	40	35	50	40	55	40
8	40	30	50	40	55	45	60	45	70	55	80	60
6	55	40	65	50	75	60	80	60	95	75	105	80
4	70	55	85	65	95	75	105	80	125	100	140	110
3	85	65	100	75	110	85	120	95	145	115	165	130
2	95	75	115	90	130	100	140	110	170	135	190	150
1	110	85	130	100	150	115	165	130	195	155	220	175
0	125	100	150	120	170	135	195	150	230	180	260	205
00	145	115	175	135	195	150	225	175	265	210	300	235
000	165	130	200	155	225	175	260	200	310	240	350	275
0000	195	150	230	180	260	205	300	235	360	280	405	315
250	215	170	255	205	390	230	450	265	405	315	455	355
300	240	190	285	230	320	255	375	290	455	350	505	395
350	260	210	310	250	350	280	420	330	505	395	570	445
500	320	260	380	310	430	350	515	405	620	485	700	545
750	400	320	475	385	535	435	655	515	785	620	885	700
INSULATION TYPES	TW, UF		THW, RH, THWN, XHHW		SIS, FEP, MI, RHH, RHW-2, THHN, THHW, THW-2, THWN-2, USE-2, XHH, XHHW-2		TW, UF		THW, RH, THWN, XHHW		SIS, FEP, MI, RHH, RHW-2, THHN, THHW, THW-2, THWN-2, USE-2, XHH, XHHW-2	

Figure 4-37 Wire ampacities for aluminum or copper conductors given wire size, voltage rating, and ambient temperature.

SIZE (AWG or kCMIL)	THW (SQ. IN.)	THHN, TFN, THWN (SQ. IN.)	XHHW (SQ. IN.)	BARE (SQ. IN.)	
18	0.0088	0.0062	0.0062	0.001	SOLID
16	0.0109	0.0079	0.0079	0.002	SOLID
14	0.0206	0.0087	0.0131	0.003	SOLID
12	0.0172	0.0117	0.0167	0.005	SOLID
10	0.0311	0.0184	0.0216	0.008	SOLID
8	0.0598	0.0373	0.0456	0.017	STRANDED
6	0.0819	0.0519	0.0625	0.027	STRANDED
4	0.1087	0.0845	0.0845	0.042	STRANDED
3	0.1263	0.0995	0.0995	0.053	STRANDED
2	0.1473	0.1182	0.1182	0.067	STRANDED
1	0.2027	0.1590	0.1590	0.087	STRANDED
0	0.2367	0.1893	0.1893	0.109	STRANDED
00	0.2781	0.2265	0.2265	0.138	STRANDED
000	0.3288	0.2715	0.2715	0.173	STRANDED
0000	0.3904	0.3278	0.3278	0.219	STRANDED
250	0.4877	0.4877	0.4026	0.26	STRANDED
300	0.5581	0.5581	0.4669	0.312	STRANDED
350	0.6291	0.6291	0.5307	0.364	STRANDED
500	0.8316	0.8316	0.7163	0.519	STRANDED
750	1.2252	1.2252	1.0936	0.782	STRANDED

Figure 4-38 Cross-sectional areas and insulations of the most commonly used wires.

CABLE TYPE	CABLE DESCRIPTION	USES
MV	MEDIUM VOLTAGE SINGLE-CONDUCTOR OR MULTI-CONDUCTOR	IN WET OR DRY LOCATIONS UP TO 35 KV IN CABLE TRAYS IN RACEWAYS DIRECTLY BURIED SUPPORTED BY MESSENGER CABLE
MI	MINERAL INSULATED, METAL SHEATHED,	EVERYWHERE, INCLUDING CLASS I DIVISION 1 LOCATIONS. NORMALLY LESS THAN 600 VOLTS.
AC	ARMORED CABLE, COMMONLY CALLED "BX"	NORMALLY USED FOR INTERIOR BRANCH CIRCUITS BRANCH CIRCUITS AND FEEDERS EXPOSED AND CONCEALED CAN BE IN CABLE TRAYS MUST BE IN DRY LOCATIONS CAN BE EMBEDDED IN PLASTER OR MASONRY CAN BE FISHED IN DRY CMU VOIDS NOT IN PLACES OF ASSEMBLY NOT IN HAZARDOUS LOCATIONS
MC	METAL CLAD	SIMILAR TO AC USES, WITH SOME FEW ADDITIONS
NM	NON-METALLIC SHEATHED COMMONLY CALLED ROMEX	RESIDENTIAL BRANCH CIRCUITS EXPOSED OR CONCEALED NOT IN BUILDINGS OVER 3 FLOORS TALL NOT IN DAMP LOCATIONS OR HAZARDOUS LOCATIONS
SE OR USE	SERVICE ENTRANCE OR UNDERGROUND SERVICE ENTRANCE	NORMALLY USED FOR RESIDENTIAL SERVICE FEEDERS
TC	TRAY CABLE	INDUSTRIAL USE IN CABLE TRAY

Figure 4-39 Commonly used cables and their applications.

WIRING METHOD	NORMAL USE
Aerial conductor	Outdoor distribution, overhead on poles
Direct burial cable	Outdoor distribution, underground in trench
Embedded PVC conduit	Underground or in concrete slab
Embedded Steel Conduit	Underground or in concrete slab
EMT	Indoors above ground, less commonly outdoors above ground
Rigid Steel Conduit	In all locations, including hazardous locations
IMC Conduit	In all locations, including hazardous locations
Aluminum Rigid Conduit	Above ground in all locations, but not embedded in concrete
ENT	Indoors above ground
Cables in Cable Tray	Outdoors in industrial plants, less commonly indoors
Nonmetallic Cable	Indoors above ground, most commonly for residential installations
Armored "AC" Cable ("BX")	Indoors above ground, normally for commercial installations
Armored "MC" Cable	Specific types of this cable are used in all locations

Figure 4.40 Common wiring methods and locations where each method is normally implemented.

hard and brittle and simply cracks with movement. This determination requires recognition that all conductors are subject to movement from three sources:

1. Outside physical vibration
2. Electromagnetic forces
3. Changes in dimension with temperature change

Figure 4-29 shows many of the most commonly used insulation types, along with locations in which their use is suitable. Since each type of insulation requires a different thickness for a given voltage rating, each wire size of each insulation type has a different cross-sectional area, and these are shown in Fig. 4-38.

Figure 4-39 is a listing of commonly used cable types, their descriptions, and locations where they can be used, and Fig. 4-40 is a listing of common wiring methods, including cables, with a listing of the locations where each type of wiring method is implemented.

NOTES

NOTES

Chapter

5

Short-Circuit Calculations

When the current flow path is directed correctly, the pressure of the source voltage forces normal current magnitudes to flow through the load impedances. During this time, the insulation surrounding the energized conductors prevents current from flowing through any path other than through the load impedance. In this situation, the load impedance is large enough to limit the current flow to "normal" low values in accordance with Ohm's law:

$$E = I \times Z$$

Problems arise, however, when the conductor insulation fails, permitting a shortened path for electron flow than through the load impedance. If the shortened path, or *short circuit* (also known as a *fault*), permits contact between a phase conductor and an equipment grounding conductor, this is known as a *ground fault,* or a *phase-to-ground fault.* If, however, the shortened path instead permits contact between two or three phase conductors, then it is known as a *phase-to-phase fault.*

If a solid connection is made between the faulted phase conductor and the other phase wire or the equipment grounding conductor, then the short circuit is identified as a *bolted fault.*

In bolted faults, little or no arcing exists, the voltage drop across the very low impedance of the almost-nonexistent arc is very low, and the fault current is of high magnitude.

If an arcing connection is made between the faulted phase conductor and the other phase wire or the equipment grounding conductor, then the short circuit is identified as an *arcing fault,* with its associated lowered fault current flow. In certain systems, this arcing fault current can be so low that it is not recognizable as a problem to the upstream overcurrent device. During such events, excessive heat buildup around the arc can occur, causing further damage to the otherwise sound electrical system or starting fires in nearby structures or processes. Even though the fault may be small, heat energy flows from it over time to the surroundings, eventually causing high temperatures. High temperatures applied to most wire insulation systems cause plastic deformation and failure, frequently resulting in an even more severe short circuit than the original one. It is because of this consideration that everything possible must be done to interrupt current flow to a fault as soon as possible after the beginning of the fault or to limit fault current flow in some way, such as with transformers or grounding resistors.

Considering that fault current can reach several hundred thousand amperes, interrupting fault current flow can be a formidable task, so one rating of circuit breakers and fuses intended to interrupt fault currents is the ampere rating that each can interrupt. These device ampere ratings frequently are given in terms of symmetrical current values (providing for no direct-current offset) or in terms of asymmetrical current values that include provision for direct-current (dc) offset. The relative amount of dc offset and resulting asymmetrical current value multiplier used to multiply by the symmetrical current value are related to the X/R value of the system at the point of the fault. With a highly resistive circuit having little reactance, the difference between the symmetrical and asymmetrical current values are minimal (there could be no difference at all), but with a highly reactive system containing only a small resistance value, the asymmetrical current value could be a multiple of three to five times the symmetrical value, or higher.

Large values of current cause large magnetic forces. When fault current flows through switchgear or a circuit breaker, the bus bars therein are attracted to one another and then are forced apart from one another at a frequency of 50 or 60 times per second depending on the frequency of the power source. This equipment must be braced to withstand these forces, and this bracing is rated and is called the *withstand rating* of the equipment. Thus a circuit breaker must be rated in full-load amperes for normal operation, it must have a withstand rating, and it must have an interrupting rating.

The *interrupting rating* of a circuit breaker depends largely on how fast its contacts can operate to begin to interrupt the current flow in the circuit. A breaker that is extremely fast, with a contact parting time of, say, one cycle, must be able to interrupt much more current in a system with a large X/R value than would a slower breaker. Fuses, on the other hand, are able to open and interrupt fault current in less than $^{1}/_{4}$ cycle, so fuses must carry accordingly greater ratings.

Although the calculation of the asymmetrical current values from the symmetrical current values is important, most of the short-circuit calculation work goes into determining the symmetrical short current availability at the prospective fault point in the electrical power system. This symmetrical current calculation normally is done for a phase-to-phase fault, with the knowledge that this is normally the most demanding case. Rarely (only where the fault occurs near very large electrical machines having solidly grounded wye connections) does the phase-to-ground fault exceed the phase-to-phase fault in current availability. Therefore, concentration is on the phase-to-phase fault for the electrical power system with the largest expected quantity of rotating machines that will exist on the system in the foreseeable future.

Short-circuit currents present a huge amount of destructive energy that can be released through electrical systems under fault conditions. These currents can cause serious damage to electrical systems and to equipment or nearby persons. Protecting persons and electrical systems against damage during short-circuit conditions is required by the *National Electrical Code* (Secs. 110-9 and 110-10).

Not only should short-circuit studies be performed when a facility electrical system is first designed, but they also should be updated when a major modification or renovation takes place and no less frequently than every 5 years. Major changes would include a change by the electrical utility, a change in the primary or secondary system configuration within the facility, a change in transformer size or impedance, a change in conductor lengths or sizes, or a change in the motors that are energized by a system.

When modifications to the electrical system increase the value of available short-circuit current, a review of overcurrent protection device interrupting ratings and equipment withstand ratings should be made. This may entail replacing overcurrent protection devices or installing current-limiting devices such as current-limiting fuses, current-limiting circuit breakers, or current-limiting reactors. The key is to know, as accurately as possible, how much short-circuit current is available at every point within the electrical power system.

Sources of Short-Circuit Current

Every electrical system confines electric current flow to selected paths by surrounding the conductors with insulators of various types. Short-circuit current is the flow of electrical energy that results when the insulation barrier fails and allows current to flow in a shorter path than the intended circuit. In normal operation, the impedances of the electrical appliance loads limit the current flow to relatively small values, but a short-circuit path bypasses the normal current-limiting load impedance. The result is excessively high current values that are limited only by the limitations of the power source itself and by the small impedances of the conductive elements that still remain in the path between the power source and the short-circuit point. Short-circuit calculations are used to determine how much current can flow at certain points in the electrical system so that the electrical equipment can be selected to withstand and interrupt that magnitude of fault current. In short-circuit calculations, the contribution of current sources is first determined, and then the current-limiting effects of impedances in the system are considered in determining how much current can flow in a particular system part.

There are three basic sources of short-circuit currents:

- The electrical utility
- Motors
- On-site generators

There are two types of motors that contribute short-circuit current:

- Induction motors
- Synchronous motors

Between these sources of short-circuit current and the point of the short circuit, various impedances act to limit (impede) the flow of current and thus reduce the actual amount of short-circuit current "available" to flow into a short circuit. Naturally, the value of these impedances is different at every point within an electrical system; therefore, the magnitudes of short-circuit currents available to flow into a short circuit at different places within the electrical system vary as well.

Several calculation methods are used to determine short-circuit currents, and reasonably accurate results can be derived by system simplifications prior to actually performing the calculations. For example, it is common to ignore the impedance effect of cables except for locations where the cables are very long and represent a large part of the overall short-circuit current path impedance. Accordingly, in the most common form of short-circuit calculations, short-circuit current is considered to be produced by generators and motors, and its flow is considered to be impeded only by transformers and reactors.

The Ability of the Electrical Utility System to Produce Short-Circuit Current

By definition, the source-fault capacity is the maximum output capability the utility can produce at system voltage.

Generally, this value can be gotten from the electrical utility company by a simple request and is most often given in amperes or kilovoltamperes.

Suppose that the utility company electrical system interface data are given as

$$\begin{aligned} \text{MVA}_{\text{SC}} &= 2500 \text{ at } 138 \text{ kilovolts (kV) with an } X/R \\ &= 7 \text{ at the interface point} \end{aligned}$$

For this system, the utility can deliver 2,500,000 kilovoltamperes (kVA) ÷ [138 kV($\sqrt{3}$)], or a total of 10,459 symmetrical amperes (A) of short-circuit current.

The short-circuit value from the electrical utility company will be "added to" by virtue of contributions from the on-site generator and motor loads within the plant or building electrical power system. That is, the short-circuit value at the interface point with the electrical utility will be greater than just the value of the utility contribution alone.

Short-Circuit Contributions of On-Site Generators

The nameplate of each on-site generator is marked with its subtransient reactance X_d'' *like this*. This subtransient value occurs immediately after a short circuit and only continues for a few cycles. For short-circuit current calculations, the subtransient reactance value is used because it produces the most short-circuit current.

Determining how many kilovoltamperes an on-site generator can contribute to the short-circuit current of an electrical power system is a simple one-step process, solved as follows:

$$\text{Short-circuit kVA} = \frac{\text{generator kVA rating}}{X_d'' \textit{ rating}}$$

For example, a typical synchronous generator connected to a 5000 shaft horsepower (shp) gas turbine engine is rated at 7265 kVA, and its subtransient reactance X_d'' is 0.17. The problem is to determine the short-circuit output capabilities of this generator. Thus

$$\text{Short-circuit kVA} = \frac{7265\ \text{kVA}}{0.17} = 42{,}735\ \text{kVA}$$

Note that when more than one source of short-circuit current is present in a system, the resulting amount of short-circuit kilovoltamperes available to flow through a short circuit is simply the arithmetic sum of the kilovoltamperes from the various sources, and the total *equivalent kilovoltamperes* from all these sources is simply the arithmetic sum of the individual equivalent kilovoltampere values.

For example, if two of the preceding 7625-kVA, 138-kV synchronous generators are cogenerating into an electrical power system that is also supplied from an electrical utility whose capabilities on the generator side of the utility transformer are 300 million voltamperes (MVA), how many total fault kilovoltamperes are available at the terminals of the generator? The answer is

$$\text{Total kVA} = 2\ (42{,}735\ \text{kVA}) + 300{,}000\ \text{kVA} = 385{,}470\ \text{kVA}$$

It is worthwhile to note that the kilowatt rating of the generator set is *not* used in this calculation. Instead, the kilovoltampere rating of the electrical dynamo in the generator set is used because it (not the engine) determines how many short-circuit kilovoltamperes can be delivered momentarily into a fault. This is so because most of the fault current is quadrature-component current having a very lagging power factor. Another way of saying this is that the fault current does little work because of its poor power factor, and the size of the engine determines the kilowatt value of real work the generator set can do.

Short-Circuit Contributions of Motors

The nameplate of each motor is marked with its locked-rotor code letter as well as with its rated continuous full-load current and horsepower. The locked-rotor value occurs immediately on motor energization and also immediately after a short circuit. After a short circuit, this large amount of current flow through the motor only continues for a few cycles (even less time with induction motors). Unless specific information about the individual motor locked-rotor current characteristics is known, normally, a value of six times the motor full-load current is assumed. *Dividing 100 percent current by 600 percent current produces the normally assumed X″ motor value of 0.17*. It is this value that is used to determine the short-circuit power contribution of a motor or group of motors. The characteristics of small induction motors limit current flow more than those of larger motors; therefore, *for calculations involving motors smaller than 50 horsepower (hp), the 0.17 value must be increased to 0.20*. In addition, *the approximation that 1 hp equals 1 kVA* normally can be made with negligible error, particularly with motors of less than 200 hp.

Determining the short-circuit contribution from a motor is a simple one-step process solved as follows:

$$\text{Short-circuit power} = \frac{\text{motor horsepower rating}}{0.17}$$

(*Note:* 1 hp = 1 kVA for this calculation.)

Unless specific values of subtransient reactance or motor locked-rotor code letter are known, for motors of 50 hp and less, use 0.20 for the subtransient reactance. For individual motors or groups of motors producing 50 hp and more, use 0.17 for the subtransient reactance. Always assume that 1 hp = 1 kVA.

For example, a typical 460-volt (V) motor is rated at 40 hp. Determine the short-circuit contribution of this motor. Thus

$$\text{Short-circuit power} = \frac{40 \text{ kVA}}{0.20} = 200 \text{ kVA}$$

If there were 10 identical motors of this same rating connected to one bus, then their total short-circuit contribution at the bus would be 10 × 200 kVA = 2000 kVA.

Note that for motors of 50 hp or larger, the X_d'' rating of 0.17 should be used. For example, a typical 4.16-kV motor is rated at 2000 hp. Determine the short-circuit power contribution of this motor.

Setting 1 hp equal to 1 kVA, we get

$$\text{Short-circuit power} = \frac{2000 \text{ kVA}}{0.17} = 11{,}764 \text{ kVA}$$

If all the short-circuit current contributors were at the same voltage, then short-circuit currents simply could be added to one another to determine the total amount available to flow in a system. However, because transformers greatly lessen the flow of fault current and also change voltages, the impact of transformers must be considered, and fault values transported through transformers must be calculated in terms of power instead of amperes.

Let-Through Values of Transformers

For any transformer, the maximum amount of power (the fault capacity) that the transformer will permit to flow from one side of the transformer to the other side, in kilovoltamperes, is calculated as

$$\text{Let-through power} = \frac{\text{transformer kilovoltampere rating}}{\%Z/100}$$

For example, a 2000-kVA transformer has a nameplate impedance of 6.75 percent, and its voltage ratings are 13.8 kV-277/480 V, three phase. Find the maximum amount of power that this transformer can let through if it is energized from an infinite power source. The answer is

$$\text{Let-through power} = \frac{\text{transformer kilovoltampere rating}}{\%Z/100}$$

$$= \frac{2000}{0.0675} = 29{,}630 \text{ kVA}$$

If the power source upstream of the transformer is not infinite, then the amount of power that would be available on the load side of the transformer would be less, and it is calculated using admittances as follows:

- Utility short-circuit power is *UP*, and its admittance is 1/*UP*.
- Maximum transformer let-through is *T*, and its admittance is 1/*T*.
- Net power from the utility let through the transformer *P* is calculated as

$$P = \frac{1}{(1/UP) + (1/T)}$$

Where the utility can supply 385,000 kVA, the amount that the preceding 2000-kVA, 6.75 percent impedance transformer can let through, or admit, is calculated as

$$P = \frac{1}{(1/385{,}000) + (1/29{,}630)} = 27{,}512 \text{ kVA}$$

Let-Through Values of Reactors

Similar to a transformer coil in its current-limiting characteristics, a reactor is a series impedance used to limit fault current. In a three-phase circuit, there are normally three identical reactor coils, each connected in series with the phase conductors between the source and the electrical load. The amount of power that a reactor will let through from an infinite power source to a short circuit on the output terminals of the reactor is calculated as

$$\text{Let-through power} = \frac{1000\ (\text{phase-to-phase kilovolt circuit rating})^2}{\text{reactor impedance in ohms per phase}}$$

For example, in a three-phase reactor operating within a 4.16-kV circuit, the impedance of each of the three reactor coils is 0.125 ohms (Ω). Assuming that an infinite power source is connected to the line terminals of the three-phase reactor, how much short-circuit power would the reactor let through to a "bolted" short circuit on the load terminals of the reactor? The answer is

$$\text{Let-through power} = \frac{1000\ (4.16)^2}{0.125} = 138{,}444\ \text{kVA}$$

Let-Through Power Values of Cables

A length of cable is a series impedance that limits short-circuit current. As with reactors, there is a maximum amount of power any cable will let through to a short circuit from an infinite power source at system voltage. Each size and configuration of cable has unique impedance characteristics (found within the manufacturer's cable catalog), and typical values are shown in Fig. 4-11 for 600-V cable, for 5-kV cable, and for 15-kV cable. The impedance values in this table contain the resistance, inductive reactance, and overall impedance of typical cables in units of ohms per thousand feet.

The amount of short-circuit power that a cable would let through to a short circuit if an infinite power source were connected to one end of the cable and the short circuit is at the other end of the cable is calculated as

$$\text{Short-circuit power} = \frac{1000\ (\text{phase-to-phase kilovolt circuit rating})^2}{\text{cable impedance in ohms per phase}}$$

For example, a 3/c 500-kCMIL copper cable that is 650 feet (ft) long is operated on a 480-V three-phase system. How much power would this cable let through to a short circuit if the impedance of the cable is given as 0.0268 Ω

per thousand feet? The short-circuit capacity of this cable is calculated as

Let-through power

$$= \frac{1000\ (\text{phase-to-phase kilovolt circuit rating})^2}{\text{cable impedance in ohms per phase}}$$

$$= \frac{1000\ (0.480)^2}{(0.0268\ \Omega)\ 0.650}$$

As mentioned earlier, the effect of cable is often neglected in short-circuit calculations. However, the amount of fault current that can flow through a cable, together with the time duration during which it can flow, can be considered in sizing medium-voltage cable using a cable damage curve. A detailed explanation of this thermal damage cable calculation is provided along with cable thermal damage curves in Figs. 4-27 and 4-28.

Sample Short-Circuit Calculation

It is desired to determine the short-circuit value available at the 13.8-kV bus in the electrical system shown in Fig. 5-1. The equivalent drawing of the electrical system is given in Fig. 5-2, along with the calculations and calculation results of the short-circuit values available at different points in the system. Note the arrows showing the direction of power flow into the fault point.

In this calculation, the fault point is the 13.8-kV bus. Fault power is available to flow into this bus from

1. The electrical utility source
2. The 24-megawatt (MW), 30-MVA generator
3. The 5000-hp motor
4. The 200-hp motor
5. The 40-hp motor

If all these loads had been connected directly to the 13.8-kV bus, the calculation would have been quite straightfor-

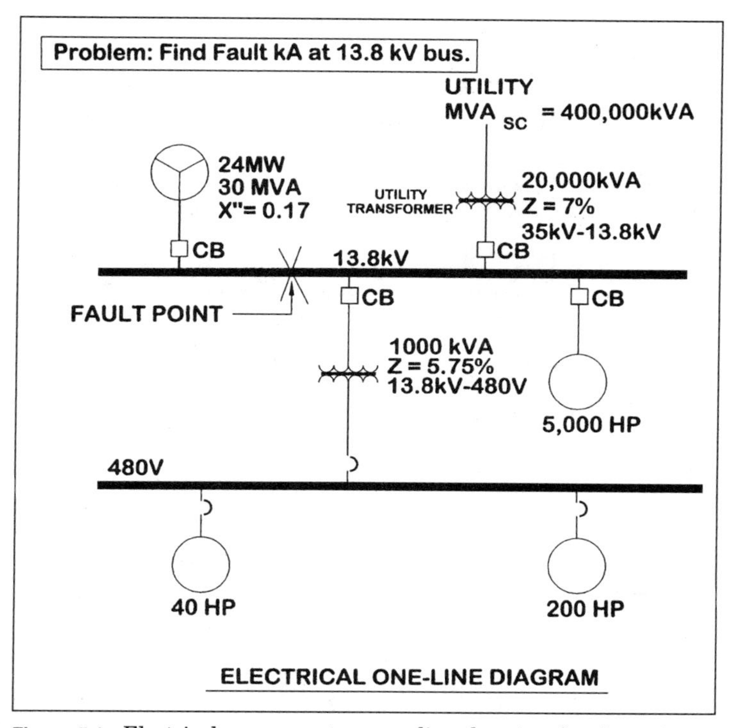

Figure 5-1 Electrical power system one-line drawing for short-circuit calculation example.

ward, but the system contains two transformers that change current values and present large oppositions to fault current flow. Therefore, measures must be used in the calculation procedure to accommodate these transformations and their impedances.

In this example problem, the following steps are taken and are shown in numerical form in Fig. 5-2. First, however, note that electrical short-circuit power can come from four sources in this circuit:

1. The 400,000-kVA electrical utility service
2. The 24-MW, 30-MVA generator
3. The 5000-hp motor
4. The motors connected to the 480-V bus

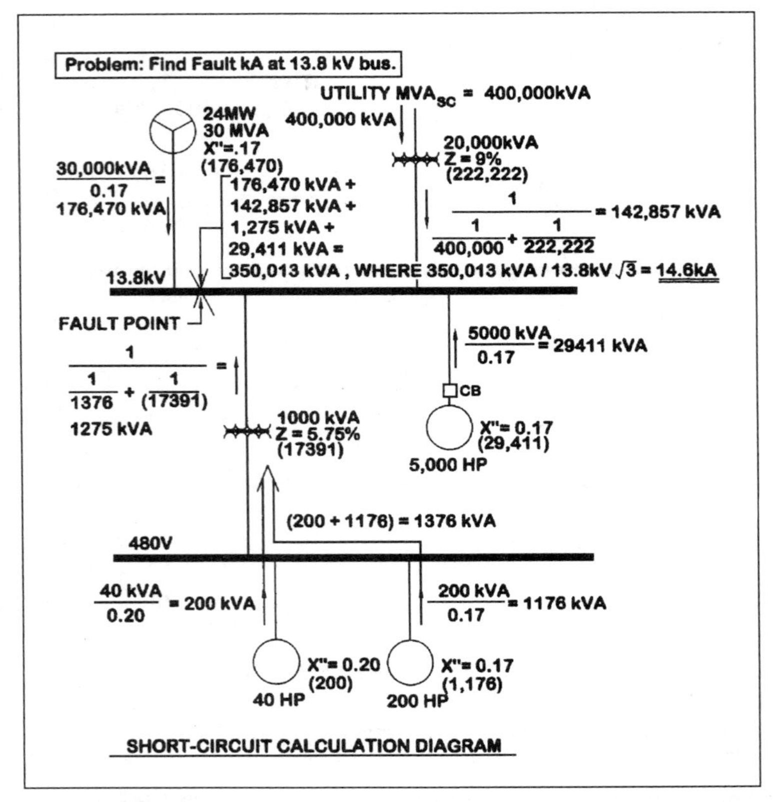

Figure 5-2 Solve for the short-circuit current in the electrical power system of Fig. 5-1.

Step 1. Power from the 400,000-kVA electrical utility service is restrained from flowing entirely into the fault point at the 13.8-kV bus by the impedance of the transformer. As is shown in the calculation in Fig. 5-2, only 142,857 kVA of the original 400,000 kVA of electrical utility power can get through the transformer to the fault point.

Step 2. As shown in the calculation beside the generator in Fig. 5-2, the subtransient reactance of the 30-MVA generator limits its contribution into the fault to 176,470 kVA.

Step 3. The contribution of the 5000-hp motor into the fault is limited by its impedance to 29,411 kVA.

Step 4. The small contributions from the 200- and 40-hp 480-V motors, 1376 kVA, are reduced to an even smaller value by the 1000-kVA transformer to 1275 kVA.

Step 5. The sum of these power contributions into this fault point is 350,013 kVA, as shown in Fig. 5-2. When this value is divided by 13.8 kV multiplied by the square root of 3, the result is the actual fault current at the fault point, which is 14.6 kA.

NOTES

Chapter

6

Generator Sizing Calculations

A generator set consists of an engine driver and an electrical dynamo known as an *alternator* or *generator*. The prime mover delivers the mechanical power to generate the electricity, and the dynamo creates the voltamperes of electrical apparent power. Ignoring inefficiencies or losses, the kilowatt output of the dynamo is equal to the kilowatt power capability of the prime mover, the engine. On the other hand, the power factor of the load may cause the kilovoltampere flow to greatly exceed the numeric value of kilowatts. Accordingly, the kilowatt rating of a generator set is limited by the output of the engine, whereas the kilovoltampere rating of the generator set is limited by the electric current output of the dynamo connected to the engine.

Engine horsepower output is determined by the burning of fuel and air and is almost always power limited by air. The larger the number of pounds of air the engine can take in over a certain time interval, the more oxygen is available to burn the fuel, and the more power the engine can produce. Air is heavier and contains more oxygen molecules when it is cold and when its pressure is high. Therefore, an engine operating on a cold day at sea level can produce more horsepower than an engine operating on a warm day on a mountaintop. It is little wonder that engine manufacturers go to great lengths to provide the largest “charges” of air possible for their engines. In gas-turbine engines this can take the form of a zero-stage turbine compressor section or an inlet air chiller, and in reciprocating engines it can take the form of an exhaust-driven turbocharger set with intercoolers.

In unusual cases, engine horsepower is limited by the heat content of the fuel. For example, in dual-fuel reciprocating engines, the engine can produce about 40 percent more power from diesel fuel than it can from methane fuel, and this is true strictly because of the limited amount of heat energy contained in the methane fuel for each intake stroke charge of fuel-air mixture compared with the amount of heat energy contained in an equivalent volume of diesel fuel–air intake stroke charge.

Sizing a Gas-Turbine Generator Set for a Known Kilowatt Load

The gas-turbine generator set consists of two distinct parts, and each must be sized with a separate calculation. The engine must be rated with sufficient horsepower (this can be measured in kilowatts) to drive the watt load, and the electrical dynamo must be sized to allow selected motors to start while other selected electrical loads are in operation. Except in very unusual electrical systems where there is one very large motor load, the manufacturer of the generator set matches an electrical dynamo with the engine that is kilowatt rated at or above the maximum engine kilowatts and is rated in kilovoltamperes at 125 percent of its kilowatt rating. (This is usually stated by saying that the dynamo is rated at 80 percent power factor.) The method of calculating the generator size is as follows:

A1. Sum all running loads, including approximately an 11 percent motor inefficiency factor for motor windage and friction.

A2. Add the largest nonsimultaneous load that must be started while its counterpart motor is still running. For example, if there is a pump *A* and a backup pump *B,* then only pump *A* is used in the A1 calculation. In this A2 calculation, pump *B* is added into the load sum because both pumps *A* and *B* can be running simultaneously.

A3. Call the load sum determined in step A2 *load A.*

B1. Select the engine-generator set whose International Standards Organization (ISO) kilowatt rating is slightly greater than load *A*. Note that the ISO rating of a gas-turbine engine is at 59°F at mean sea level.

B2. Derate the ISO rating of the engine-generator set for the maximum historical ambient temperature recorded for the site. Do *not* use the maximum design temperature stated for the facility, since there will be days that are hotter than the design temperature, and it is unacceptable for the engine-generator set to trip off on underfrequency (the result of the engine not being able to deliver enough power for the load) just because a day occurs that is hotter than the design temperature. Call this resulting temperature rating *rating B*. This value is determined either from the engine manufacturer's performance curves for the machine or from derating formulas for other than ISO conditions. Either of these normally requires specific performance information for each machine. In the absence of specific performance data on the gas-turbine engine used, an approximation of the reduction in power can be made by assuming that engine shaft horsepower will be reduced by 1 percent for every 2.7°F above the ISO rating of 59°F.

B3. Derate rating *B* by 4 percent for dirt buildup on the turbine blades, and call the resulting rating *rating C*.

B4. Derate rating *C* by 3 percent for blade erosion losses after 25,000 hours of operation (this is immediately before a scheduled hot-section overhaul). Call the resulting rating *rating D*.

C1. Make certain that rating *D* is greater than load *A*. If it is not, then incorporate additional generator sets into the system or select a machine with a greater kilowatt rating. The amount by which rating *D* is greater than load *A* is the percent spare capacity for future loads. Generally, the amount of this spare capacity is within a 15 to 30 percent window, depending on the project and the certainty of the known loads at the time of the generator sizing calculation. Additional spare capacity occasionally can be provided through load shedding.

C2. So that electrical power can still be provided while one generator set is down for maintenance, common practice is to add a second generator that is of the same kilowatt rating as the first generator and normally to operate both generator sets simultaneously at approximately 50 percent load. Another commonly used possibility is to provide three generator sets, two of which can operate the load, while the third is the running spare.

An example of sizing the gas-turbine generators for an industrial plant is given in Fig. 6-1.

Sizing a Reciprocating Engine-Driven Generator Set for a Known Kilowatt Load

Sizing a reciprocating engine-driven generator set for a known kilowatt load is similar to sizing a gas-turbine generator, as described earlier, except that the derating of the engine due to blade-tip burning, dirt buildup, and turbine blade fouling need not be considered. In addition, responses to the altitude of operation are different for the two types of engines, as are their responses to ambient intake air temperature. This is so because most reciprocating engine-driven generator sets incorporate turbocharging and intercooling to effectively negate much of the density altitude characteristics of low-pressure "thin" air and high-temperature "thin" air. Accordingly, as long as the largest motor that must be started by the reciprocating engine-driven generator set is less than 10 percent of the overall rating of the generator set, the problem simply becomes one of determining the kilowatt value of the load, selecting the next larger commercially available reciprocating engine-driven generator set, and making certain that the kilovoltampere rating of the electrical dynamo bolted to the engine is large enough to provide current to the running and starting loads without exceeding the allowable voltage dip for the system. Normally, the allowable voltage dip is around 80 percent of normal sys-

Problem: An coastal industrial plant has a brake horsepower load of 49 MW. Historical checks reveal that the hottest day at the plant site occurred 18 years ago, and that temperature was recorded as 87 degrees F. There largest motor in the plant is 300 horsepower, one of the gas turbine starter motors. And the largest back-up motor that must be started while the "normal-duty" motor is running is 160kW (BHP). The initial selection of LM 2500 gas turbine generator sets has been made, and these units are iso rated at 59 degrees F at sea level at 22.8MW. One running spare generator is to be installed. Determine the quantity of generator sets that must be installed.

A1.	TOTAL RUNNING LOAD = 52 MW (BHP)	49 MW
	ADD 11% FOR MOTOR LOSSES	5.39
A2.	THE LARGEST NON-SIMULTANEOUS LOAD THAT MUST BE STARTED: 160 kW X 111%	0.1776 MW
A3.	**LOAD A:**	**54.5676 MW**
B1.	ISO-RATING OF LM-2500 GENSET (MFGR DATA)	22.8 MW
B2.	DERATION OF ENGINE FOR HOTTEST DAY:	
	87 DEG HOTTEST DAY MINUS 59 DEG ISO RATING= 28 DEG CHANGE	
	28 DEG/2.7 DEG = 10.37	
	(10.37) X 1% = 10.27% DERATING	
	(22.8 MW) X (100% - 10.37%) = 20.43MW CLEAN, NEW CAPABILITY	
	RATING B:	**20.43 MW**
B3.	DERATING OF 4% FOR DIRT BUILD-UP BEFORE WATER WASH:	
	(20.43MW) X (0.96) = 19.612 MW NEW AND DIRTY CAPABILITY	
	RATING C:	**19.612 MW**
B4.	DERATING OF 3% FOR BLADE EROSION:	
	(20.43 MW) X (1.0 - .03) = 19.02 MW DIRTY NEEDING OVERHAUL	
	THE SITE RATING OF ONE GENSET IS 19.02 MW.	
	RATING D:	**19.02 MW**
C1.	POWER AVAILABLE FROM GENSET #1	19.02 MW
	POWER AVAILABLE FROM GENSET #2	19.02 MW
	POWER AVAILABLE FROM GENSET #3	19.02 MW
	CAPABILITY OF THESE 3 GENSETS	**57.06 MW**
	SPARE CAPACITY IN THE 3 GENSETS: (57.06 MW) - (54.5676 MW) =	2.49 MW
C2	IF ONE GENERATOR IS "OFF LINE" FOR MAINTENANCE THEN THE REMAINING TWO GENERATORS COULD NOT SUPPLY THE PLANT LOAD. THEREFORE A FOURTH GENERATOR SET IS SUPPLIED	19.02 MW
	TOTAL SITE RATING OF ALL FOUR GENSETS:	**76.08 MW**

Figure 6-1 Solve for generator site rating given load, temperature, and altitude.

tem voltage at the terminals of the motor being started and around 90 percent of the normal system voltage at the generator switchgear bus. A good rule of thumb to achieve this is to assume that the generator has a subtransient reactance of approximately 15 percent (normally this is a valid assumption) and then to select a generator set whose kilowatt rating exceeds the total kilowattage of the summed loads [setting 1 kilowatt (kW) equal to 1 horsepower (hp) for this calculation] and whose kilovoltampere rating is equal to or greater than 50

Gas-turbine generators arrive at the installation site prepackaged on skids.

percent of the total of the running kilowattage plus six times the kilowatt rating of the largest motor to be started while the other loads are running. For example, if the running load is 1000 kW and a 150-hp motor is to be started meanwhile, the minimum kilowatt rating of the generator would be 1000 kW + 150 kW, or 1150 kW, and the minimum kilovoltampere rating of the generator would be $0.5 \times [1000\text{ kW} + 6(150)]$, or 950 kilovoltamperes (kVA). Select a generator set that meets or exceeds both the 1150-kW and the 950-kVA requirements and that has a voltage regulator that can offset approximately a 15 percent internal voltage dip (15 percent regulation) while the large motor is starting. In this case, the usual offering of a generator set manufacturer would be a set rated at 1150 kW that carries a kilovoltampere rating of 1150/0.8, or 1437 kVA. If a larger motor is to be started, then the kilovoltampere calculation would have more largely determined the size of the generator set than would the running kilowatt load.

Sizing of Generator Feeder Conductors

As is shown in Fig. 6-2, solving for the required ampacity of generator power conductors and generator overcurrent pro-

PROBLEM: A 250 kW GENERATOR HAS A NAMEPLATE RATING OF 312.5 kVA AT 480 VOLTS, 3-PHASE. DETERMINE THE MAXIMUM OVERCURRENT DEVICE SETTING TO PROTECT THE GENERATOR AND THE MINIMUM WIRE AMPACITY CONNECTING THE GENERATOR WITH THE LOAD.

GENERATOR FULL-LOAD AMPERES

$$\frac{312{,}500 \text{ VA}}{(480\text{V})(1.73)} = 376 \text{ AMPERES}$$

MAXIMUM GENERATOR PROTECTIVE OVERCURRENT DEVICE: 376 AMPERES

IF THE GENERATOR IS FITTED WITH OVERCURRENT SENSORS OTHER THAN THE MAIN CIRCUIT BREAKER, THEN THE MAIN CIRCUIT BREAKER CAN BE INCREASED IN RATING UP TO THE AMPACITY OF THE CONDUCTORS IT PROTECTS, 115% OF 376 AMPERES.

MINIMUM AMPACITY OF CONDUCTORS FROM GENERATOR TO THE FIRST OVERCURRENT DEVICE:
115% OF 376 AMPERES = 432 AMPERES

Figure 6-2 Solve for generator overcurrent protection and the wire ampacity of generator conductors.

Diesel engine generator set.

tection devices requires unique methodologies. The generator kilowatt rating is set by the engine power, but the kilovoltampere and current-carrying capabilities of the electrical dynamo bolted to the engine depend on the size of the dynamo and its reactance values. Almost universally, the generator set kilovoltampere rating is designed to be 125 percent of the generator kilowatt rating; and this is done to accommodate the lagging power factor of the loads, since quadrature-component currents do not require engine power because they do no work. Figure 6-1 shows how to protect the electrical dynamo in the generator set from overcurrent and how to size and protect the conductors that extend from the generator set toward the loads.

NOTES

Chapter

7

Grounding

The Functions of Grounding

The work of grounding systems is probably one of the best kept set of secrets in the electrical industry. At first glance, the deceptively simple passive elements of grounding systems obviously could not do very much, or could they? The answer is that grounding systems come in many shapes, forms, and sizes and do many duties, many of which are absolutely essential. If they are designed and constructed well, then the systems they support have a good chance of working well. However, if the grounding system is flawed in design or installation, or if it is damaged by impact or chemical attack, the related systems are negatively affected.

Consider the case of a static grounding grid with its variety of grounding electrode shapes in an industrial plant that is energized through a high-voltage utility substation. This almost completely hidden grounding system performs all these tasks:

- It minimizes the ground potential rise and coincident step and touch potentials that occur from high-voltage system zero sequence current flowing through the earth during utility system ground faults, such as insulator-string arc-over.

- It equalizes the direct-current (dc) potentials within the plant that build up from process flows.
- It limits the system-to-frame voltage for human safety and prevents overstress in phase-to-ground voltage.
- For all practical purposes, it provides an equipotential plane on which humans can stand and not be harmed during times of ground fault within the plant. That is, it equalizes the potential of, say, a motor stator that a maintenance person might be touching during a ground fault and the surrounding earth on which the person would be standing. With no potential across the person's body, no harmful current can flow through the body.
- It provides a ground reference plane to which all the instruments in the plant control system can be referenced.
- It provides a secondary path through which ground-fault current can flow back to the last transformer (or generator) ground point in the event of loss of the equipment grounding conductor path, thereby providing increased assurance of tripping of overcurrent devices on ground fault and providing enhanced personnel safety from stray current flow and from flash burns and induced fires.
- It provides an earthing point for lightning protection or for lightning-avoidance systems.
- It provides a catholic protection current return path.

While providing all these functions, the grounding system also controls the stress on the system insulation during times of ground fault. In ungrounded systems, such as a 460-volt (V), three-phase, three-wire delta system, an arcing fault that repeatedly restrikes can cause voltage "jacking" of up to six times the normal system phase-to-neutral voltage. In these systems, even where restrike does not occur, the power system insulation must withstand 173 percent of the normal phase-to-ground system voltage because the potential difference between the point of the first ground fault and the opposite phase conductor is full phase-to-phase voltage.

In a typical high-rise commercial building, the grounding-electrode system functions in a manner similar to that in an industrial plant, except that the shapes of the grounding electrodes and grounding-electrode conductors are different, and additional functions are accomplished. The structural steel columns within the building are suitable for lightning-protection "down" conductors, and they are also suitable for use as the grounding-electrode system for each local transformer secondary on upper floors. These steel members assist in the attenuation of magnetic noise emanating from outside the building by forming a sort of "Faraday cage" around the building contents. In buildings containing radio transmitters with rooftop antennas, however, these same steel columns form a part of the radiating-element–ground-plane system that tends to cause high-frequency noise within the building systems rather than attenuate it. In almost every type of structural steel building design, however, building steel forms a very good low-impedance path for ground-fault current to promote rapid overcurrent device tripping and enhance personnel safety and to eliminate arcing noise of the type that comes from arcing faults in high-impedance equipment grounding conductor paths.

If the building contains specialty systems, such as Article 645 information technology equipment or Article 517 anesthetizing locations, the grounding system is relied on to perform all the functions just listed *and* perform the additional duties of minimizing even low-voltage potential differences between any two conductive points in locations where delicate biologic tissues or semiconductor devices are located. Part of the methodology used to perform these functions rests in the ability of equipment grounding-conductor forms, such as conduits, to absorb transmitted energy by transforming electromagnetic (emf) waves into eddy currents and heat instead of letting the emf "cut" system wires, thereby inducing noise within the wires. In fact, the entire functionality of digital systems requires this elimination of noise voltages that the equipment could erroneously interpret as valid information. It is toward this goal that specific modifications of grounded-cable shields (terminate and ground on only one end), conduits (install insulating section with internal equipment grounding conductor), or cables (provide many concentric wraps per foot of cable) are done. And if these and similar steps are insufficient to guard against voltage transients, then the grounding-electrode system provides the equipotential plane to which one side of transient surge suppressors can be connected to "short to ground" these unwanted voltages at wire terminations.

For verification that the grounding system is really extremely important to the normal operation of a facility, just remember what has happened in locations where the grounding systems have been impaired: Fires have started from arcing faults, computers have crashed from spurious noise data, data-control systems have shut down processes in error, human hearts have been defibbed and stopped, motors have burned up and arc-type lamps have turned off due to voltage imbalances caused by bad service bonding jumper terminations, voltage jacking has occurred on ungrounded power systems or systems that have lost their grounding connection, and pipes have sprung leaks as a result of catholic erosion. Truly, although they are generally hidden from view, grounding systems and their many actions are extremely important.

Calculating the Resistance to Remote Earth of Ground Rods

The *National Electrical Code* requires a minimum resistance to remote earth of 25 ohms (Ω) for a grounding electrode made of a rod of which at least 8 feet (ft) is buried. When its measured resistance exceeds the 25 Ω, the code requires the installation of *one* additional ground rod at least 6 ft away from the first. The rods must have a minimum outside diameter (OD) of $^3/_4$ inch (in) if they are made of galvanized pipe, a minimum of $^5/_8$ in OD if they are solid iron or steel, and a minimum of $^1/_2$ in OD if they are listed and made of nonferrous material such as copper or stainless steel. Due to the sacrificial nature of aluminum, the code prohibits the use of ground rods made of aluminum.

The function of the grounding-electrode system is to keep the entire grounding system at earth potential during lightning and other transients. Its function is not principally for conducting ground-fault current, even though some zero-sequence current could flow through the grounding electrode during a ground fault. However, where served from overhead lines where the fault current return path(s) could break and become an open circuit, grounding system designs also should be aimed at reducing the potential gradients in the vicinity of the ground rods. This will help to achieve safe step and touch potentials under ground-fault conditions in the electrical power supply system.

Earth-electrode resistance is the number of ohms of resistance measured between the ground rod and a distant point on the earth called *remote earth.* Remote earth is the point where the earth-electrode resistance no longer increases appreciably when the distance measured from the grounding electrode is increased, which is typically about 25 ft for a 10-ft ground rod. Earth-electrode resistance equals the sum of the resistance of the metal ground rod and the contact resistance between the electrode and the soil plus the resistance of the soil itself. The relative resistances of commercially available metal rods are as follows:

Copper	100 percent relative conductance
Stainless steel	2.4 percent relative conductance
Zinc-coated steel	8.5 percent relative conductance
Copper-clad steel	40 percent relative conductance

Since the resistance values of the ground rod and the soil contact resistance are very low, for all practical purposes, earth-electrode resistance equals the *resistance of the soil surrounding the rod.* Except for corrosion considerations, the type of metal of which the ground rod is made has *almost no effect* on its earth-electrode resistance because this resistance value is almost entirely determined by the soil. Evidence of this is shown in the following formula for calculating the resistance of a ground rod to remote earth, where the type of metal in the rod is not even in the formula. The resistance R of a ground rod can be approximated as

$$R = \frac{\rho}{1.915L}\left(\ln \frac{96L}{d} - 1\right)$$

where ρ = soil resistivity, ohm-meters ($\Omega \cdot$ m)
L = rod length, ft
d = rod diameter, in

For example, if the soil resistivity averages 100 Ω·m, then the resistance of one 0.75-in × 10-ft electrode is calculated to be 32.1 Ω.

The values of some typical soil resistivities, given in ohm-meters, are as follows:

Loam	25
Clay	33
Sandy clay	43
Slate or shale	55
Silty sand	300
Gravel-sand mixture	800
Granite	1000
Gravel with stones	2585
Limestone	5000

Variables other than soil resistivity are rod length and diameter. Experimenting with a series of calculations of rods having differing lengths shows that the diameter of the rods also makes very little difference in the ultimate resistance to remote earth. It follows that unless they penetrate the local water table, ground rods that are longer than 10 ft often provide only insignificant additional reductions in resistance to remote earth, assuming uniformity of soil resistivity. For example, the resistance of a $^3/_4$-in rod in a loam soil only decreases from 8.2 Ω for a 10-ft rod to 3.2 Ω for a 30-ft rod. This is a relatively small improvement when compared with the reduction from 52 Ω for a 1-ft rod to 8.2 Ω for a 10-ft rod. Improving the soil resistivity characteristics immediately surrounding the rod can do this same job and normally do it much more easily and cost-effectively.

The two exceptions to this rule are (1) in a very dry soil, extending the ground rod down into the permanent groundwater dramatically improves the resistance value, and (2) during the winter, having the ground rod extended to the deep nonfrozen soil greatly improves its resistance value over what it would have been in frozen soil or ice.

Soil resistance is nonlinear. Most of the earth-electrode resistance is contained within a few feet of the ground rod and is concentrated within a horizontal distance that is 1.1 times the length of the ground rod. Therefore, ground rods that are installed too close together are essentially trying to flow current in the same earth volume, so their parallel resistance to remote earth is less than would be expected for parallel resistances in a normal electric circuit. For maximum effectiveness, each rod must be provided with its own volume of earth having a diameter that is approximately 2.2 times the rod length. Figure 7-1 presents a calculation for the resistance to remote earth of a $^3/_4$-in × 10-ft copperweld ground rod driven into soil having a resistivity of 200 $\Omega \cdot m$.

Grounding-Electrode Conductors

Connecting an electrical system to a grounding electrode requires a grounding-electrode conductor. The minimum size of grounding-electrode conductor is shown in the *National Electrical Code* in Table 250-66. The size of the grounding-electrode conductor is based on the amount of fault current that it might be called on to carry, and this is measured by the size of the largest phase conductor in the service feeder. See Fig. 7-2 for an example problem in sizing the grounding-electrode conductor.

Equipment-Grounding Conductors

When there is a ground fault, a low-impedance path must be provided from the point of fault to the neutral of the supply transformer or to the generator. This low-impedance path is provided by the equipment-grounding conductor. This conductor can take several forms, such as different

PROBLEM:
A 3/4" x 10' copperweld ground rod is driven into soil having a resistivity of 200 ohm-meters. Calculate the resistance of the ground rod to remote earth.

where ρ = soil resistivity in Ohm-meters
L = Rod length in feet
d = rod diameter in inches

$$R = \frac{\rho}{1.915L}\left[\ln \frac{96L}{d} - 1\right]$$

$$R = \frac{200}{(1.915)(10)}\left[\ln \frac{(96)(10)}{0.75} - 1\right]$$

$$R = 10.443\left[\ln \frac{960}{0.75} - 1\right]$$

$$R = 10.443\left[(\ln\ 1280) - 1\right]$$

$$R = 10.443\left[(7.1546) - 1\right]$$

$$R = \underline{\underline{64.27}}\ \text{ohms}$$

Figure 7-1 Solve for the resistance of a ground rod to remote earth given rod and soil characteristics.

types of conduit or wire, any of which must be installed in close proximity to the phase conductors. When the conductor is wire, the minimum size of equipment-grounding conductor that must be installed is based on the ampere rating of the overcurrent device immediately upstream of the feeder or branch circuit. The minimum size of this conductor is shown in Table 250-122 of the *National Electrical Code*. See Fig. 7-3 for an example problem in sizing the equipment-grounding conductor.

When feeders other than service feeders are installed in parallel using wires for the equipment-grounding conductors,

GROUNDING ELECTRODE CONDUCTOR FOR AC SYSTEMS (This table is based upon Table 250-66 of the National Electrical Code)	
SIZE OF LARGEST PHASE CONDUCTOR (COPPER)	**REQUIRED MINIMUM SIZE OF GROUNDING ELECTRODE CONDUCTOR**
#2 OR SMALLER	8
#1 - 0	6
00 - 000	4
000 - 350 kCMIL	2
351 kCMIL - 600 kCMIL	0
601 kCMIL - 1100 kCMIL	00
over 1100 kCMIL	000

Figure 7-2 Solve for the grounding-electrode conductor size given the size of the largest phase conductor.

EQUIPMENT GROUNDING CONDUCTOR FOR AC SYSTEMS (This table is based upon Table 250-122 of the National Electrical Code)	
AMPERE SETTING OF UPSTREAM OVERCURRENT DEVICE	**REQUIRED MINIMUM SIZE OF COPPER EQUIPMENT GROUNDING CONDUCTOR**
15	14
20	12
30	10
40	10
60	10
100	8
200	6
300	4
400	3
500	2
600	1
800	0
1000	00
1200	000
1600	0000
2000	250
2500	350
3000	400
4000	500

Figure 7-3 Solve for the equipment-grounding conductor size given the ampacity rating of the overcurrent device.

although the phase and grounded (neutral) conductors can be reduced in size down to a minimum of 1/0 American Wire Gauge (AWG), load ampacity permitting, the fully sized equipment-grounding conductor must be installed within each and every parallel raceway. This can become an issue when the parallel phase and neutral conductors are installed in the form of a multiconductor cable. Standard equipment-grounding conductor sizes for nonparalleled cables are predetermined and inserted into standard cables. The net result of this is that when using factory-standard cables for parallel circuits, slightly oversized phase conductors must be selected to provide large enough equipment-grounding conductors, as shown in Fig. 7-4.

Since system grounding is done at the service-disconnecting means, the equipment-grounding conductor and the neutral ("grounded") conductor are two separate conductors that are insulated from one another at every point downstream of the service-disconnecting means, where they are bonded together with the bonding jumper. Upstream of the service-disconnecting means, however, the neutral conductor is both

EQUIPMENT GROUNDING CONDUCTOR IN CABLE FOR AC SYSTEMS

AMPERE SETTING OF UPSTREAM OVERCURRENT DEVICE	REQUIRED MINIMUM SIZE OF COPPER EQUIPMENT GROUNDING CONDUCTOR	STANDARD TC CABLE LAYUP FOR 1 CABLE PHASE	STANDARD TC CABLE LAYUP FOR 1 CABLE GROUND
150	6	0	6
200	6	000	4
300	4	350	3
400	3	500	2

FOR A 400 AMPERE CIRCUIT, TWO 3/0 3/C + #4g CABLES ARE CONSIDERED; BUT THE #4 IS TOO SMALL TO BE IN COMPLIANCE WITH NEC TABLE 250-122 THAT REQUIRES A #3. TO OBTAIN A LARGE ENOUGH EQUIPMENT GROUNDING CONDUCTOR FOR THIS PARALLEL CABLE INSTALLATION, A CABLE WITH AT LEAST A #3 EQUIPMENT GROUNDING CONDUCTOR MUST BE SELECTED FOR EACH OF THE TWO PARALLEL CABLES. THEREFORE, SELECTING FROM THE STANDARD TC CABLE LAYUP, TWO PARALLEL TC CABLES THAT ARE 300 kCMIL + #3 G MUST BE USED. THIS MEANS THAT CABLE RATED AT THE SUM OF 285 + 285*, OR 570 AMPERES MUST BE USED DOWNSTREAM OF THIS 400 OVERCURRENT DEVICE SO THAT A LARGE ENOUGH EQUIPMENT GROUNDING CONDUCTOR CAN BE OBTAINED WITHIN EACH CABLE.

* *(NOTE THAT EACH 300 kCMIL CABLE IS RATED AT 285 AMPERES AT 75 DEGREES C.)*

Figure 7-4 Solve for cable size required in standard cable layup to obtain large enough equipment-grounding conductors with parallel cables.

NEUTRAL SERVICE CONDUCTOR FOR AC SYSTEMS	
SIZE OF LARGEST PHASE CONDUCTOR (COPPER)	**REQUIRED MINIMUM SIZE OF NEUTRAL SERVICE CONDUCTOR**
#2 OR SMALLER	8
#1 - 0	6
00 - 000	4
000 - 350 kCMIL	2
351 kCMIL - 600 kCMIL	0
601 kCMIL - 1100 kCMIL	00
over 1100 kCMIL	12.5% OF THE AREA OF THE LARGEST PHASE CONDUCTOR

Figure 7-5 Solve for the minimum size of neutral service conductor size given the size of the largest phase conductor.

the grounded neutral conductor and the equipment-grounding conductor, so raceways, metering enclosures, and similar conductive equipment are made safely grounded upstream of the service-disconnecting means by being bonded to the neutral conductor there. The neutral conductor in the service feeder must be large enough to carry the fault current back to the transformer neutral point, so it must be at least as large as the grounding-electrode conductor, sized in accordance with Table 250-66 of the *National Electrical Code,* and it must be a minimum of 12.5 percent of the size of the largest phase conductor as well, as is shown in Fig. 7-5.

Methods of Grounding Systems

In residential and commercial establishments, the most common way of grounding an electrical power system is by *solidly grounding* it. This means that a conductive path is installed between the system grounding point (most commonly, this is the "neutral" common center point of a three-phase wye system or the "neutral" common center point of a single-phase system) and the grounding-electrode conductor system. Where this solidly grounded method is installed, the idea is to facilitate current flow during times of phase-to-ground fault so that the overcurrent device can rapidly operate and "clear" the fault. Once a fault occurs in such systems, the tripping of the overcurrent device deenergizes that portion of the electrical power system within only a few cycles.

In many industrial establishments, however, the preferred method of generator grounding or transformer grounding is through an impedance (*reactor* grounding used to limit fault current in a generator to 25 to 100 percent of three-phase fault-current levels) or a resistance. The value of this system grounding method is that even after the first ground fault, the electrical equipment can remain in service instead of being tripped off-line, as is done in a solidly grounded system. Grounding resistors or reactors are inserted in series with the grounding-electrode conductor from the center-tap point of a wye transformer or generator. The advantages of resistor grounding of electrical power systems include

- Reduced magnitudes of transient overvoltages
- Simplification of ground-fault protection
- Improved system and equipment protection from ground faults
- Improved service reliability
- Reduced fault frequency
- Greater personnel safety

There are several things to keep in mind when designing this type of a grounding system:

1. The type of grounding system and resistor is normally decided by the magnitude of capacitive charging current of the system. If the system is small (i.e., the total length of low-voltage cable or shielded medium-voltage cable is short), the charging current is small. An example of this would be a low-voltage distribution system. If the charging current is less than 5 amperes (A), the system could be operated as *high-resistance grounded,* wherein the adjustable grounding resistance is selected to limit the ground current to approximately

5 to 7 A. Ground fault on this type of system is normally alarmed only and not tripped. The grounding resistor must be sized for 10 A, continuous. Although the resistance-grounded system limits fault current to low values, it still effectively limits transient overvoltages.

2. If the system is larger, the charging current is more because of the larger size of electrical equipment (generators, transformers, motors, etc.) and the use of longer and larger cable sizes or shielded cable. If the charging current is about 10 to 15 A, the system could be operated using a *low-resistance grounded system,* where the fault current is limited to between 200 and 1000 A. Most often, 400 A is selected as the relay pickup point. With this type of system, during a ground-fault condition, the faulted feeder and connected equipment are isolated by the tripping of associated protective devices. Where the high-resistance grounding system of item 1 permitted the faulted equipment to remain in operation, the low-resistance grounded system does not. For this system, the grounding resistor must be rated for 400 A, 10 seconds, and the overcurrent device must be set to trip open the circuit within 10 seconds to prevent resistor damage.

3. If the charging current is more than 15 A, the system is normally designed and operated as solidly grounded. Under such conditions, the faulted feeder and associated equipment are isolated immediately by tripping.

4. Reactance grounding is not considered to be an alternative to resistance grounding because it still allows a high percentage (25 to 100 percent) of the fault current to flow. It is generally used only to limit ground-fault current through a generator to a value that is no greater than the three-phase fault current contributed by the generator.

Obtaining the System Grounding Point

The point at which an electrical power system is grounded must have an infinite impedance to ground-fault current while effectively being a short circuit to ground-fault current. This point is most commonly the neutral point of a

CONSIDERATION	UNGROUNDED SYSTEM	SOLID-GROUNDED SYSTEM	REACTANCE-GROUNDED SYSTEM LOW REACTANCE	REACTANCE-GROUNDED SYSTEM HIGH REACTANCE	LOW-RESISTANCE RESISTOR-GROUNDED SYSTEM	HIGH-RESISTANCE RESISTOR-GROUNDED SYSTEM
CURRENT FOR PHASE-TO-GROUND FAULT IN PERCENT OF 3-PHASE FAULT CURRENT	LESS THAN 1%	VARIES, IS APPROXIMATELY 100%	60% - 100%	5% - 25%	5% - 20% 5 - 20%	LESS THAN 1%
TRANSFORMER OVERVOLTAGES	VERY HIGH	NEGLIGIBLE	NEGLIGIBLE	VERY HIGH	NEGLIGIBLE	NEGLIGIBLE
AUTOMATIC SEGREGATION OF FAULTED ZONE	NO	YES	YES	YES	YES	NO
REMARKS	NOT RECOMMENDED DUE TO OVER-VOLTAGES AND NONSEGREGATION OF FAULT	GENERALLY USED ON SYSTEMS OF LESS THAN 600V AND OVER 15kV	GENERALLY USED ON SYSTEMS OF LESS THAN 600V AND OVER 15kV	NOT USED DUE TO EXCESSIVE OVERVOLTAGES AND HIGH VALUES OF FAULT CURRENT	GENERALLY USED ON INDUSTRIAL SYSTEMS OF 2.4kV TO 15 kV	GENERALLY USED ON SYSTEMS OF LESS THAN 5 kV

Figure 7-6 Summary of grounding methods, their characteristics, and their results.

wye-connected transformer or generator. However, where a neutral point of a wye-connected machine or transformer is not available, a fully functional alternative grounding point can be made at a convenient point in the electrical power system through the use of a wye-delta grounding transformer or a zigzag grounding autotransformer. Figure 7-6 presents a summary of grounding methods for electrical power systems along with their characteristics and results.

NOTES

Chapter 8

Lighting

The Lumen Method

It is always valuable to keep in mind the definition of a *footcandle:* It is the quantity of lumens per square foot of illuminated work surface. This definition alone provides the lighting designer a quick rough estimate of the quantity of lumens required to illuminate (fall on) a certain area: *Take the footcandle value and multiply it by the area in square feet:*

Approximate lumen quantity falling on the area

$$= \text{footcandles} \times \text{area in square feet}$$

A more accurate and easy approximation of the actual lamp lumen requirement (how many lumens must be emitted by the lamps *within* the luminaires) is also possible. Note that some of the lamp lumens are trapped within the fixture and do not reach the area to be illuminated. To provide for this inaccuracy, a determination is made of the rough approximate total quantity of lumens required to illuminate the area; dividing the total lumen requirement by the lumen output from each luminaire (typically one-half the lamp lumens) provides a good estimate of the quantity of luminaires required to illuminate the area. One-half the lamp lumens is a good estimate of the light emitted by each fixture after the combination of luminaire coefficient of use (CU) and light losses is considered:

$$\text{Approximate lumen quantity emitted by lamps} = \frac{\text{footcandles} \times \text{area in square feet}}{0.50}$$

or, stated in another way,

$$\text{Approximate lumen quantity emitted by lamps} = 2 \times (\text{footcandles} \times \text{area in square feet})$$

This quick calculation method is also a good method of checking intensive manual calculations or computer calculations to see if their results are reasonable.

From this basic logic, the following lumen method of calculation formulas are derived:

$$\text{Luminaire quantity} = \frac{\text{footcandles} \times \text{area in square feet}}{(\text{lumens/lamp})\ (\text{lamps/fixture})\ (\text{fixture CU})\ (\text{maintenance factor})}$$

The footcandle illuminance determination can be made from a known area and known lighting layout in this way:

$$\text{Footcandles} = \frac{(\text{no. of fixtures})\ (\text{lumens/lamp})\ (\text{lamps/fixture})\ (\text{fixture CU})\ (\text{maintenance factor})}{\text{area in square feet}}$$

For example, if a 10,000-square-foot (ft^2) area is to be illuminated by lamps whose lumen output is 2500 lumens per lamp and the type of luminaire and maintenance factors are unknown (except that it is known that one luminaire will contain one lamp), the approximate quantity of luminaires required to achieve an illuminance of 5 fc will be

$$\text{Luminaire quantity} = \frac{(5 \text{ fc}) \times (10{,}000 \text{ ft}^2)}{(2500 \text{ lumens per lamp})\,(1)\,(0.5)}$$

$$= 40 \text{ single-lamp luminaires}$$

The definitions and finer points of maintenance factors and fixture CUs are discussed in detail later in this chapter and in Chap. 10.

The lumen method does not give an indication of what the footcandle level will be at any one specific point or workstation. For this, it is necessary to use the point-by-point method.

The Point-by-Point Method

How to make point-by-point calculations

Later in this chapter, lighting calculations within areas having reflective surfaces, such as interior walls, are shown and

High-mast lighting can provide even illuminance levels.

explained. However, where no reflective surfaces exist, all the light falling on the work surface must be provided directly from the luminaires shining on the work surface, and the method of manual lighting calculation that is most accurate for points on the work surface is known as the *point-by-point method of lighting calculation* (also known as the *point-to-point method*). This method is also used most often when light at specific locations on the work surface must be known and for floodlighting calculations.

Direct lighting diminishes inversely as the distance squared. This relationship can be used to determine the illumination level, or footcandle level, at a specific point, for with this relationship the footcandle value can be calculated from the candlepower directed toward that point, the distance from that point to the light source, and the angle of incidence the light rays make with the lighted surface.

When the light rays are not falling perpendicularly onto the lighted surface, the full impact of the light is not available to illuminate the surface. Exactly how much illumination will result is easily determined by these two relationships:

1. Footcandles measured at the work surface with the lightmeter laid flat on the work surface are equal to the candlepower (CP) intensity multiplied by the excluded angle made by the light ray and the work surface divided by the square of the distance between the luminaire and the point on the work surface:

$$\text{Footcandles} = \frac{\text{CP} \times \cos\theta}{(\text{distance in feet})^2}$$

Note that the angle θ is the excluded angle that the light ray makes with the work surface or with the face of an imaginary lightmeter laid flat on the work surface.

2. Note that for "normal" footcandle values, the face of the lightmeter is perpendicular to the light ray, so the angle θ that the light ray makes with the face of the lightmeter is zero, and the cosine of zero is 1.0 . Therefore, for footcandle values immediately below the centerline of a luminaire (known as at *nadir*), this formula simplifies to

$$\text{Footcandles} = \frac{\text{CP}}{(\text{distance in feet})^2}$$

The left side of Fig. 8-1 shows a completed sample problem solving for normal footcandle values for the case directly below (at nadir) the luminaire.

To understand how these equations are used, it is necessary to know the following definitions and concepts:

1. Distance in point-by-point calculations is the quantity of feet between the lighting fixture and the point at which an imaginary lightmeter is placed at the work surface to be illuminated.

2. Candlepower is the value of light intensity emitted by the lighting fixture in the direction formed by a line between the center of the lamp and the center of the imaginary lightmeter.

3. The lighting calculations result in footcandle illumination values that would be displayed on an imaginary footcandle lightmeter located at the illumination point on the work surface. The lightmeter would be positioned so that its photocell pickup transducer would be parallel to the plane of the work surface rather than perpendicular to the ray of light coming from the luminaire.

4. Unless specifically stated otherwise in a given problem, light illuminance is stated in horizontal footcandles. *Horizontal footcandles* are the measure of light falling perpendicularly onto a horizontal surface.

The quantity of horizontal footcandles is equal to the candlepower emitted by the luminaire in the exact direction of the point on the surface to be lighted multiplied by the cosine of the angle the light ray makes with the surface to be lighted and divided by the square of the distance between the luminaire and the point on the surface to be lighted.

It is not necessary that the lighting designer be a mathematician skilled in trigonometry, but rather that the lighting designer simply understand that the light ray is not hitting the work surface squarely. Compensation must be made for this by multiplying the lighting intensity value by

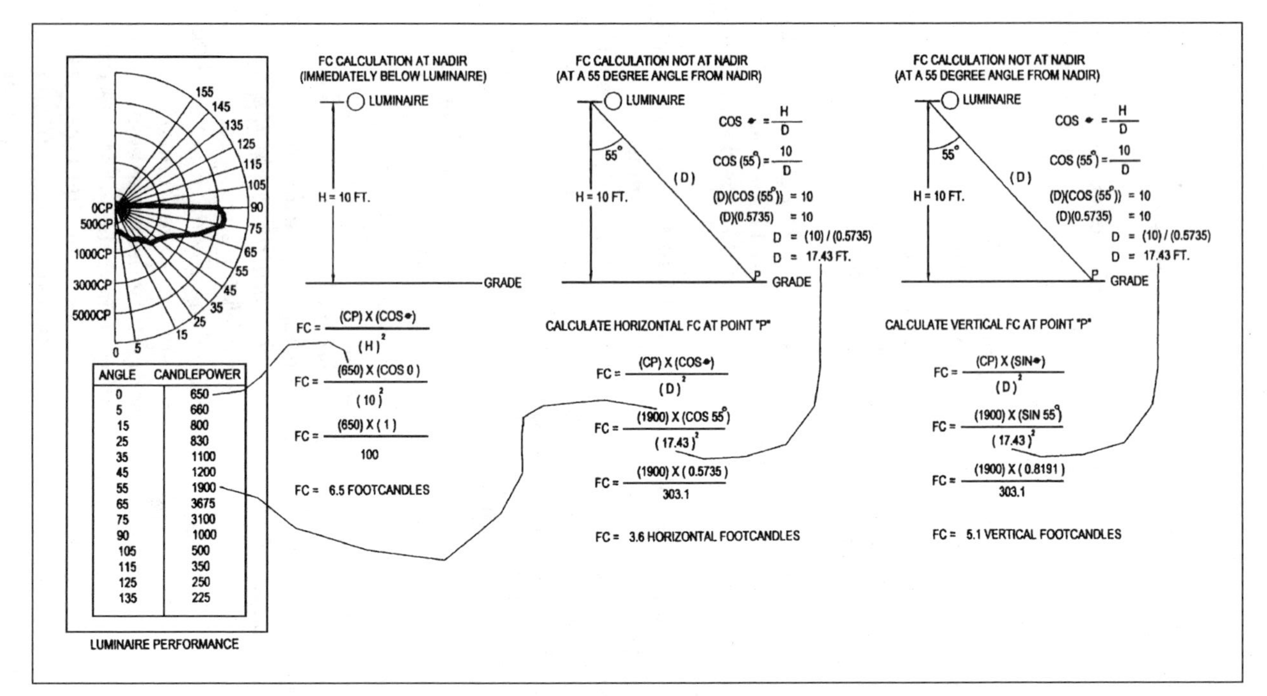

ANGLE	CANDLEPOWER
0	650
5	660
15	800
25	830
35	1100
45	1200
55	1900
65	3675
75	3100
90	1000
105	500
115	350
125	250
135	225

Figure 8-1 Solve for "normal" vertical and horizontal footcandles using the point method given candlepower intensity, height, and distance or angle.

the factor every scientific calculator will show when the angle is entered and the "COS" key is depressed.

5. When it is specifically stated within a given problem that the lighting designer is to solve for a light intensity hitting a facia, a sign, the side of a building or tank, or some other vertical surface, the light intensity must be solved in vertical footcandles. *Vertical footcandles* are the measure of light falling perpendicularly onto a vertical surface, similar to light from an automobile headlight illuminating a garage door.

The quantity of vertical footcandles is equal to the candlepower emitted by the luminaire in the exact direction of the point on the surface to be lighted multiplied by the sine of the angle the light ray makes with the surface to be lighted and divided by the square of the distance between the luminaire and the point on the surface to be lighted.

Again, it is not necessary that the lighting designer be a mathematician skilled in trigonometry, but rather that the lighting designer simply understand that the light ray is not hitting the work surface squarely. Compensation must be made for this by multiplying the lighting intensity value by the factor that every scientific calculator will show when the angle is entered and the "SIN" key is depressed.

6. The point directly under a lighting fixture that is aimed directly downward is known as *nadir.*

7. Candlepower emitted from a fixture at any one angle from nadir does *not* represent the candlepower emitted at any other angle.

8. All frequencies of light follow the same intensity and formula calculations. Therefore, lighting intensity calculations simply ignore light color.

9. The angle θ is the angle formed between straight down, known as *nadir,* and the line formed between the center of the lamp and the center of the imaginary lightmeter that is centered on the beam of light.

10. The candlepower values shown in luminaire photometric data already include the fixture CU and efficiency; therefore, in point-by-point calculations using candlepower data, the fixture CU and efficiency can be ignored.

And the last concept is the one on which all area lighting calculations are based, so it can be considered to be possibly the most important point of all.

11. If several luminaires contribute to the illumination at a point, the resulting illumination is determined by making an individual calculation of the horizontal footcandle contribution by each individual luminaire and then summing these contributions in a normal algebraic manner. For example, if two luminaires shine on one certain point, the total horizontal footcandle level at that point would be equal to the sum of the horizontal footcandles from the first luminaire plus the horizontal footcandles from the second luminaire.

Sample point-by-point lighting calculations

Refer to Fig. 8-1 for sample point-by-point lighting calculations for both the nadir point and a point that is not directly below the fixture aimed in a downward direction where the horizontal and vertical footcandle values must be determined.

The first sample calculation is as follows: What is the footcandle intensity immediately below the fixture? To solve this

High-mount fixtures are the most cost-effective way of lighting an outdoor area.

problem, it is first necessary to note that the question has not specifically asked for vertical footcandle intensity, and therefore, the final answer should be in horizontal footcandles.

Next, an inspection of the figure shows that the photometric curves of the fixture have already been read, and their values at key angles have been placed in table form beside the sketch of the lighting fixture and the surface to be illuminated. The formula incorporating the cosine function provides horizontal footcandles (important for lighting a walkway), whereas the formula incorporating the sine function provides vertical footcandles (important for lighting a wall).

Indoor Lighting

Zonal cavity method for indoor lighting calculations

To be able to properly design lighting systems for indoor locations, the lighting designer needs to understand the *zonal cavity method* of calculations and all the factors that enter into them. This section details the zonal cavity method of calculations, describes how these calculations are made, and provides reference material such as reflectance values for different colors and textures of interior surfaces.

Lighting is provided by two components:

- Direct light
- Reflected light

In all point-to-point calculations, only the direct-light component is considered, and this is acceptable for use outdoors, where few reflective surfaces exist.

When lighting calculations are made for indoor areas, consideration of the reflected light is frequently needed for more accuracy because of the large amount of reflected light from the surfaces of rooms. The zonal cavity method of lighting calculations provides a way of calculating the sum of the direct light and the reflected light, thus calculating all the light that will shine on a work surface. The reflectances (in percent reflected lumens) of different colors and textures of wall, ceiling, floor, and furniture surfaces (painted with flat paint) are shown below:

Light red	70%
Dark red	21%
Light orange	68%
Dark orange	35%
Light yellow	82%
Light green	73%
Dark green	7%
Light blue	68%
Dark blue	8%
Light gray	65%
Dark gray	14%

As its name implies, the zonal cavity calculation method suggests that there are certain cavities that are affected by light or which affect the lighting within them. In this calculation methodology, there are three cavities in a room:

- The room cavity (h_{RC})
- The ceiling cavity (h_{CC})
- The floor cavity (h_{FC})

Actually, one or more of these cavities may have no depth and thus may be neglected within the calculation. See calculation step E below for the steps to determine actual cavity ratios.

Preparation steps and related information

A. Determine the mean footcandle level desired. The footcandle level desired for a given use can be determined by referring to specifications for the area, to illumination engineering manuals, to life safety codes, or to the following abbreviated suggested mean footcandle values:

Safety egress path	1 fc
General corridor pathway illumination	20 fc
Reading and general office tasks	75 fc
Drawing on tracing paper	200 fc
Background lighting in hospital operating room	500 fc
Paint shop lighting	500 fc

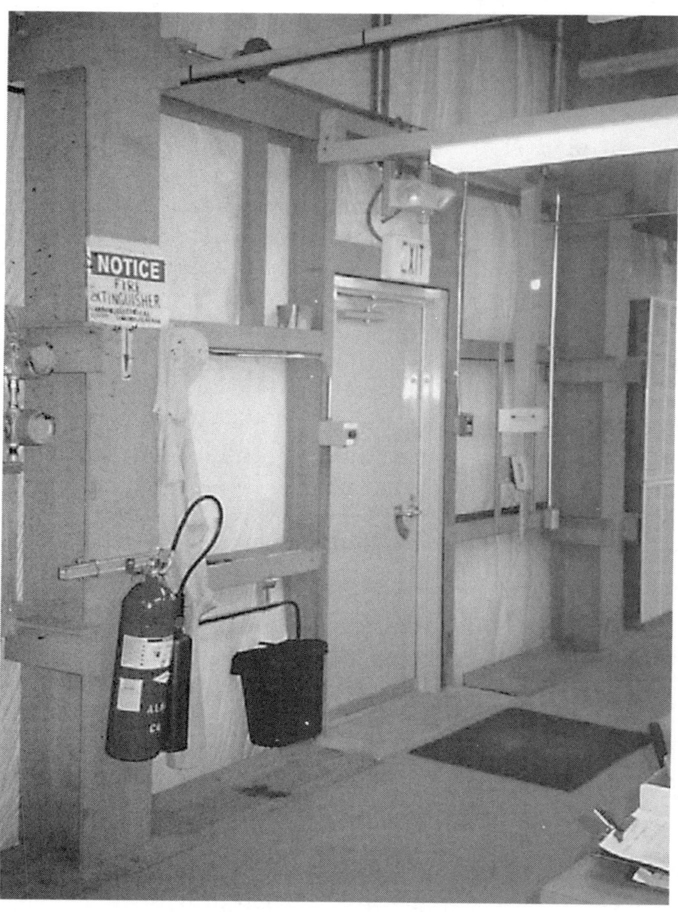

Illumination for egress paths must be provided in manned installations.

Note that the issue of veiling reflections on cathode-ray tube (CRT) screens must be considered when selecting the luminaire and the lens or louver in the luminaire. That is, every effort must be made in areas where computers will be used to prevent the computer operator from seeing the reflection of the luminaire in the CRT screen. Also, in certain offices, shiny glass desk tops are used. In these unusual locations, luminaires whose photometric data resemble a "bat wing" are required. With bat-wing lenses, almost all the light is emitted at large angles from nadir, whereas almost no light is emitted straight downward. A knowledge of the planned use of the space to be lighted is necessary to make a good lighting design.

B. Select the type of fixture and lamp to be used. This is done from a vendor catalog or from operator-client details and specifications. For example, frequently the lighting fixtures used within an office space measure 2 × 4 ft, mount into an inverted-T lay-in ceiling, and are equipped with flat prismatic acrylic lenses and two, three, or four F32T8 lamps.

C. Determine initial lumens per lamp. This is done most easily from lamp manufacturer catalog data. Lamp catalogs show a wealth of information about each lamp, including

- The catalog number of the lamp
- The energy use of the lamp in watts
- The quantity of lamps that are packaged by the factory in a standard case
- The nominal length in inches and in millimeters
- The initial light output in lumens
- The mean light output in lumens
- The average rated life in hours
- The color temperature in degrees Kelvin
- The color rendering index information in percent
- Additional information about special phosphors or ambient temperatures

- The type of lamp (e.g., fluorescent rapid-start medium bipin)
- The type of socket base into which the lamp is intended to mount

A summary of data for some of the most frequently used lamps is given in Fig. 8-2.

D. Calculate the light-loss factors. The initial lumen values published within a catalog for a given lamp are based on certain ambient temperature levels and lamp aging criteria. Also, not all the lumens emitted from the lamps escape from the lighting fixture. Finally, dirt accumulation on the fixture and on the lamps, as well as on the surfaces of the room, absorbs some of the light. These must all be considered when designing a lighting system:

1. Ambient temperature

- Does not significantly affect high-intensity-discharge (HID) output levels.
- Does not affect incandescent output levels.
- Affects fluorescent output levels when the ambient temperature is warmer or colder than 77°F. For example, at 20°F, the output of fluorescent lamps is reduced to 40 percent.
- See the second bulleted item under "Ballast Factor" below for the results of extreme overtemperature.

2. Ballast factor. There are several issues concerning ballasts that must be considered, but most of them can be summarized as follows:

- A "poor" fluorescent ballast causes decreased overall luminaire performance, but this is most often ignored in lighting calculations. If ambient temperature can exceed 135°F during the hot summer months, then the thermal element within Class P ballasts will open, initially deenergizing the fixture. Ultimately (after approximately four deenergization cycles), however, this will destroy the bal-

LAMP WATTAGE	LAMP TYPE	DESCRIPTION	MEAN LIFE	INITIAL LUMENS	MEAN LUMENS
100	IF	100 WATT INCANDESCENT, INSIDE FROSTED, ARBITRARY SHAPE	750	1750	1750
150	PAR/SP	150 WATT INCANDESCENT, REFLECTOR SPOT, THICK LENS	800	2175	2175
150	PAR/FL	150 WATT INCANDESCENT, REFLECTOR FLOOD, THICK LENS	750	2175	2175
400	BT-37	400 WATT METAL HALIDE, CLEAR	20000	36000	25000
1000	BT-56	1000 WATT METAL HALIDE, CLEAR	11000	110000	88000
50	E-17	50 WATT HIGH PRESSURE SODIUM	24000	4000	3600
70	E-17	70 WATT HIGH PRESSURE SODIUM	24000	6300	5350
100	E-17	100 WATT HIGH PRESSURE SODIUM	24000	9500	8000
150	E-17	150 WATT HIGH PRESSURE SODIUM	24000	15800	13400
250	ET-18	250 WATT HIGH PRESSURE SODIUM	24000	29000	26000
400	ET-18	400 WATT HIGH PRESSURE SODIUM	24000	50000	45000
1000	E-25	1000 WATT HIGH PRESSURE SODIUM	24000	130000	121000
32	T-8	32 WATT, 48-INCH, OCTRON FLUORESCENT,	20000	2800	1875
32	U-6	FB40 FLUORESCENT, U-SHAPED,	18000	2730	1825
40	T-12	F40 FLUORESCENT, 48-INCH, BIPIN	20000	3300	2200
60	T-12	F96 SLIMLINE FLUORESCENT, 96-INCH, SINGLE PIN	12000	5800	3900

Figure 8-2 Common lamps and their characteristics.

last unless system voltage can be reduced sufficiently to offset the increased ambient temperature. If this is not possible, then an anticipated quantity of ballasts that will be deenergized at one time must be factored into the overall calculation. One item that affects the value of this factor is the type of maintenance program planned for the facility.

- The type of ballast used for an HID lamp greatly affects lumen maintenance.
- Base the design on standard voltage. If a nonstandard voltage is to be used, then the output of the lamps must be reduced. For example, fluorescent lamp output should be reduced 5 percent for every 10 percent reduction in line voltage.

3. Room surface dirt depreciation. Room surface dirt depreciation accounts for decreased wall reflectance when the walls become aged or dirty. The following dirt depreciation factors should be *subtracted* from 100 percent to obtain the dirt depreciation factor (DDF) that is used in the zonal cavity calculation formula:

0–12 percent loss	Very clean
13–24 percent loss	Clean
25–36 percent loss	Medium
37–48 percent loss	Dirty
49–60 percent loss	Very dirty

4. Lamp lumen depreciation. This is one of the key parts of the lighting calculation because it determines the output of the lamps as they burn over time. This part of the calculation recognizes that

- Aged lamps emit less light. For example, if fluorescent lamps are not replaced after approximately three-quarters of their rated life, their output is reduced to 66 percent of initial lumen output. This reduction is even more severe (reduction to 40 percent) when certain HID lamps are used.

Parking lot lighting using HID lamps.

- As an alternative to entering the initial lamp lumen rating and then the lamp lumen depreciation value into the calculation, the lighting designer can instead enter the mean lumen rating from the catalog data.
- This section of the calculation must include a consideration of whether "burned out" lamps will be replaced as they

die, or whether the passage of time will discover that at the end of the mean lamp life curve, half the lamps in the facility will not be burning. If they will not be replaced as they burn out, then a factor of 0.5 must be entered to accommodate this fact.

5. Luminaire dirt depreciation. If dirt is allowed to build up on lighting fixture lenses and on the lamps, or if airflow is directed over the lamps in the planned luminaire, then less

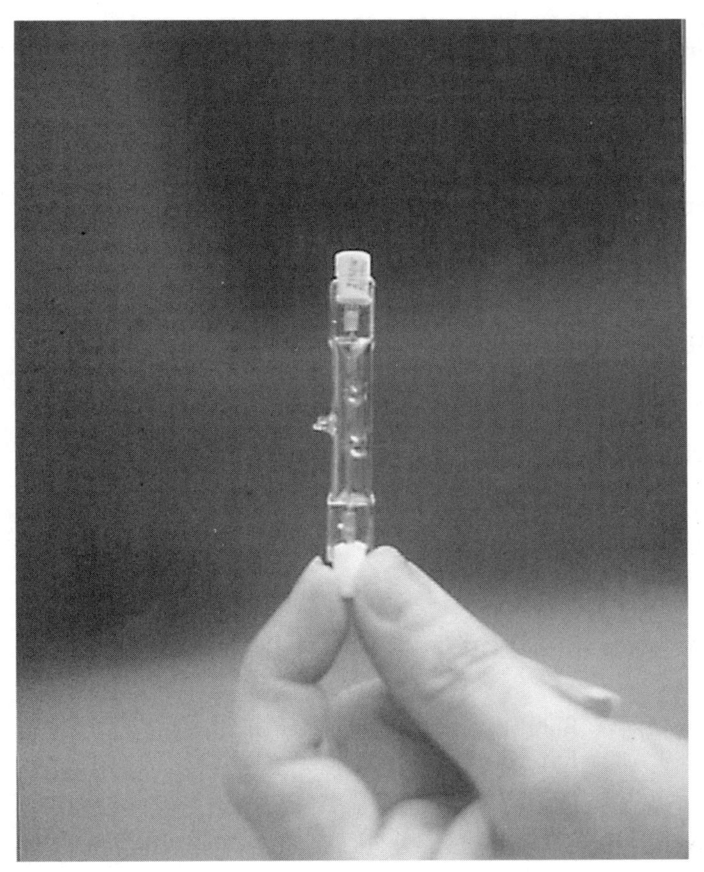

Quartz tungsten lamps are useful where stroboscopic effect is not permitted.

light will be emitted, and this must be accommodated within the lighting calculation. Light emitted from the fixture is reduced by 10 percent over time in "clean" environments, by 20 percent in typical industrial areas, and by 30 percent in very dirty areas.

The total light-loss factor (often called the maintenance factor) is the product of the factors shown above in items 1 through 5.

E. Calculate the cavity ratios. An individual calculation must be made of the ceiling cavity, the floor cavity, and the room cavity. The cavity ratios for each are calculated as follows:

$$\text{Room cavity ratio: RCR} = \frac{5h_{RC}\,(L + W)}{L \times W}$$

where h_{RC} = is the height of the room cavity, feet (ft)
L = the length of the room cavity, ft
W = width of the room cavity, ft

$$\text{Ceiling cavity ratio: CCR} = \frac{5h_{CC}\,(L + W)}{L \times W}$$

where h_{CC} = height of the ceiling cavity, ft
L = length of the ceiling cavity, ft
W = width of the ceiling cavity, ft

Note that if the fixtures are recessed, the ceiling cavity has a height of zero, and RCR = 0.

$$\text{Floor cavity ratio: FCR} = \frac{5h_{FC}\,(L + W)}{L \times W}$$

where h_{FC} = height of the floor cavity, ft
L = length of the floor cavity, ft
W = width of the floor cavity, ft

Note that if the work plane is at the floor, then the floor cavity has a height of zero, and FCR = 0.

When FCR = 0, the actual reflectance of the floor can be used by ignoring further consideration of the FCR in the lighting calculation. Similarly, when CCR = 0, the actual reflectance of the ceiling can be used by ignoring further consideration of the CCR in the lighting calculation. These are done simply by looking up the CU of the fixture within the manufacturer's data for the room cavity and using it in the calculation for quantity of luminaires or in the footcandle formula. That is, if no ceiling cavity or floor cavity exists, then the CU of the nonexistent floor cavity or ceiling cavity can be ignored.

If, however, the luminaires are suspended below the ceiling, then there is a ceiling cavity. And if the work plane is some distance above the floor, typically 30 inches (in), then there is also a floor cavity. In this case, one would calculate the ceiling cavity and look up the CU for the ceiling cavity in the luminaire table. Then one would calculate the floor cavity and again look up the CU for the floor cavity in the luminaire table. Then one would arrive at the proper CU for the luminaire by interpolating between the individual CU values for the ceiling, the room, and the floor cavities.

Find the CU for the planned luminaire from the calculated RCR by referring to the specific catalog data for the luminaire. Enter the table with the calculated RCR, then find the effective ceiling reflectance, and then find the effective wall reflectance. The resulting number is the CU for the luminaire.

F. Make the actual zonal cavity calculation

1. Refer to Fig. 8-3 for the basic calculation form to be filled in with each calculation. Fill in the values for reflectances from the data provided at the beginning of this chapter and by referring to Fig. 8-4, and then fill in room dimensions to match the problem at hand.

2. Calculate the cavity ratios as shown in step E above, and record them in the calculation form of Fig. 8-3.

3. Refer to the manufacturer's lighting fixture data (refer to Fig. 8-5 for an example of some typical data that

ROOM DATA

ROOM DIM.	LENGTH	1	______	FT
	WIDTH	2	______	FT
	FLOOR AREA	3	______	SQ. FT.
	CEILING HT.	4	______	FT
SURF. REFLECT	CEILING	5	______	%
	WALL	6	______	%
	FLOOR	7	______	%
FIXTURE MOUNTING HT.		8	______	FT

CAVITY DATA

ROOM CAVITY	HEIGHT	9	______	FT
	RATIO	10	______	
CEILING CAVITY	HEIGHT	11	______	FT
	RATIO	12	______	
	EFF.REFL.	13	______	%
FLOOR CAVITY	HEIGHT	14	______	FT
	RATIO	15	______	
	EFF.REFL.	16	______	%

FIXTURE DATA

MFGR.	______	17
CAT. NO.	______	18
LAMP TYPE	______	30
	______	
LAMPS/FIXT.	______	19
LUMENS/LAMP	______	20
C.U.	______	21
M.F.	______	22

FOOTCANDLES

QUANTITY OF LUMINAIRES REQUIRED TO PRODUCE REQUIRED FOOTCANDLES (FC)

DESIRED ILLUMINANCE: 23 ______ FC

QUANTITY OF LUMINAIRES 24 ______ FIXT.

QUANTITY OF FOOTCANDLES PRODUCED BY A GIVEN QUANTITY OF LUMINAIRES:

______ 25 FIXTURES

______ 28 FOOTCANDLES

INTERPOLATE CAVITY RATIO HERE:

CALCULATING CAVITY RATIOS

$$\text{CAVITY RATIO} = \frac{5 \times \text{CAVITY HEIGHT} \times (\text{LENGTH} + \text{WIDTH})}{(\text{LENGTH}) \times (\text{WIDTH})}$$

$$\text{ROOM CAVITY RATIO} = \frac{5 \times \text{LINE 9} \times (\text{LINE 1} + \text{LINE 2})}{(\text{LINE 1}) \times (\text{LINE 2})} = \_\_\_\_\_\_ \text{LINE 10}$$

$$\text{CEILING CAVITY RATIO} = \frac{5 \times \text{LINE 11} \times (\text{LINE 1} + \text{LINE 2})}{(\text{LINE 1}) \times (\text{LINE 2})} = \_\_\_\_\_\_ \text{LINE 12}$$

$$\text{FLOOR CAVITY RATIO} = \frac{5 \times \text{LINE 14} \times (\text{LINE 1} + \text{LINE 2})}{(\text{LINE 1}) \times (\text{LINE 2})} = \_\_\_\_\_\_ \text{LINE 15}$$

CALCULATING THE REQUIRED QUANTITY OF LUMINAIRES

$$\text{QUANTITY OF LUMINAIRES} = \frac{\text{FLOOR AREA} \times \text{DESIRED FOOTCANDLES}}{\text{LAMPS/FIXTURE} \times \text{LUMENS/LAMP} \times \text{COEFFICIENT OF UTILIZATION} \times \text{MAINTENANCE FACTOR}}$$

$$\text{QUANTITY OF LUMINAIRES} = \frac{\text{LINE 3} \times \text{LINE 23}}{\text{LINE 19} \times \text{LINE 20} \times \text{LINE 21} \times \text{LINE 22}} = \text{\_\_\_\_\_\_\_\_ FIXTURES. LINE 24}$$

CALCULATING FOOTCANDLES (LUMENS PER SQUARE FOOT.)

$$\text{FOOTCANDLES} = \frac{\text{NO. OF FIXTURES} \times \text{LAMPS/FIXT.} \times \text{LUMENS/LAMP} \times \text{CU} \times \text{MF}}{\text{FLOOR AREA}}$$

$$\text{FOOTCANDLES} = \frac{\text{LINE 25} \times \text{LINE 19} \times \text{LINE 20} \times \text{LINE 21} \times \text{LINE 22}}{\text{LINE 3}} = \text{\_\_\_\_\_\_\_\_ FC}$$

MAINTENANCE FACTOR

0.88 - BEST, CLEAN
0.8, - GOOD, CLEAN
0.75, - MEDIUM, CLEAN
0.65,- FAIR, DIRTY
0.5, - POOR, DIRTY

Figure 8-3 Zonal cavity calculation worksheet.

EFFECTIVE CAVITY REFLECTANCE

(TO BE USED WHEN CEILING, FLOOR, AND WALL REFLECTANCE ARE UNEQUAL)

% CEILING OR FLOOR REFLECTANCE	80	80	80	80	70	70	70	70	70	50	50	50	30	30	30	10	10	10
% WALL REFLECTANCE	**80**	**50**	**30**	**10**	**80**	**70**	**50**	**30**	**10**	**50**	**30**	**10**	**50**	**30**	**10**	**50**	**30**	**10**
CEILING OR FLOOR CAVITY RATIO																		
0	80	80	80	80	70	70	70	70	70	50	50	50	30	30	30	10	10	10
0.2	79	77	76	74	69	70	67	66	65	48	47	46	29	29	28	10	10	9
0.4	78	74	72	70	68	68	65	63	61	47	45	44	29	28	26	11	10	9
0.6	77	71	68	65	67	65	63	59	57	45	43	41	28	26	25	11	10	8
0.8	75	69	65	61	66	64	60	56	53	44	40	38	28	25	23	11	10	8
1.0	74	67	62	57	64	62	58	53	50	43	38	36	27	24	22	12	10	8
1.2	73	64	58	53	65	61	57	50	46	41	36	34	26	23	21	12	10	7
1.4	72	62	55	50	64	60	55	47	44	40	35	32	25	22	19	12	10	7
1.6	71	60	53	47	63	59	53	45	41	39	33	30	25	22	18	12	9	7
1.8	70	58	50	44	62	58	51	42	38	38	31	28	24	21	17	13	9	6
2.0	69	56	48	41	61	56	49	40	36	37	30	26	24	20	16	13	9	6
2.2	68	54	45	38	61	55	48	38	34	36	29	24	24	19	15	13	9	6
2.4	67	52	43	36	60	54	46	37	32	35	27	23	24	19	14	13	9	6
2.6	66	50	41	34	60	54	45	35	30	34	26	21	23	18	14	13	9	5
2.8	66	48	39	32	59	53	43	33	28	33	25	20	23	17	13	13	9	5
3.0	65	47	37	30	58	52	42	32	27	32	24	19	22	17	12	13	9	5
3.2	64	45	35	28	58	51	40	31	25	31	23	18	22	16	12	13	9	5
3.4	63	44	34	27	58	50	39	29	24	30	22	17	22	16	11	13	9	5
3.6	62	43	32	25	56	49	38	28	22	29	21	16	21	15	10	13	9	4
3.8	62	41	31	24	56	49	37	27	21	29	21	15	21	15	10	14	9	4
4.0	61	40	30	22	55	48	36	26	20	28	20	15	21	14	9	14	9	4
4.2	60	39	29	21	55	47	35	25	19	28	20	14	20	14	9	14	9	4
4.4	59	38	28	20	54	46	34	24	18	27	19	13	20	14	9	14	8	4
4.6	59	37	27	19	54	45	33	24	17	26	18	13	20	13	8	14	8	4
4.8	58	36	26	18	53	45	32	23	16	26	18	12	20	13	8	14	8	4
5.0	57	35	25	18	53	44	31	22	16	25	17	12	19	13	8	14	8	4
5.2	57	34	24	17	52	43	31	22	15	25	17	11	19	12	7	14	8	4
5.4	57	33	23	16	52	43	30	21	15	24	16	11	19	12	7	14	8	3
5.6	56	32	23	16	51	42	29	20	14	24	16	10	19	12	7	14	8	3
5.8	56	31	22	15	51	41	29	20	14	23	16	10	18	12	6	14	8	3
6.0	55	31	22	15	50	41	28	19	13	23	15	10	18	11	6	14	8	3
6.2	55	30	21	14	50	40	28	19	13	23	15	9	18	11	6	14	8	3
6.4	54	29	20	14	49	39	27	18	12	22	15	9	18	11	6	14	8	3
6.6	54	29	20	13	49	39	26	18	12	22	14	9	18	11	6	14	8	3
6.8	54	28	19	13	49	38	26	18	12	21	14	9	17	11	5	14	8	3
7.0	53	28	19	12	48	38	26	17	11	21	14	8	17	10	5	13	8	3
7.2	53	27	18	12	48	37	25	17	11	21	13	8	17	10	5	13	8	3
7.4	52	27	18	12	47	37	25	16	11	20	13	8	17	10	5	13	7	3
7.6	51	26	18	11	47	36	24	16	10	20	13	8	16	10	5	13	7	3
7.8	51	26	17	11	47	36	24	16	10	20	13	8	16	10	5	13	7	3
8.0	50	25	17	11	46	35	23	15	10	19	12	7	16	9	4	13	7	3
8.2	50	25	16	11	46	35	23	15	9	19	12	7	16	9	4	13	7	3
8.4	50	24	16	10	46	34	22	15	9	19	12	7	16	9	4	13	7	2
8.6	50	24	16	10	45	34	22	14	9	19	12	7	16	9	4	13	7	2
8.8	49	23	15	10	45	33	22	14	9	18	11	7	15	9	4	13	7	2
9.0	49	23	15	10	45	33	21	14	9	18	11	7	15	9	4	13	7	2
9.2	48	22	15	9	44	32	21	13	8	18	11	6	15	8	4	13	7	2
9.4	48	22	14	9	44	32	20	13	8	17	11	6	15	8	4	13	7	2
9.6	47	22	14	9	44	32	20	13	8	17	10	6	15	8	4	13	7	2
9.8	47	21	14	9	43	31	20	13	8	17	10	6	15	8	3	12	7	2
10.0	46	21	14	8	43	31	19	12	8	17	10	6	14	8	3	12	7	2

Figure 8-4 Effective cavity reflectances.

SUGGESTED COEFFICIENTS OF UTILIZATION FOR TYPICAL LUMINAIRES

(FOR USE WITH ZONAL CAVITY CALCULATIONS)

(FOR SPECIFIC LUMINAIRES, SEE MANUFACTURER'S SPECIFIC CU DATA)

	CEILING REFLECTANCE	80%	80%	50%	50%
	WALL REFLECTANCE	50%	30%	50%	30%
	FLOOR REFLECTANCE	10%	10%	10%	10%
LUMINAIRE TYPE	ROOM RATIO	COEFFICIENTS OF UTILIZATION (CU)			
RECESS-MOUNTED					
1-LAMP FLUORESCENT	1	0.43	0.4	0.42	0.37
1-LAMP FLUORESCENT	1.5	0.5	0.47	0.49	0.44
1-LAMP FLUORESCENT	2.5	0.55	0.53	0.54	0.5
1-LAMP FLUORESCENT	4	0.59	0.57	0.56	0.54
2-LAMP FLUORESCENT	1	0.4	0.36	0.38	0.35
2-LAMP FLUORESCENT	1.5	0.47	0.45	0.46	0.39
2-LAMP FLUORESCENT	2.5	0.53	0.5	0.52	0.47
2-LAMP FLUORESCENT	4	0.57	0.55	0.55	0.52
3-LAMP FLUORESCENT	1	0.37	0.31	0.35	0.27
3-LAMP FLUORESCENT	1.5	0.45	0.4	0.43	0.35
3-LAMP FLUORESCENT	2.5	0.52	0.48	0.49	0.43
3-LAMP FLUORESCENT	4	0.56	0.54	0.54	0.5
SURFACE-MOUNTED					
2-LAMP FLUORESCENT	1	0.76	0.73	0.64	0.62
2-LAMP FLUORESCENT	1.5	0.67	0.63	0.57	0.54
2-LAMP FLUORESCENT	2.5	0.6	0.55	0.51	0.48
2-LAMP FLUORESCENT	4	0.54	0.48	0.46	0.42
PENDANT-MOUNTED					
2-LAMP FLUORESCENT	1	0.76	0.73	0.64	0.62
2-LAMP FLUORESCENT	1.5	0.67	0.63	0.57	0.54
2-LAMP FLUORESCENT	2.5	0.6	0.55	0.51	0.48
2-LAMP FLUORESCENT	4	0.54	0.48	0.46	0.42
LUMINOUS CEILING					
TRANSPARENT LENSES	1	0.31	0.26	—	—
TRANSPARENT LENSES	1.5	0.4	0.35	—	—
TRANSPARENT LENSES	2.5	0.49	0.46	—	—
TRANSPARENT LENSES	4	0.56	0.54	—	—
OPAL WHITE LOUVERS	1	0.25	0.22	—	—
OPAL WHITE LOUVERS	1.5	0.3	0.26	—	—
OPAL WHITE LOUVERS	2.5	0.33	0.31	—	—
OPAL WHITE LOUVERS	4	0.35	0.34	—	—
INDIRECT LIGHTING					
	1	0.18	0.15	0.1	0.07
	1.5	0.25	0.21	0.15	0.11
	2.5	0.33	0.3	0.2	0.17
	4	0.36	0.34	0.22	0.2

NOTE: ABOVE CU VALUES ARE BASED ON 10% FLOOR REFLECTANCE.
INCREASE ALL CU VALUES BY THE FOLLOWING PERCENTAGES FOR FLOORS HAVING 30% REFLECTANCE

1	5%	3%	4%	1%
1.5	7%	6%	5%	2%
2.5	10%	8%	7%	4%
4	14%	12%	8%	6%

NOTE: INTERPOLATE FOR ALL ROOM RATIOS NOT SHOWN, OR SEE MFGR DATA FOR SPECIFIC LUMINAIRES

Figure 8-5 Solve for the coefficient of utilization for a typical luminaire given fixture type, room ratio, and surface reflectances.

are similar to the manufacturer's data for a recessed fluorescent lighting fixture), and from this data select the CU for the floor cavity based on the h_{FC}. Then select the CU for the room cavity based on the h_{RC}, and then interpolate between these CUs to arrive at the overall CU for the overall calculation. Then record the overall CU on the calculation form.

4. Calculate the resulting maintained footcandle value by the following formula:

$$\text{Footcandles} = \frac{\text{lumens/lamp} \times \text{lamps/fixture} \times \text{(no. of fixtures)} \times \text{LLF} \times \text{CU}}{\text{area to be illuminated in square feet}}$$

where LLF = the combined light loss factor from step D above, and CU = the coefficient of utilization from step F.3 above.

5. Check to make certain that the spacing-to-mounting height ratio of the installed luminaires is not greater than is shown in the photometric data for the specific luminaire. If this number is higher, then uneven lighting distribution is likely, and pools of light immediately under fixtures with dark areas between the fixtures are likely to result. Note that almost no light emanates from the end of fluorescent fixtures, while most of the light from the lamp is emitted from the side of the lamp. Accordingly, to avoid "spotty" lighting, placing fluorescent fixtures lamp end pointing to lamp end with no more then one mounting height between fixture ends is to be expected, while side-to-side spacing can be increased 1.5 to 2 times the mounting height without causing excessive unevenness of light on the work surface. When the spacing criteria require that more luminaires be installed than was calculated for the required footcandle value, then a new footcandle value can be calculated based on the increased quantity of luminaires using the preceding formula.

Refer to Fig. 8-6 for a completed sample problem showing the calculation for lighting in a classroom using the zonal cavity method.

G. Prepare the lighting layout. Draw the lighting layout to scale, and look for obvious "dark" spots, filling them in with extra lighting fixtures or rotating fixtures or relocating fixtures to eliminate the dark spots. In determining exact lighting fixture locations while providing even, continuous illumination over the entire work plane, it is necessary to consider the photometric data of the luminaires to be used. This is particularly important when using luminaires having nonsymmetrical lighting output, such as fluorescent luminaires.

Lighting Rules of Thumb

To estimate footcandle values or required quantities of fixtures, some rules of thumb are invaluable, such as

- In outdoor situations, the number of initial lamp lumens coming from all lamps within the system multiplied by a 50 percent factor for combined fixture CU, beam CU, object reflectances, and all maintenance factors will approximate the lumens per square foot footcandle value found in rigorous solutions.
- In indoor situations, some of the light is reflected onto the work surface instead of being lost as "spillover" light; therefore, the resulting illuminance levels are anticipated to be greater than the quantity of initial lamp lumens divided by a 50 percent factor calculation would indicate.
- In indoor situations, approximately one four-lamp recessed 4-ft fluorescent luminaire per 75 ft^2 will provide 75 maintained footcandles at the work surface 5 ft below the ceiling in a typical office space environment.
- In many instances, the quantity of fluorescent luminaires is determined by the office or warehouse furniture or storage rack and aisle layout. That is, rows of book shelves, file cabinets, or warehouse bins often prevent contributions of light from more than one luminaire at any one point. In these locations, calculations below and surrounding one luminaire dictate the anticipated illuminance level on both

ZONAL CAVITY CALCULATIONS

Problem: A classroom measures 20' x 30' x 10' high. The ceiling and wall are painted with white paint of 80% reflectance, and the floor is covered with tile having a reflectance of 20%. The luminaires that will be used will be 3-lamp fluorescent four-foot lay-in *recessed fixtures, each fitted with three F34T8 lamps and a clear, prismatic, acrylie lens. Maintenance in the school is very good.* How many luminaires will be required to achieve an overall mean footcandle illuminance value of 30 footcandles at 36" above the finished floor?

ROOM DATA

		Line	Value	Unit
ROOM DIM.	LENGTH	1	20	FT
	WIDTH	2	30	FT
	FLOOR AREA	3	600	SQ. FT.
	CEILING HT.	4	10	FT
SURF. REFLECT	CEILING	5	80	%
	WALL	6	80	%
	FLOOR AREA	7	20	%
FIXTURE MOUNTING HT.		8	10	FT

CAVITY DATA

		Line	Value	Unit
ROOM CAVITY	HEIGHT	9	7	FT
	RATIO	10	2.91	
CEILING CAVITY	HEIGHT	11	0	FT
	RATIO	12	N.A.	
	EFF.REFL.	13	80	%
FLOOR CAVITY	HEIGHT	14	3	FT
	RATIO	15	1.25	
	EFF.REFL.	16	20	%

FIXTURE DATA

	Value	Line
MFGR.	SAMPLE LAY-IN	17
CAT. NO.	SPECIAL WITH LENS	18
LAMP TYPE	F34T8	30
LAMPS/FIXT.	THREE	19
LUMENS/LAMP	3250 INITIAL	20
C.U.	0.523 (SEE NOTE***)	21
M.F.	0.8	22

FOOTCANDLES

QUANTITY OF LUMINAIRES REQUIRED TO PRODUCE REQUIRED FOOTCANDLES (FC)

DESIRED ILLUMINANCE: 23 30 FC

QUANTITY OF LUMINAIRES 24 **5** FIXT.

QUANTITY OF FOOTCANDLES PRODUCED BY A GIVEN QUANTITY OF LUMINAIRES:

$$\frac{\text{25 FIXTURES}}{\text{28 FOOTCANDLES}}$$

CALCULATING CAVITY RATIOS

$$\text{CAVITY RATIO} = \frac{5 \times \text{CAVITY HEIGHT} \times (\text{LENGTH} + \text{WIDTH})}{(\text{LENGTH})(\text{WIDTH})}$$

$$\text{ROOM CAVITY RATIO} = \frac{5 \times (7) \times (20 + 30)}{(20)(30)} = \underline{\quad 2.91 \quad} \text{ LINE 10}$$

NOTE * FOR CALCULATION OF CU FOLLOWS:**
CU FOR RCR OF 2.91 IS 0.55
CU FOR FCR OF 1.25 IS 0.45 + 2% FOR EXTRA FLOOR REFLECTANCE OVER 10% THUS MAKING FLOOR CAV. CU=0.46
OVERALL CU IS (.7)(.55)+(.3)(.46) = 0.523
(7' OF 10' ARE AT 0.55, & 3' OF10' ARE AT 0.46)

CEILING CAVITY RATIO = $\frac{5 \times \text{LINE 11} \times (\text{LINE 1} + \text{LINE 2})}{(\text{LENGTH})(\text{WIDTH})}$ = _N.A.______LINE 12

FLOOR CAVITY RATIO = $\frac{5 \times 3 \times (20 + 30)}{(20)(30)}$ = __ **1.25**____LINE 15

CALCULATING THE REQUIRED QUANTITY OF LUMINAIRES

QUANTITY OF LUMINAIRES = $\frac{\text{FLOOR AREA} \times \text{DESIRED FOOTCANDLES}}{\text{LAMPS/FIXTURE} \times \text{LUMENS/LAMP} \times \text{COEFFICIENT OF UTILIZATION} \times \text{MAINTENANCE FACTOR}}$

QUANTITY OF LUMINAIRES = $\frac{(600)(30)}{(3)(3250)(0.523)(0.8)}$ ' = ___4.41 **(USE 5)**______ FIXTURES. LINE 24

CALCULATING FOOTCANDLES (LUMENS PER SQUARE FOOT.)

FOOTCANDLES = $\frac{\text{NO. OF FIXTURES} \times \text{LAMPS/FIXT.} \times \text{LUMENS/LAMP} \times \text{CU} \times \text{MF}}{\text{FLOOR AREA}}$

FOOTCANDLES = $\frac{\text{LINE 24} \times \text{LINE 19} \times \text{LINE 20} \times \text{LINE 21} \times \text{LINE 22}}{\text{LINE 3}}$

FOOTCANDLES = 5 x 3 x 3250 x 0.523 x 0.8 / 600 = 34.0 F.C.

MAINTENANCE FACTOR

0.88 - BEST, CLEAN
0.8, - GOOD, CLEAN
0.75, - MEDIUM, CLEAN
0.65,- FAIR, DIRTY
0.5, - POOR, DIRTY

Figure 8-6 Solve for required quantity of luminaires given room and luminaire characteristics.

horizontal and vertical surfaces, and luminaires are either placed end to end or close enough that the light patterns from adjacent luminaires overlap slightly on the surface(s) to be lighted. Note that the lighting pattern at the end of a fluorescent fixture is significantly reduced in intensity and angle from nadir when compared with the luminaire's lighting pattern perpendicular to the lamp.

NOTES

NOTES

Chapter

9

Transformers

While direct-current (dc) systems are essentially "stuck" with the source voltage (with only a very few exceptions), alternating-current (ac) systems offer great flexibility in voltage due to magnetic coupling in transformers. As their name implies, *transformers* are used in ac systems to transform, or change, from one voltage to another.

Since transformers are among the most common types of devices in electrical power systems, second only to wires and cables, specific attention is paid to designing electrical systems that contain these devices.

In its simplest form, a transformer consists of two coils that are so near to one another that the magnetic flux caused by exciting current in the first, or primary, coil cuts the three-dimensional space occupied by the second coil, thereby inducing a voltage in the second coil. With this action, it is essentially acting just like a generator's rotating magnetic field. The voltage imparted to the second coil can be calculated simply by the ratio

$$\frac{N_P}{N_S} = \frac{V_P}{V_S}$$

where N_P = the number of turns in the primary coil
N_S = number of turns in the secondary coil
V_P = voltage measured across the primary coil terminals
V_S = voltage generated in the secondary coil

Figure 9-1 is a sample calculation showing how to determine what the voltage will be out of the secondary terminals of a transformer with a given input voltage connected to the primary coil terminals.

Some transformers are more robust than others, and the amount of electrical abuse that a transformer can withstand is closely related to the method of heat removal employed within the transformer. A given transformer that can carry load x when cooled by convection air can carry more than x when cooled by an auxiliary fan. Further, when the transformer coils are immersed in an insulating liquid such as mineral oil, internal heat is dispersed and hot spots are minimized. Thus liquid cooling permits given sizes of transformer coils to carry much more load without damage. Moreover, some transformers are insulated with material that can remain viable under much hotter temperatures than others. All these things increase the load-carrying capability of transformers:

- Convective air circulation
- Forced fan air circulation
- Coil immersion in an insulating liquid
- Convective air circulation around oil cooling fins
- Forced fan air circulation around oil cooling fins
- The addition of two or three sets of oil cooling fins
- Forced pumping of insulating liquid through cooling fins
- Coil insulation of a higher temperature rating

Liquid-filled transformers are always base rated according to their load-carrying capability by convective air circulation around the transformer and around the first set of

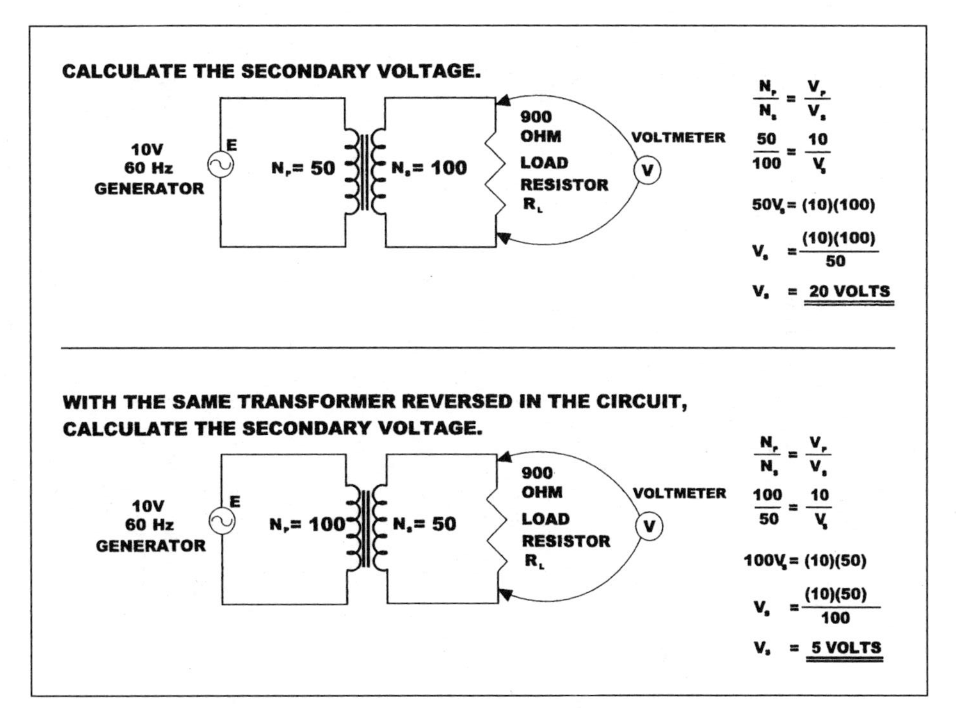

Figure 9-1 Solve for transformer output voltage given input voltage and turns ratio.

cooling fins if the transformer normally is equipped with these cooling fins as standard equipment. Normally, liquid-filled transformers are rated at the highest temperature that the insulation system can withstand over a long period without degrading prematurely.

Figure 9-2 is a sample calculation showing how much additional load-carrying capability a transformer of a given size can gain when some of the more usual auxiliary cooling methods are applied. A transformer that is rated OA 55°C/FA 65°C can carry 12 percent more load when permitted to rise to 65°C, even without the cooling fans in operation. How much each of the more usual insulation systems and auxiliary cooling methods can increase transformer load capabilities is shown in Fig. 9-2, and the resulting transformer kilovoltampere ratings and full-load current ratings are shown in Fig. 9-3. Note that the percentage increase is different for very large transformers when compared with transformers in the 1000-kilovoltampere (kVA) range. Also note that the transformer rating is the 24-hour average load rating and that it can be exceeded somewhat for short periods without deleterious effects.

There are a great many types of transformer ratings, and some are more usual than others. A summary of these ratings is given at the top of Fig. 9-2.

All the things just stated about transformers are predicated on the transformer being in operation with a sinusoidal voltage of the exact frequency for which the transformer is designed and at approximately the voltage for which the transformer is designed. If the voltage is reduced, maintaining the kilovoltampere level requires increased current flow, thus tending to overheat the transformer. If the voltage is increased too much, too much exciting current flows, and core magnetic saturation occurs. This also causes transformer overheating. Operating a transformer in an electrical system having a large value of voltage distortion and/or current distortion also causes transformer overheating due to increased eddy current flow and greatly increased hysteresis losses.

In electrical systems having nonsinusoidal currents and voltages, either the use of greatly oversized transformers or special transformers with K-ratings is required to handle all the extra heat generated within the transformers. *A K-rating of 8 on a transformer nameplate means that the transformer can safely carry a specific nonsinusoidal kilovoltampere load that would heat a non-K-rated transformer to the same temperature as if it were carrying a load that was eight times larger.* This is due to additional eddy current core losses and conductor heating losses due to skin-effect current flow at the higher frequencies. Figure 3-27 shows how to calculate the transformer K-rating requirements for a given load containing harmonics.

In calculating the required K-rating of a transformer, the first thing that is necessary is to determine the magnitudes and frequencies of the currents that the transformer must carry. These are normally stated in terms of amperes at each harmonic or multiple of the first-harmonic base frequency, but sometimes the currents are stated as a percentage of the fundamental frequency. The first harmonic (i.e., the fundamental frequency) is 60 in a 60-hertz (Hz) system, and it is 50 in a 50-Hz system.

Three-Phase Transformers

Most of the transformers in operation in the electrical power systems of the world today are three-phase transformers. This is so largely because of economics and partly because of the innate rotating flux provided by three-phase systems in electrical motors. Given the correct coil-winding equipment and design software, almost any voltage can be created with three-phase transformers, but there are only a few standard transformer connections that are used frequently, and they are summarized here for American National Standards Institute/National Electrical Manufacturers Association (ANSI/NEMA) installations as well as for International Electrotechnical Commission (IEC) installations and Australian designs.

THE MOST COMMON CLASSES OF TRANSFORMER COOLING SYSTEMS

CLASS	REMEMBER BY	COOLING METHOD
OA	OUTSIDE-AIR	SELF-COOLED (BY CONVECTION)
OA/FA	OUTSIDE-AIR/FAN-AIR	SELF-COOLED OR FAN COOLED
OA/FA/FA	OUTSIDE-AIR W/ 2 FAN COOLING SETS	SELF-COOLED/FAN-COOLED
OA/FA/FOA	OUTSIDE-AIR/FAN-AIR/FORCED (PUMPED) OIL	SELF-COOLED, FAN-COOLED, PUMPED OIL
FOA	FORCED OIL/FAN COOLED	PUMPED OIL WITH FANS
AA	AIR-AIR DRY TYPE (OR CAST INSULATION)	SELF-COOLED (BY CONVECTION)
AA/FA	AIR-AIR/FAN-AIR IN DRY TYPE	SELF-COOLED WITH FANS

TRANSFORMERS CAN HAVE VARIOUS BUILT-IN OVERLOAD CAPABILITIES. FOR EXAMPLE, A 750 KVA OIL-FILLED TRANSFORMER THAT IS BASE RATED AT 55°C WITH 65°C INSULATION CAN ACTUALLY CARRY OVER A LONG TERM THE FOLLOWING LOAD:

FINAL 65° C RATING = 55° C BASE RATING X 1.12
FINAL 65° C RATING = 750 X 1.12 = 840 kVA

ADDITIONAL LOAD CAPABILITY IS ACHIEVED BY ADDING RADIATORS WITH FANS AND PUMPING OIL. THIS TABLE IS A SUMMARY OF THE PROPER USE OF THESE FACTORS:

THREE-PHASE OIL-FILLED TRANSFORMER KVA (PROX) RATINGS				EXTRA-CAPACITY FACTORS		
55°OA	65°OA	65°FA	65°FOA	55°-TO-65°OA	65°OA-TO-65°FA	65°FA-TO-65°FOA
750	840	966		1.12	1.15	
1000	1120	1288		1.12	1.15	
1500	1680	1932		1.12	1.15	
2000	2240	2576		1.12	1.15	
2500	2800	3500		1.12	1.25	
3750	4200	5250		1.12	1.25	
5000	5600	7000		1.12	1.25	
7500	8400	10500		1.12	1.25	
10000	11200	14000		1.12	1.25	
12000	13440	17875	22344	1.12	1.33	1.25
60000	67200	89376	111720	1.12	1.33	1.25

NOTE THAT THE FACTORS INCREASE IN LARGER TRANSFORMER SIZES ABOVE 2000 kVA AND ABOVE 10 MVA.

EXAMPLE PROBLEM: A 3750 KVA 55°C OA/65°C OA IS OVERLOADED. WHAT CONTINUOUS LOAD CAN IT SUPPORT WITHOUT DAMAGING THE TRANSFORMER IF A RADIATOR COOLING STAGE WITH FANS (FA) IS INSTALLED?

FINAL FA65° C RATING = 55 C BASE RATING X 1.12 X 1.33

FINAL FA65° C RATING = (3750) X 1.12 X 1.33

FINAL FA65° C RATING = 5250 kVA

Figure 9-2 Solve for oil-filled transformer kilovoltampere capability given increased insulation temperature capability and with added cooling systems.

55 Deg C OA Rating kVA	FULL LOAD AMPS @ 480V	FULL LOAD AMPS @ 2400 VOLTS	FULL LOAD AMPS @ 4160 VOLTS	FULL LOAD AMPS @ 13.8 kV	65 Deg C OA Rating kVA	FULL LOAD AMPS @ 480V	FULL LOAD AMPS @ 2400 VOLTS	FULL LOAD AMPS @ 4160 VOLTS	FULL LOAD AMPS @ 13.8 kV	65 Deg C FA Rating kVA	FULL LOAD AMPS @ 480V	FULL LOAD AMPS @ 2400 VOLTS	FULL LOAD AMPS @ 4160 VOLTS	FULL LOAD AMPS @ 13.8 kV
750	902	180	103	31	**840**	1010	202	116	35	**966**	1162	232	133	40
1000	1203	241	138	42	**1120**	1347	269	154	47	**1288**	1549	310	177	54
1500	1804	361	207	63	**1680**	2021	404	232	70	**1932**	2324	465	266	81
2000	2406	481	276	84	**2240**	2694	539	309	94	**2576**	3098	620	355	108
2500	3007	601	345	105	**2800**	3368	673	386	117	**3500**	4210	842	482	146
3750	4511	902	517	157	**4200**	5052	1010	579	176	**5250**	6315	1263	723	219
5000	6014	1203	689	209	**5600**	6736	1347	772	234	**7000**	8420	1684	965	293
7500	9021	1804	1034	314	**8400**	10104	2020	1158	351	**10500**	12629	2525	1447	439
10000	12028	2405	1378	418	**11200**	13471	2694	1543	468	**14000**	16839	3367	1929	585
12000	14434	2886	1654	502	**13440**	16166	3232	1852	562	**17875**	21500	4299	2463	747
15000	18042	3608	2067	628	**16800**	20207	4040	2315	702	**22344**	26875	5374	3079	934
20000	24056	4810	2756	837	**22400**	26943	5387	3087	936	**29792**	35834	7165	4105	1245
25000	30070	6013	3445	1046	**28000**	33678	6734	3858	1170	**37240**	44792	8956	5132	1557
30000	36084	7215	4134	1255	**33600**	40414	8081	4630	1404	**44688**	53751	10747	6158	1868
45000	54126	10823	6201	1883	**50400**	60621	12121	6945	2107	**67032**	80626	16121	9237	2802
50000	60140	12025	6890	2092	**56000**	67357	13468	7717	2341	**74480**	89585	17912	10263	3113
60000	72168	14430	8268	2510	**67200**	80828	16162	9260	2809	**89376**	107501	21495	12316	3736

Note: Each row represents one transformer with progressive stages of cooling: 55 Deg. C OA, 65 Deg. C OA, 65 Deg. C FA

Figure 9-3 Solve for transformer full-load current values for common kilovoltampere transformer ratings at common system voltage values.

Liquid-filled transformers can carry extra power when fitted with cooling stages.

Three-phase delta

Figure 9-4 shows the connections of the three individual coils of a generator connected as single-phase units. It also shows an improvement on the single-phase connection by adding jumpers at the generator that connect the three single-phase coils into a *delta* configuration. In the delta configuration, each phase appears to be an individual single-phase system, while together the three single-phase systems combine to provide three times the load capability while eliminating three circuit conductors and reducing the size of the remaining wires to 70.7 percent of the size of the former single-phase conductors. In the delta connection, the phase-to-phase voltage is also the coil voltage.

An identical connection is made at a three-phase transformer, where all three coils are connected end to end, with one "phase" wire brought out at every end-to-end joint, and the 120 electrical degree voltage displacement is faithfully displayed in vector form on graph paper in the shape of the Greek letter delta. Figure 9-5 shows the wiring connections of the three phases at the generator and at three-phase motors and single-phase loads. There are two basic problems with delta systems:

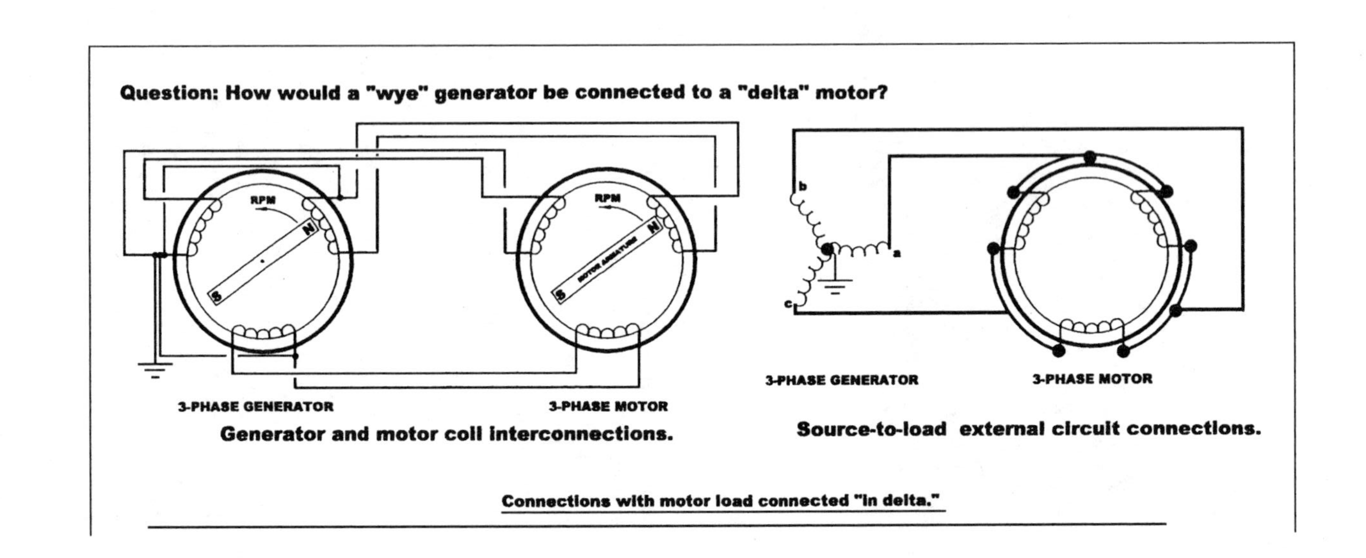
Question: How would a "wye" generator be connected to a "delta" motor?
RPM
N
S
RPM
MOTOR ARMATURE
3-PHASE GENERATOR
3-PHASE MOTOR
Generator and motor coil interconnections.
b
a
c
3-PHASE GENERATOR
3-PHASE MOTOR
Source-to-load external circuit connections.
Connections with motor load connected "in delta."

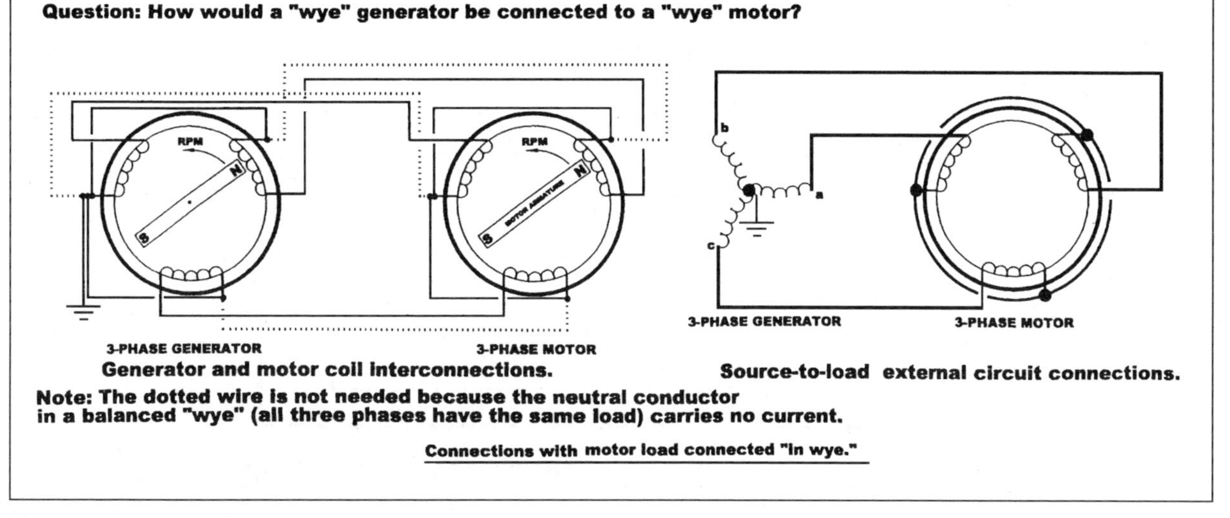

Figure 9-4 Solve for wiring connections from a wye-connected generator to a wye-connected or delta-connected motor.

Determine the voltage across each delta-connected motor coil if a delta-connected generator coil voltage is 230 volts.

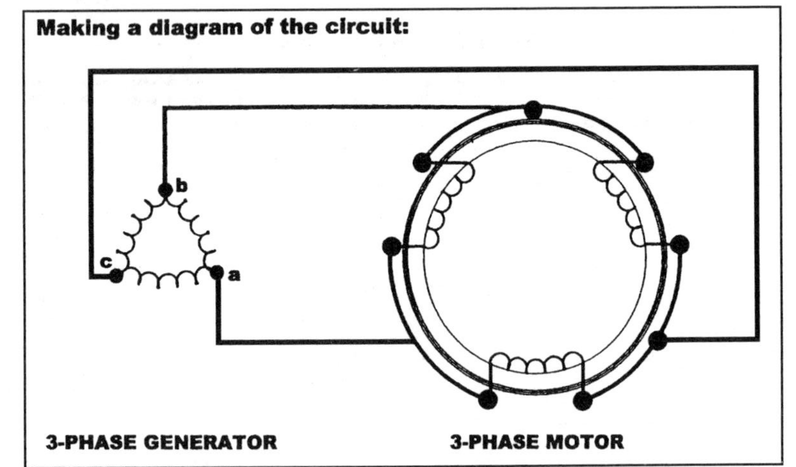

Connection diagram of the system under analysis.

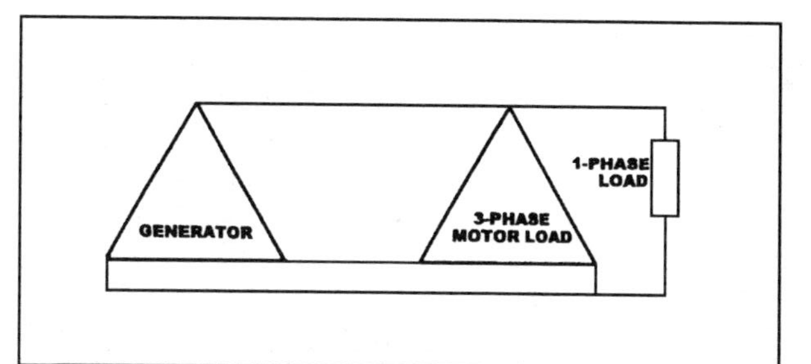

Overall connection diagram of generator and load.

By inspection, the phase-to-phase voltage in a "delta" equals coil voltage, 230 volts.

Figure 9-5 Solve for motor coil voltage given delta-connected generator coil voltage and wiring connection diagram.

- They offer no lower voltage for smaller loads.
- Grounding, if done, must be done at one phase, thus increasing voltage stress on the insulation of the other two phases.

For these reasons, wye connections are often used.

Figure 9-6 shows how to solve for the motor coil voltage from a wye-connected generator whose coil voltage is 120 volts (V). The steps are to diagram the system under analysis first and then to sketch the generator coil voltages. Next, sketch the voltage vectors to be added, and then transform the voltage vectors to the rectangular coordinate form so that they can be added. Finally, the resulting voltage sum is converted back into polar form, showing that in a wye-connected system the phase-to-phase voltage is equal to the coil voltage multiplied by the square root of 3.

Three-phase wye

Figure 9-4 showed the connections of the three individual generator coils connected as single-phase units. Figure 9-7 improves on the single-phase connection by adding jumpers at the generator that connect the three single-phase coils into a *wye* configuration. In the wye configuration, as was true in the delta configuration, each phase appears to be an individual single-phase system, while together the three single-phase systems combine to provide three times the load capability while eliminating three circuit conductors and reducing the size of the remaining wires to approximately two-thirds of the size of the former single-phase conductors. In addition, the wye system offers a "neutral" point at which grounding of the system is convenient and functional without voltage overstressing anywhere in the system, and the "neutral" grounded conductor provides a path for imbalance current to return to the source while providing a phase-to-neutral coil voltage source applicable for use with smaller loads at lower voltage.

As at a wye-connected generator, an identical connection is made at a three-phase wye-connected transformer, where all three coils are connected at one end to form the neutral grounding point, and the 120 electrical degree voltage displacement is faithfully displayed in vector form beside the transformer symbol on the electrical one-line drawing in the shape of the letter Y. This is shown on Fig. 9-7, which also shows the connections of the three phases at both a three-phase motor and at a single-phase load, as well as at a line-to-neutral load.

Figure 9-8 shows many of the most common transformer connections, along with their voltages, for both 50- and 60-Hz systems around the world.

Frequently, slight modification of the voltage is needed for proper operation of load appliances. Changing voltage can be done easily by using multiple "taps" at the transformer to increase or decrease the output voltage. Generally, when taps are provided, there are two 2.5 percent taps above and below the center voltage. Figure 9-9 shows how to change the output voltage by simply changing taps at the transformer coils.

Overcurrent Protection of Transformers

All electrical equipment must be protected against the effects of both short-circuit current and long-time overload current, and transformers are no exception. Although transformers are quite tolerant to short-time overloads because of their large thermal mass (since it takes a long time for the transformer to heat when subjected to long-time overloads), specific rules regarding the maximum overcurrent device settings for most electrical power transformers are set out in Tables 450-3(a) and (b) of the *National Electrical Code.* These tables and the rules that refer to them apply to a bank of single-phase transformers connected to operate as a single unit, as well as to individual single-phase or three-phase units operating alone.

Where an overcurrent device on the transformer secondary is required by these rules, it can consist of not more than six circuit breakers or sets of fuses grouped in one location. Where multiple overcurrent devices are used, the total

Calculate the voltage across each delta-connected motor coil if wye-connected generator coil voltage is 120 volts.

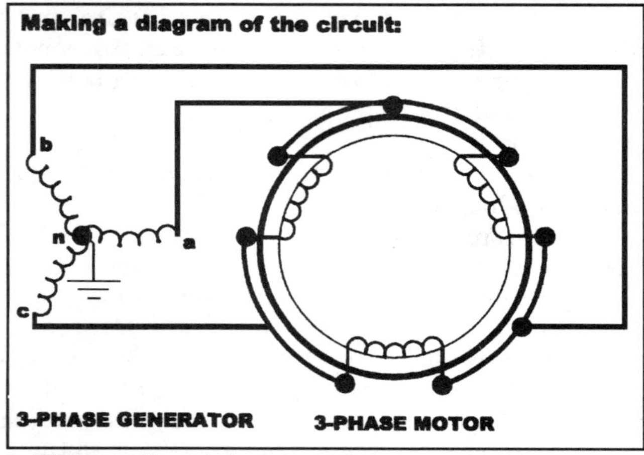

1. Diagram the system under analysis.

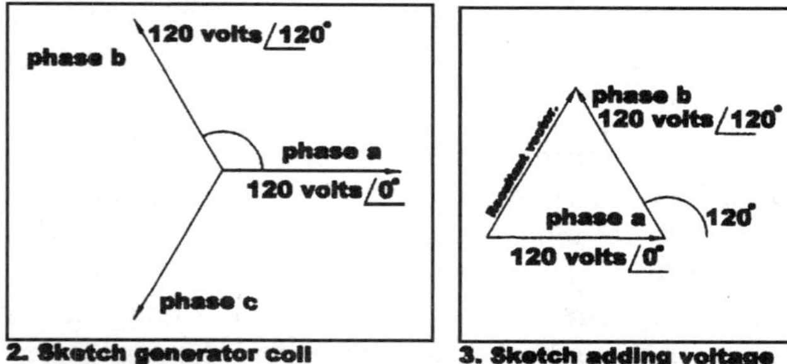

Applying this general finding in the problem stated above:

Phase-to-phase voltage = delta-connected motor coil voltage = Wye coil voltage times $\sqrt{3}$

Phase-to-phase voltage in a wye =(120) (1.732) = 208 Volts

Figure 9-6 Solve for motor coil voltage in a delta-connected motor given the source is a wye-connected generator with a 120-V coil voltage.

SUM THE VOLTAGE VECTORS IN RECTANGULAR FORM.

Phase a

120 ∠0 °

= 120 COS 0 + j 120 SIN 0

= [(120)(1) +j (120)(0)]

= 120 + j 0.0

Phase b

120 ∠120° = 120 ∠-60 °

= -120 COS 60 +j 120 SIN 60

= [(-120)(5) +j (120)(.866)]

= -60 +j 103.92

= 60 +j 103.92

THEN CHANGE BACK TO POLAR FORM

-60 +j 103.92 = $\sqrt{(60)^2 + (103.92)^2}$ **∠ARCTAN (103.92/-60)**

$\sqrt{14399}$ **∠ARCTAN (1.732)**

120 ∠-60 ° VOLTS

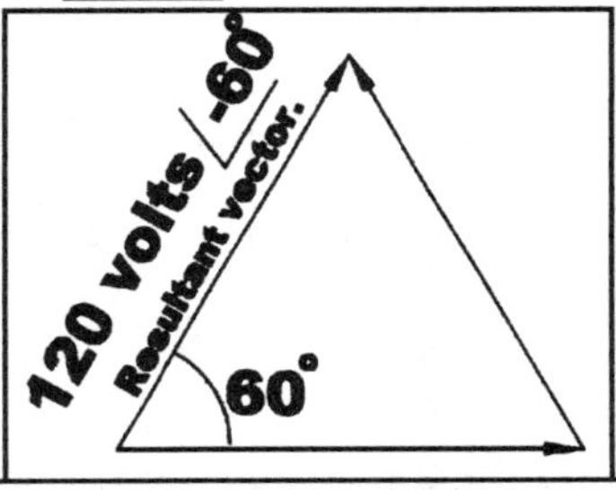

4. Add voltage vectors.

5. Change voltage vector form to incorporate the 60° angle into the scalar phase-to-phase voltage value.

Note that the Tan of 60 also equals the $\sqrt{3}$, which is 1.732

Therefore the phase-to-phase voltage in a "wye" equals coil voltage times $\sqrt{3}$.

Stated in another way, the phase-to-phase voltage equals the phase-to-neutral voltage times $\sqrt{3}$.

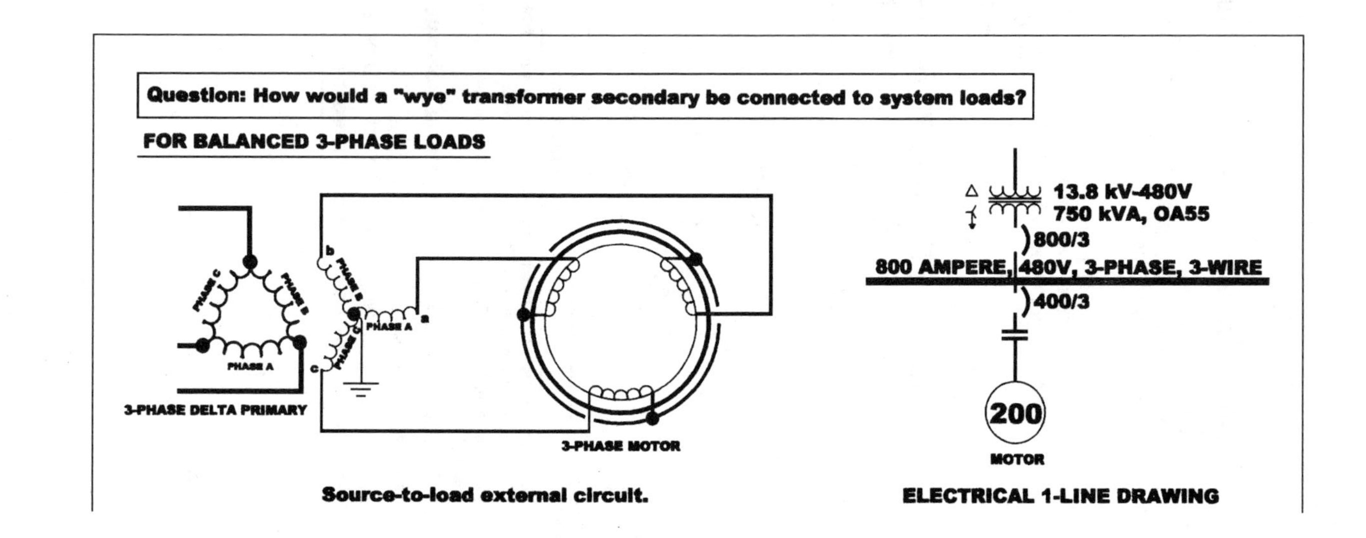

Source-to-load external circuit.

ELECTRICAL 1-LINE DRAWING

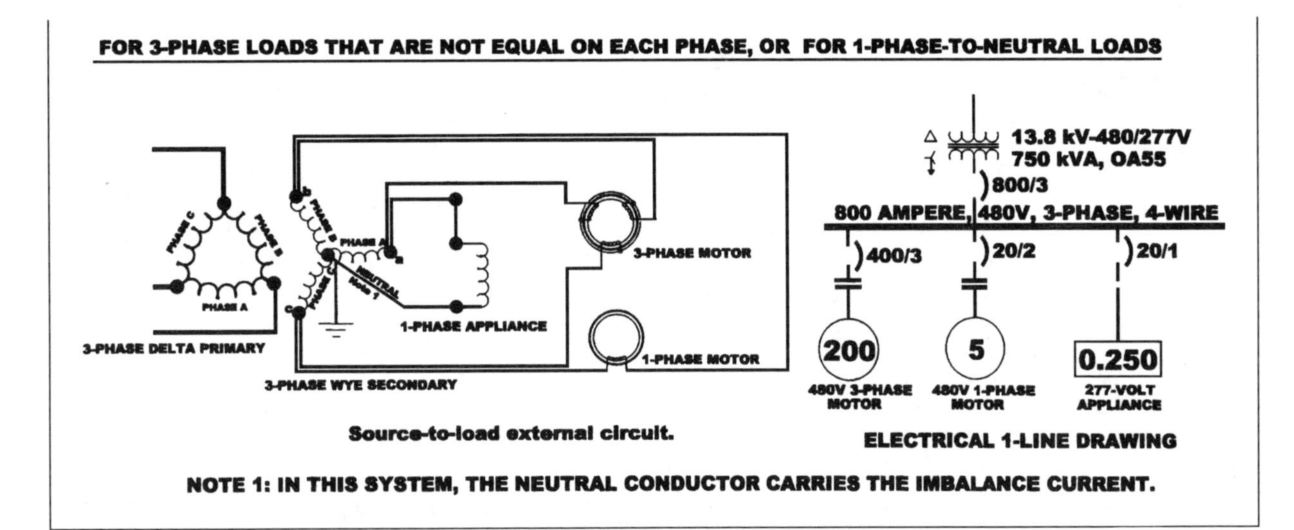

Figure 9-7 Solve for the connection diagram of a wye-connected transformer secondary to one-phase and to balanced and unbalanced three-phase loads.

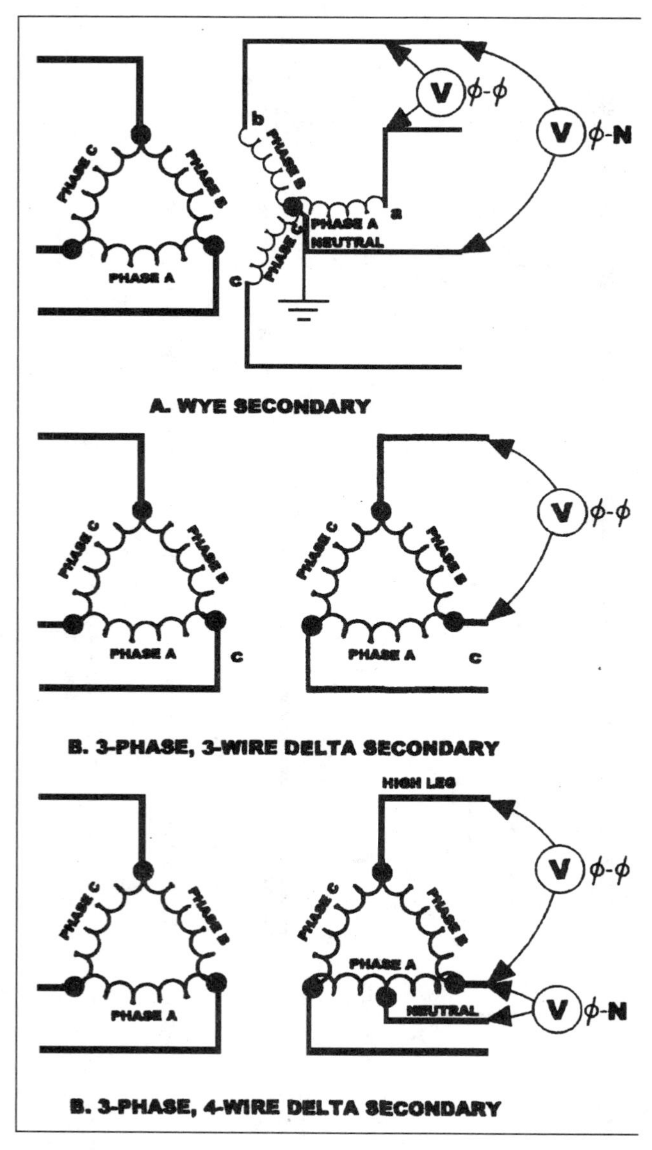

Figure 9-8 Solve for the correct voltage and matching transformer connection configuration for common 50- and 60-Hz systems.

SYSTEM VOLTAGES

60 HERTZ

WYE			DELTA		
DIAGRAM	ϕ-ϕ	ϕ-N	DIAGRAM	ϕ-ϕ	ϕ-N
A	208	120	B	240	NA
A	480	277	B	240	NA
A	595	343	C	230	115
A	4160	2400	B	460	NA
A	12460	7200	B	2400	NA
A	25000	14400			
A	34500	19920			

50 HERTZ

WYE		
DIAGRAM	ϕ-ϕ	ϕ-N
A	380	220
A	3300	1900
A	6600	3800
A	11000	6350

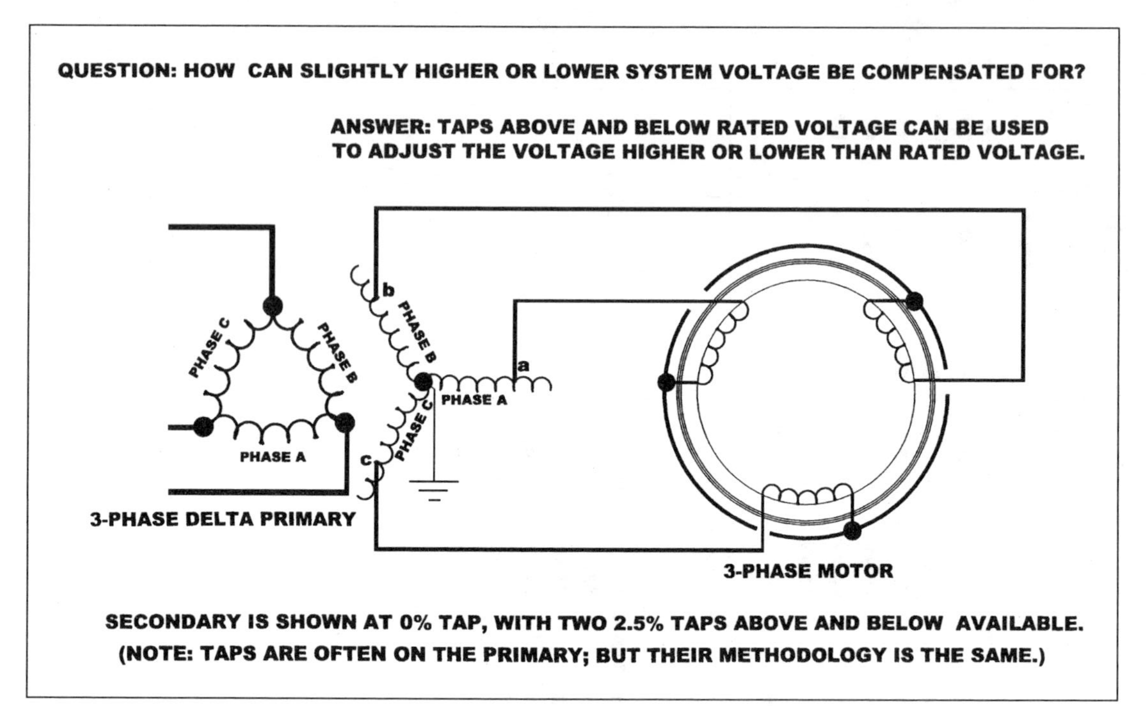

Figure 9-9 The transformer output voltage can be adjusted by switching to different "taps" of the transformer coil.

of all the devices ratings must not exceed the allowed value of a single overcurrent device. If both circuit breakers and fuses are used as the overcurrent device, the total of the device ratings must not exceed that allowed for fuses.

Note that these rules are only for the protection of the transformer and do not apply to protection of the conductors to or from the transformer. For protection of the transformer feeder conductors, compliance with the rules for conductor protection found in Article 240 of the *National Electrical Code* is required.

For transformers having at least one coil operating at over 600 V

Transformers that have at least one coil operating at over 600 V must have overcurrent protection on both their primary and secondary. The rating of each overcurrent device is provided in Table 450-3(a) of the *National Electrical Code,* and for the reader's convenience, it is replicated in Fig. 9-10. The general rule is that when the required overcurrent device rating does not correspond to a standard rating, use of the next-higher standard rating is permitted. The secondary overcurrent device can be one to six overcurrent devices, but the sum of their ratings must not exceed the value shown in the table. Other specific cases are mentioned in the code where these rating rules are relaxed somewhat, but these are left to the reader to explore in the code. A sample calculation showing the application of these rules is shown in Fig. 9-11.

For transformers operating at below 600 V

When all coil voltages are below 600 V, the basic rule in transformer overcurrent protection is for the overcurrent protective device on the primary of the transformer to be rated at 125 percent of the rated full-load primary transformer current. There are three minor exceptions to this rule, and all four rules are shown in Table 450-3(b) of the *National Electrical Code,* replicated in Fig. 9-12 on p. 278. A sample calculation using this table is shown in Fig. 9-13 on p. 279.

MAXIMUM RATING OF OVERCURRENT DEVICE PROTECTING TRANSFORMERS OVER 600 VOLTS
(STATED VALUES ARE PERCENTAGE OF TRANSFORMER CURRENT)

LOCATION	TRANSFORMER	PRIMARY PROTECTION	PRIMARY PROTECTION	SECONDARY PROTECTION		
LIMITATIONS	IMPEDANCE	(OVER 600 VOLTS)	(OVER 600 VOLTS)	OVER 600V		UNDER 600V
		CIRCUIT BREAKER	FUSE	CIRCUIT BREAKER	FUSE	
ANYWHERE	6% OR LESS	600%	300%	300%	250%	125%
ANYWHERE	6.01 - 10%	400%	300%	250%	225%	125%
SUPERVISED	ANY	300%	250%	NOT REQUIRED	NOT REQUIRED	NOT REQUIRED
SUPERVISED	6% OR LESS	600%	300%	300%	250%	125%
SUPERVISED	6.01 - 10%	400%	300%	250%	225%	125%

THE NOTES TO NEC TABLE 450-3(a) ARE NOT SHOWN IN THIS PARTIAL TABLE REPLICATION

Figure 9-10 Table replicating part of NEC Table 450-3(a), overcurrent protection of transformers over 600 V.

In many calculations, such as circuit breaker selection and harmonic resonance scans, the reactance/resistance ratio, or *X/R* ratio, of a transformer is required. The *X/R* value of a transformer can be determined either from a graph or from a unique calculation. See Fig. 9-14 on p. 280 for both a specific calculation method to determine the exact *X/R* ratio of a certain transformer and a graph and typical curve to approximate the general *X/R* value of different sizes of transformers.

Buck-Boost Autotransformers

A single-phase two-winding transformer normally has two separate windings, primary and secondary, that are connected one to the other only through flux coupling, as shown in Fig. 9-15*a* on p. 281. In this circuit, the primary winding carries the exciting current, and its 240-V connection to the incoming power circuit normally creates 24 V in the secondary coil because there is a 10:1 turns ratio between the primary and the secondary.

It is possible and operable to make one solidly conductive connection between the primary and secondary of this transformer, connecting the two coils as shown in Fig. 9-15*b*. Note that the output voltage is either 240 V + 24 V, or 264 V, or 240 V − 24 V, or 216 V, depending on whether the secondary is connected in additive polarity or subtractive polarity. This is where the name *buck-boost* originates. The same transformer, when connected as an autotransformer (part of the primary winding is electrically connected to the secondary winding), can either *buck* (reduce) or *boost* (increase) the incoming voltage.

Besides the ability of the one autotransformer to provide a range of output voltages, its kilovoltampere rating (as a transformer) can be significantly less than its kilovoltampere rating as an autotransformer. Tracing current through the circuit in Fig. 9-15*b*, it is apparent that the majority of the load current is simply conducted through the autotransformer. Recognize that the 24-V side of the transformer consists of conductors that are large enough to carry 10 times

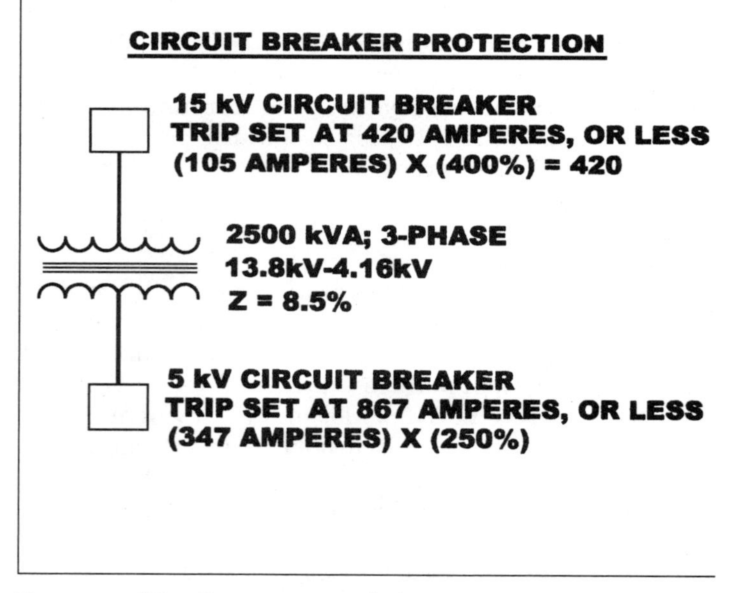

Figure 9-11 Solve for overcurrent device protecting a transformer operating at over 600 V.

the line current of the primary coil; therefore, the load current can be conducted safely straight through the secondary winding. The only flux coupling (power transfer through magnetics) done within the autotransformer is that creating the voltage change. Therefore, if the voltage changes 10 percent, then a transformer rated at x kVA will be rated at $x/10$ kVA as an autotransformer. That is, a 10 percent buck-boost

TRANSFORMER FULL-LOAD AMPERES

PRIMARY

$$\frac{2{,}500{,}000\text{VA}}{(13800\text{V})(1.73)} = 105 \text{ AMPERES}$$

SECONDARY

$$\frac{2{,}500{,}000\text{VA}}{(4160\text{V})(1.73)} = 347 \text{ AMPERES}$$

FUSE PROTECTION

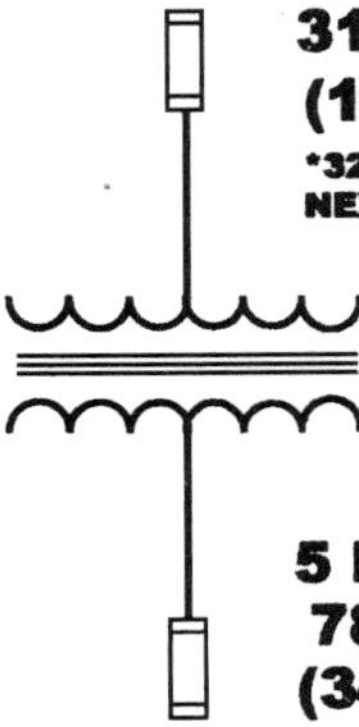

15 kV FUSE
315 AMPERE FUSE, OR LESS*
(105 AMPERES) X (300%) = 315

*325AMP FUSE CAN BE USED AS THE NEXT STANDARD SIZE

2500 kVA; 3-PHASE
13.8kV-4.16kV
Z = 8.5%

5 kV FUSE
781 AMPERE FUSE, OR LESS**
(347 AMPERES) X (225%) = 781

**800 AMP FUSE CAN BE USED AS THE NEXT STANDARD SIZE

transformer need only be rated at 1/10 the load kilovoltampere value.

If it is necessary to boost or buck the voltages in a three-phase system, three individual buck-boost transformers can be connected as shown in Fig. 9-16 on p. 282 to serve the three-phase load. In this service, each of the three-phase 10 percent buck-boost transformers need only be rated at 10 percent of the kilovoltampere load for one phase.

MAXIMUM RATING OF OVERCURRENT DEVICE PROTECTING TRANSFORMERS OF 600 VOLTS AND LESS
(STATED VALUES ARE PERCENTAGE OF TRANSFORMER CURRENT)

PROTECTION METHOD	PRIMARY PROTECTION (OVER 600 VOLTS)	PRIMARY PROTECTION (OVER 600 VOLTS)	PRIMARY PROTECTION (OVER 600 VOLTS)	SECONDARY PROTECTION	SECONDARY PROTECTION
	CURRENT OF 9 AMP OR MORE	CURRENT LESS THAN 9 AMP	CURRENT LESS THAN 2 AMP	CURRENT OF 9 AMP OR MORE	CURRENT LESS THAN 9 AMP
PRIMARY ONLY	125%	167%	300%	NOT REQUIRED	NOT REQUIRED
PRIMARY & SECONDARY	250%	250%	250%	125%	167%

Figure 9-12 Table replicating part of NEC Table 450-3(b), overcurrent protection of transformers less than 600 V.

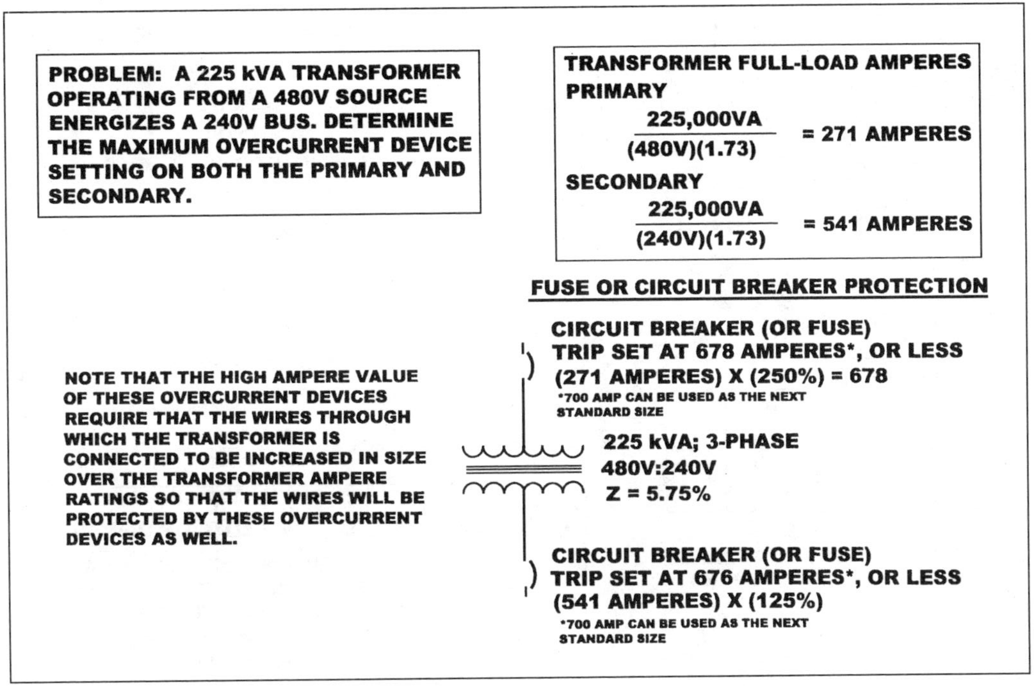

PROBLEM: A 225 kVA TRANSFORMER OPERATING FROM A 480V SOURCE ENERGIZES A 240V BUS. DETERMINE THE MAXIMUM OVERCURRENT DEVICE SETTING ON BOTH THE PRIMARY AND SECONDARY.

TRANSFORMER FULL-LOAD AMPERES

PRIMARY

$$\frac{225{,}000\text{VA}}{(480\text{V})(1.73)} = 271 \text{ AMPERES}$$

SECONDARY

$$\frac{225{,}000\text{VA}}{(240\text{V})(1.73)} = 541 \text{ AMPERES}$$

FUSE OR CIRCUIT BREAKER PROTECTION

CIRCUIT BREAKER (OR FUSE)
TRIP SET AT 678 AMPERES*, OR LESS
(271 AMPERES) X (250%) = 678
*700 AMP CAN BE USED AS THE NEXT STANDARD SIZE

225 kVA; 3-PHASE
480V:240V
Z = 5.75%

CIRCUIT BREAKER (OR FUSE)
TRIP SET AT 676 AMPERES*, OR LESS
(541 AMPERES) X (125%)
*700 AMP CAN BE USED AS THE NEXT STANDARD SIZE

NOTE THAT THE HIGH AMPERE VALUE OF THESE OVERCURRENT DEVICES REQUIRE THAT THE WIRES THROUGH WHICH THE TRANSFORMER IS CONNECTED TO BE INCREASED IN SIZE OVER THE TRANSFORMER AMPERE RATINGS SO THAT THE WIRES WILL BE PROTECTED BY THESE OVERCURRENT DEVICES AS WELL.

Figure 9-13 Solve for overcurrent device protecting a transformer operating at less than 600 V.

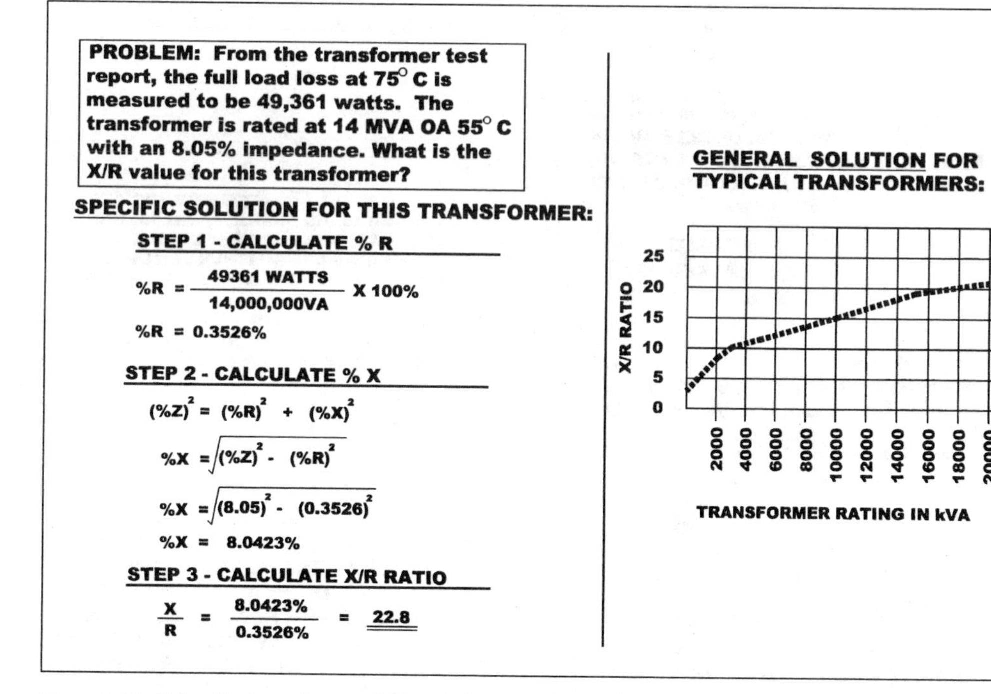

Figure 9-14 Solve for transformer *X/R* ratio by graphic means and by calculation given transformer impedance and full load loss.

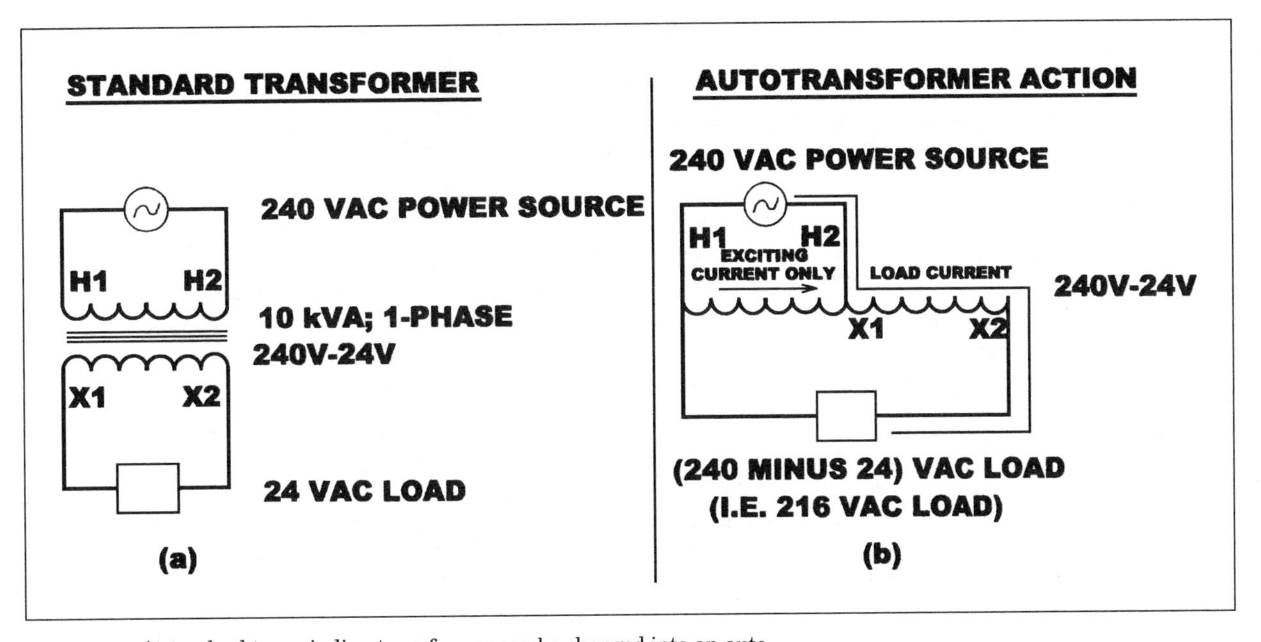

Figure 9-15 A standard two-winding transformer can be changed into an auto-transformer simply by interconnecting the primary coil with the secondary coil, as shown.

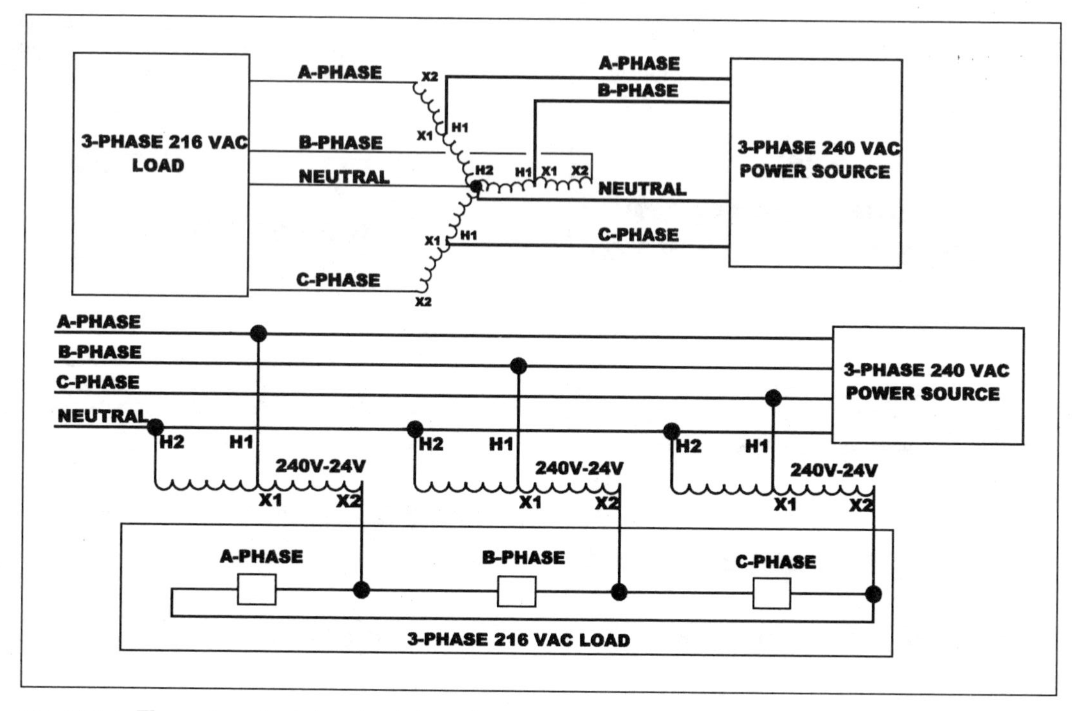

Figure 9-16 Three-phase buck-boost autotransformer connections.

NOTES

NOTES

Chapter

10

Motors

Given that a very high percentage of the electrical loads in the world are electric motors, this chapter pays specific attention to design of electrical systems for these very important loads. While there are many unique specific-duty motors, the alternating-current (ac) squirrel-cage three-phase induction motor is the primary "workhorse" of the industry.

The rotor of an ac squirrel-cage induction motor consists of a structure of steel laminations mounted on a shaft. Embedded in the rotor is the rotor winding, which is a series of copper or aluminum bars that are all short-circuited at each end by a metallic end ring. The stator consists of steel laminations mounted in a frame containing slots that hold stator windings. These stator windings can be either copper or aluminum wire coils or bars connected to the motor t-leads that are brought out to the motor junction box. Energizing the stator coils with an ac supply voltage causes current to flow in the coils. The current produces an electromagnetic field that creates magnetic fields within the stator. The magnetic fields vary in intensity, location, and polarity as the ac voltage varies, thus creating a rotating flux within the stator. The rotor conductors "cut" the stator flux, inducing current flow (and its own magnetic field) within the rotor. The magnetic field of the stator and the magnetic field of the rotor interact, causing rotation of the rotor and motor shaft. This action causes several motor characteristics, such as rotating speed (given in revolutions per minute), motor torque, motor horsepower, motor starting current, motor running current, and motor efficiency.

Selecting Motor Characteristics

Motor voltage

The power supply to motors can be either single-phase or three-phase, where single-phase is normally applied to motors having nameplate ratings of less than 1 horsepower (hp) and three-phase for larger motors.

Single-phase power is always 120 volts (V), and it is generally used to supply motors no larger than $^1/_3$ hp. Three-phase voltage sources of 208, 240, 480, and 600 V are, respectively, normally applied to motors having nameplate ratings of 200, 230, 460, and 575 V to offset voltage drop in the line. This is especially important where torque is of concern because torque is a function of the square of the voltage (decreasing the applied voltage to 90 percent decreases torque to 81 percent).

Motor speed

The speed of a motor is determined mainly by the frequency of the source voltage and the number of poles built into the structure of the winding. With a 60-hertz (Hz) power supply, the possible synchronous speeds are 3600, 1800, 1200, and 900 revolutions per minutes (rpm), and slower. Induction motors develop their torque by operating at a speed that is slightly less than synchronous speed. Therefore, full-load speeds for induction motors are, respectively, approximately 3500, 1750, 1160, and 875 rpm. Motors whose coils can be connected as two-pole, four-pole, or six-pole coils, for example, can have their speeds changed merely by switching pole wiring connections.

Let P = quantity of magnetic poles in the motor
Let f = the frequency of the voltage wave
RPM = synchronous angular velocity in revolutions per minute

$$\text{RPM} = \frac{(120) \times f}{P}$$

For example, calculating the RPM of a 4-pole synchronous motor operating at 60 Hertz:

$$\text{RPM} = \frac{(120) \times f}{P}$$

$$\text{RPM} = \frac{(120) \times (60)}{4}$$

RPM = 1800 RPM

SOLVING THIS EQUATION FOR 2P, 4P, & 6P MOTORS OPERATING FROM 50 HERTZ AND 60 HERTZ SYSTEMS YIELDS THE FOLLOWING TABLE OF SYNCHRONOUS SPEEDS:

f	P	RPM
60	2	3600
60	4	1800
60	6	1200
50	2	3000
50	4	1500
50	6	1000

For induction motors, actual RPM is less than synchronous speed by the motor "slip."
An approximate full-load slip value for a 4-pole, 60 Hertz motor is 50 RPM;
and slip increases with load increases.

Figure 10-1 Solve for motor synchronous speed given frequency, quantity of magnetic poles in the motor, and type of motor.

The speed (in rpm) at which an induction motor operates depends on the speed of the stator rotating field and is approximately equal to 120 times the frequency (f) divided by the quantity of magnetic poles (P) in the motor stator minus the rotor *slip*. Every induction motor must have some slip to permit lines of stator flux to cut the rotor bars and induce rotor current; therefore, no induction motor can operate at exactly synchronous speed ($120f/P$). The more heavily the motor is loaded, the greater the slip. Thus, the greater the voltage, the less is the slip. Figure 10-1 shows a typical motor speed calculation.

Ambient temperature and humidity

The ambient conditions must be considered in selecting the type of motor to be used in a specific location. *Ambient temperature* is the temperature of the air surrounding the motor. If it is very hot, special lubricant that does not decompose or “coke” at elevated temperatures and special wire insulation normally are required. Locations where high moisture levels or corrosive elements also exist require special motor characteristics, such as two-part epoxy paint, double-dip paint processes, and waterproof grease. Standard motors are designed to operate in an ambient temperature of up to 40°C (104°F) and normally are lubricated with high-temperature grease. At altitudes of greater than 3300 feet (ft), the lower density of the air reduces the self-cooling ability of the motor; therefore, compensation for altitude as well as ambient temperature must be made. Additional information about altitude compensation is provided below under the heading "Service Factor."

Torque

The rotating force that a motor develops is called *torque.* Due to the physical laws of inertia, where a body at rest tends to remain at rest, the amount of torque necessary to start a load (*starting torque*) is always much greater than the amount of torque required to maintain rotation of the load after it has achieved normal speed. The more quickly a load must accelerate from rest to normal rotational speed, the greater must be the torque capability of the motor driver. For very large inertia loads or loads that must be accelerated quickly, a motor having a high starting torque should be applied.

The National Electrical Manufacturers Association (NEMA) provides design letters to indicate the torque, slip, and starting characteristics of three-phase induction motors. They are as follows:

> Design A is a general-purpose design used for industrial motors. This design exhibits normal torques and full-load slip of approximately 3 percent and can be used for many types of industrial loads.
>
> Design B is another general-purpose design used for industrial motors. This design exhibits normal torques while also having low starting current and a full-load slip of approximately 3 percent. This design also can be used for many types of industrial loads.

Design C motors are characterized by high starting torque, low starting current, and low slip. Because of its high starting torque, this design is useful for loads that are hard to start, such as reciprocating air compressors without unloader kits.

Design D motors exhibit very high starting torque, very high slip of 5 to 13 percent, and low starting current. These motors are excellent in applications such as oilfield pumping jacks and punch presses with large flywheels.

Variable-torque motors exhibit a speed-torque characteristic that varies as the square of the speed. For example, a two-speed 1800/900-rpm motor that develops 10 hp at 1800 rpm produces only 2.5 hp at 900 rpm. Variable-torque motors are often a good match for loads that have a torque requirement that varies as the square of the speed, such as blowers, fans, and centrifugal pumps.

Constant-torque motors can develop the same torque at each speed; thus power output from these motors varies directly with speed. For example, a two-speed motor rated at 10 hp at 1800 rpm would produce 5 hp at 900 rpm. These motors are useful in applications with constant-torque requirements, such as mixers, conveyors, and positive-displacement compressors.

Service factor

The *service factor* shown on a motor nameplate indicates the amount of continuous overload to which the motor can be subjected at nameplate voltage and frequency without damaging the motor. The motor may be overloaded up to the horsepower found by multiplying the nameplate-rated horsepower by the service factor.

As mentioned earlier, service factor also can be used to determine if a motor can be operated continuously at altitudes higher than 3300 ft satisfactorily. At altitudes greater than 3300 ft, the lower density of air reduces the motor's cooling ability, thus causing the temperature of the motor to be higher. This higher temperature can be compensated for, in part, by reducing the effective service factor to 1.0 on motors with a 1.15 (or greater) service factor.

Motor enclosures

The two most common types of enclosures for electric motors are the totally enclosed fan-cooled (TEFC) motor and the open drip-proof (ODP) motor. The TEFC motor limits exchange of ambient air to the inside of the motor, thus keeping dirt and water out of the motor, whereas the ODP motor allows the free exchange of air from the surrounding air to the inside of the motor. Other types include the totally enclosed nonventilated (TENV), the totally enclosed air over (TEAO), and the explosionproof enclosure. Selection of the enclosure is determined by the motor environment.

Winding insulation type

The most common insulation classes used in electric motors are class B, class F, and class H. Motor frame size assignments are based on class B insulation, where, based on a 40°C ambient temperature, class B insulation is suitable for an 80°C temperature rise. Also based on a 40°C ambient temperature, class F insulation is suitable for a 105°C rise, and class H insulation is suitable for a 125°C rise. Using class F or class H insulation in a motor that is rated for a class B temperature rise is one way to increase the service factor or the motor's ability to withstand high ambient temperatures. Also, these insulations incorporate extra capability for localized "hot spot" temperatures.

Efficiency

Efficiency of an appliance is defined as the measure of the input energy to the output energy. The efficiency of an electric motor is the usable output power of the motor divided by the input power to the motor, and the differences between input and output power are losses in the motor. Smaller motors generally are less efficient than larger motors, and motors operated at less than half load usually are inefficient compared with their operation at full load. Therefore, for maximum operating efficiency, motors should be selected such that their nameplate horsepower or kilowatt rating is nearly the same as the driven load.

All the operating characteristics of a motor are interdependent, as shown in Fig. 10-2. A summary of these characteristics is provided in Fig. 10-2 to assist in expediting proper motor selection.

Motor starting current

When typical induction motors become energized, a much larger amount of current than normal operating current rushes into the motor to set up the magnetic field surrounding the motor and to overcome the lack of angular momentum of the motor and its load. As the motor increases to slip speed, the current drawn subsides to match (1) the current required at the supplied voltage to supply the load and (2) losses to windage and friction in the motor and in the load and transmission system. A motor operating at slip speed and supplying nameplate horsepower as the load should draw the current printed on the nameplate, and that current should satisfy the equation

$$\text{Horsepower} = \frac{\begin{array}{c}\text{voltage} \times \text{current} \times \\ \text{power factor} \times \text{motor efficiency} \times \sqrt{3}\end{array}}{746}$$

Typical induction motors exhibit a starting power factor of 10 to 20 percent and a full-load running power factor of 80 to 90 percent. Smaller typical induction motors exhibit an operating full-load efficiency of approximately 92 percent, whereas large typical induction motors exhibit an operating full-load efficiency of approximately 97.5 percent.

Since many types of induction motors are made, the inrush current from an individual motor is important in designing the electrical power supply system for that motor. For this purpose, the nameplate on every motor contains a code letter indicating the kilovoltampere/horsepower starting load rating of the motor. A table of these code letters and

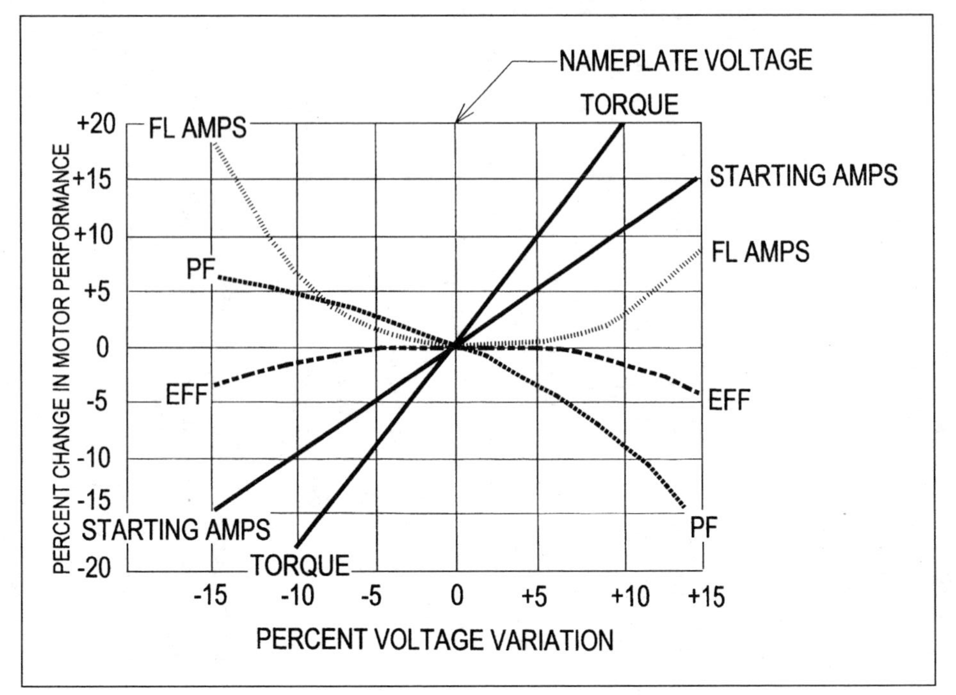

Figure 10-2 Solve for motor torque, speed, power factor, and efficiency reactions to varying voltage above and below nameplate voltage rating.

CODE LETTER ON MOTOR NAMEPLATE	KVA PER HORSEPOWER WITH LOCKED ROTOR MINIMUM	KVA PER HORSEPOWER WITH LOCKED ROTOR MEAN VALUE	KVA PER HORSEPOWER WITH LOCKED ROTOR MAXIMUM
A	0	1.57	3.14
B	3.15	3.345	3.54
C	3.55	3.77	3.99
D	4	4.245	4.49
E	4.5	4.745	4.99
F	5	5.295	5.59
G	5.6	5.945	6.29
H	6.3	6.695	7.09
J	7.1	7.545	7.99
K	8	8.495	8.99
L	9	9.495	9.99
M	10	10.595	11.19
N	11.2	11.845	12.49
P	12.5	13.245	13.99
R	14	14.995	15.99
S	16	16.995	17.99
T	18	18.995	19.99
U	20	29.2	22.39
V	22.4	NO LIMIT	

Figure 10-3 Solve for the kilovoltampere/horsepower value given motor code letter.

their meanings in approximate kilovoltamperes and horsepower is shown in Fig. 10-3. Using these values, the inrush current for a specific motor can be calculated as

$$I_{\text{inrush}} = \frac{\text{code letter value} \times \text{horsepower} \times 577}{\text{voltage}}$$

An example of this calculation for a 50-hp code letter G motor operating at 460 V is shown in Fig. 10-4.

Because of the items listed above, motors that produce constant kilovoltampere loads make demands on the electrical power system that are extraordinary compared with the demands of constant kilowatt loads. To start them, the overcurrent protection system must permit the starting current, also called the *locked-rotor current,* to flow during the normal starting period, and then the motor-running overcurrent must be limited to approximately the nameplate full-load ampere rating. If the duration of the locked-rotor

Starting currents exhibited by large induction motors are so much greater than those for smaller motors that starting voltage dip is a concern.

Problem: A 50 horsepower, 460 volt, 3-phase induction motor is designated on its nameplate as a class H motor. How much inrush current will this motor draw during starting on a 480 volt system?

Solution Step 1: Determine how many locked-rotor kVA the motor draws; and this is done with reference to the Code Letter.
A Code Letter of H designates that the motor draws between 6.3 and 7.09 kVA per nameplate horsepower. Therefore, this motor draws this starting (locked rotor) power:

kVA = (7.09kVA/HP) X (50 HP) = 354.5 kVA

Solution Step 2: Calculate inrush current:

P = E X I X 1.73
354,500 = (480) X I X (1.73)
I = 426.4 amperes

Solution:
During motor starting, the locked-rotor current drawn by the motor is up to 426.4 amperes.

Figure 10-4 Solve for inrush current of a 50-hp code letter G motor operating at 480 V, three-phase.

% of Table Full Load Current

Motor Type	Single-element Fuse	Dual-element Time Delay Fuse	Inverse Time Breaker	Instantaneous-only Magnetic Trip only Breaker
1-phase motor	300	175	250	800
3-phase squirrel cage	300	175	250	800
Design E 3-phase squirrel cage	300	175	250	1100
Synchronous	300	175	250	800
Wound Rotor	150	150	150	800
Direct Current	150	150	150	250

For example, a 50 horsepower, Design B, 460 volt, 3-phase motor has a table full load current of 65 amperes at 460 volts. The maximum rating of a thermal-magnetic circuit breaker protecting the motor branch circuit would be 65 amperes X 250%, or 162.5 amperes. The next higher standard rating is 175 ampers, so **175 amperes** is the maximum rating that can be used to protect the motor circuit.

Figure 10-5 Replication of NEC Table 430-152 of maximum overcurrent protective devices for motor circuits. Solve for overcurrent device rating for motor branch circuit given table ampere load.

current is too long, the motor will overheat due to I^2R heat buildup, and if the long-time ampere draw of the motor is too high, the motor also will overheat due to I^2R heating. The *National Electrical Code* provides limitations on both inrush current and running current, as well as providing a methodology to determine motor disconnect switch ampere and horsepower ratings.

Table 430-152 of the *National Electrical Code* provides the maximum setting of overcurrent devices upstream of the motor branch circuit, and portions of this table are replicated in Fig. 10-5. The code provides motor running current for typical three-phase induction motors in Table 430-150, portions of which are replicated in Fig. 10-6, and it provides motor disconnect switch horsepower and ampere criteria in Table 430-151, portions of which are replicated in Fig. 10-7 on pp. 298 and 299.

Calculating Motor Running Current

The following figures illustrate the calculations required by specific types of motors in the design of electric circuits to permit these loads to start and to continue to protect them during operation:

HORSEPOWER	208 VOLTS	230 VOLTS	460 VOLTS	575 VOLTS
0.5	2.4	2.2	1.1	0.9
0.75	3.5	3.2	1.6	1.3
1	4.6	4.2	2.1	1.7
1.5	6.6	6	3	2.4
2	7.5	6.8	3.4	2.7
3	10.6	9.6	4.8	3.9
5	16.7	15.2	7.6	6.1
7.5	24.2	22	11	9
10	30.8	28	14	11
15	46.2	42	21	17
20	59.4	54	27	22
25	74.8	68	34	27
30	88	80	40	32
40	114	104	52	41
50	143	130	65	52
60	169	154	77	62
75	211	192	96	77
100	273	248	124	99
125	343	312	156	125
150	396	360	180	144
200	528	480	240	192

Figure 10-6 Table of full-load currents for three-phase ac induction motors.

Figure 10-8: Continuous-duty motors driving a continuous-duty load (pp. 300 and 301)

Figure 10-9: Continuous-duty motors driving an intermittent-duty load (pp. 302 and 303)

Figure 10-10: Continuous-duty motors driving a periodic-duty load (pp. 304 and 305)

Figure 10-11: Continuous-duty motors driving a varying-duty load (pp. 306 and 307)

Calculating Motor Branch-Circuit Overcurrent Protection and Wire Size

Article 430-52 of the *National Electrical Code* specifies that the minimum motor branch-circuit size must be rated at 125 percent of the motor full-load current found in Table 430-150 for motors that operate continuously, and Section 430-32 requires that the long-time overload trip rating not be greater than 115 percent of the motor nameplate current unless the motor is marked otherwise. Note that the values of branch-circuit overcurrent trip (the long-time portion of a thermal-magnetic trip circuit breaker and the fuse melt-out curve ampacity) are changed by Table 430-22b for motors that do not operate continuously.

LOCKED-ROTOR CURRENT FOR THREE-PHASE ALTERNATING-CURRENT MOTORS

HORSEPOWER	208 VOLTS	230 VOLTS	460 VOLTS	575 VOLTS
0.5	22.1	20	10	8
0.75	27.6	25	12.5	10
1	33	30	15	12
1.5	44	40	20	16
2	55	50	25	20
3	71	64	32	25.6
5	102	92	46	36.8
7.5	140	127	63.5	50.8
10	179	162	81	64.8
15	257	232	116	93
20	321	290	145	116
25	404	365	183	146
30	481	435	218	174
40	641	580	290	232
50	802	725	363	290
60	962	870	435	348
75	1200	1085	543	434
100	1603	1450	725	580
125	2007	1815	908	726
150	2400	2170	1085	868
200	3207	2900	1450	1160

Note: Currents for Design E motors are not shown.

Example problem: Three motors are on a skid and share a disconnecting means.
They are 40 HP, 60 HP, and 75 HP 460 volt motors.
Determine the horsepower rating of the common disconnect switch.

Solution Step 1:

	Locked-rotor current	Motor HP
40 HP	290	40
60 HP	435	60
75 HP	543	75
subtotals.	1268 amperes	175

Solution Step 2:
By simple addition, the switch must be rated for at least 175 HP; but it must also be sized as follows:
The equivalent HP to 1268 amperes is 200 hp, therefore by the sum of current method, the switch must be rated for 200HP.

Final Solution:
Since the switch must be rated at the greater of 200 HP or 175 HP, the switch must be rated at **200 HP**.

Figure 10-7 Solve for the horsepower rating of motor disconnecting means using both horsepower and locked-rotor current.

PROBLEM

A 40 HORSEPOWER, 460 VOLT, 3-PHASE, CODE LETTER G, SERVICE FACTOR OF 1.0, IS PLANNED FOR OPERATION FROM A 480 VOLT, 3-PHASE SYSTEM. THE MOTOR HAS A NAMEPLATE FULL LOAD AMPERE RATING OF 50 AMPERES, AND THE TABLE AMPERE RATING FOR THIS MOTOR IS 52 AMPERES. THE MOTOR IS RATED FOR CONTINUOUS DUTY AND THE LOAD IS CONTINUOUS. SOLVE FOR THE MINIMUM SIZES OF BRANCH CIRCUIT ELEMENTS.

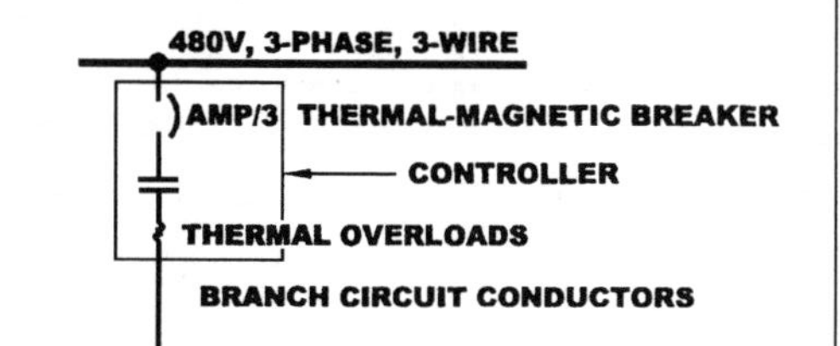

ELECTRICAL 1-LINE DRAWING OF PROBLEM

SOLUTION

STEP 1. CONFIRM TABLE FULL LOAD AMPERES FOR THIS MOTOR: 52 AMPERES
STEP 2. DETERMINE WIRE SIZE: 125% OF 52 AMPERES = 65 AMPERES
STEP 3. DETERMINE INVERSE-TIME BREAKER RATING: 250% OF 52 AMPERES = 130 AMPERES
STEP 4. DETERMINE THE RATING OF THE THERMAL OVERLOADS: 115% OF 50 AMPERES = 57.5 AMPERES
STEP 5. DETERMINE DISCONNECT SWITCH AMPERE RATING: 115% OF 52 AMPERES = 59.8 AMPERES
STEP 6. DETERMINE CONTROLLER HORSEPOWER RATING: 40 HP (SAME AS MOTOR NAMEPLATE HP RATING).

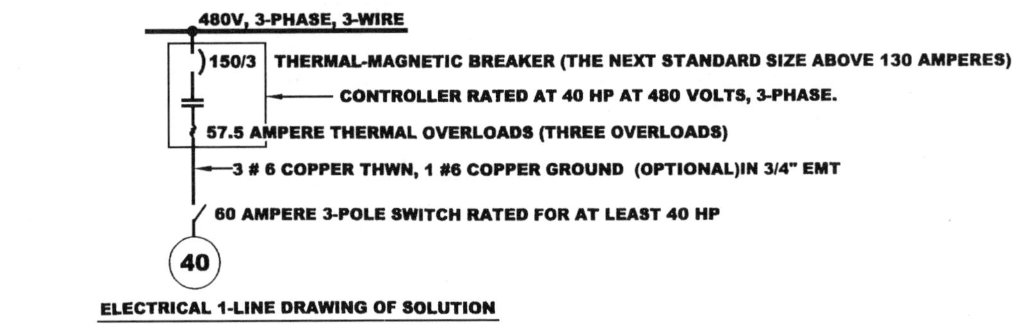

Figure 10-8 Solve for the wire ampere rating required for a continuous-duty ac motor driving a continuous load.

PROBLEM

A 40 HORSEPOWER, 460 VOLT, 3-PHASE, CODE LETTER G, SERVICE FACTOR OF 1.0, IS PLANNED FOR OPERATION FROM A 480 VOLT, 3-PHASE SYSTEM.
THE MOTOR HAS A NAMEPLATE FULL LOAD AMPERE RATING OF 50 AMPERES, AND THE TABLE AMPERE RATING FOR THIS MOTOR IS 52 AMPERES.
THE MOTOR IS RATED FOR CONTINUOUS DUTY AND THE LOAD IS INTERMITTENT.
SOLVE FOR THE MINIMUM SIZES OF BRANCH CIRCUIT ELEMENTS.

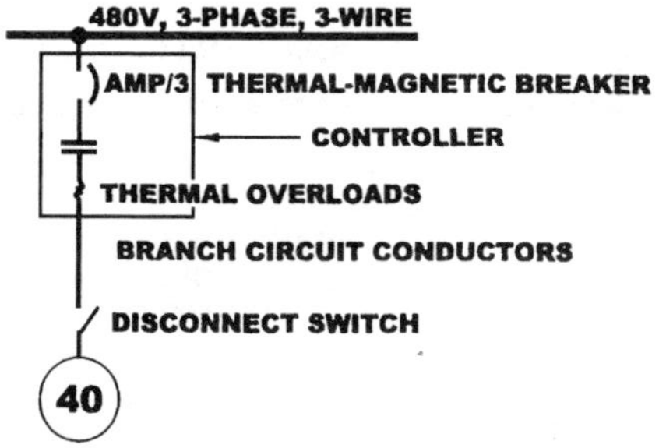

480V, 3-PHASE, 3-WIRE
AMP/3 THERMAL-MAGNETIC BREAKER
CONTROLLER
THERMAL OVERLOADS
BRANCH CIRCUIT CONDUCTORS
DISCONNECT SWITCH
40

ELECTRICAL 1-LINE DRAWING OF PROBLEM

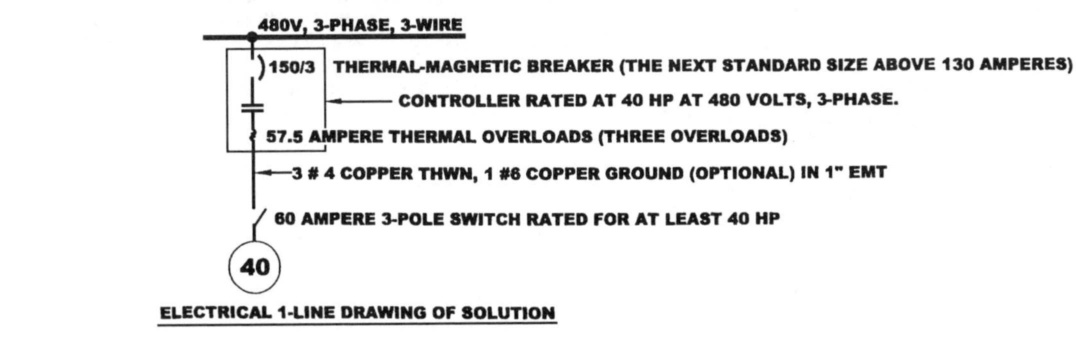

Figure 10-9 Solve for the wire ampere rating required for a continuous-duty ac motor driving an intermittent-duty load.

PROBLEM

A 40 HORSEPOWER, 460 VOLT, 3-PHASE, CODE LETTER G, SERVICE FACTOR OF 1.0, IS PLANNED FOR OPERATION FROM A 480 VOLT, 3-PHASE SYSTEM. THE MOTOR HAS A NAMEPLATE FULL LOAD AMPERE RATING OF 50 AMPERES, AND THE TABLE AMPERE RATING FOR THIS MOTOR IS 52 AMPERES. THE MOTOR IS RATED FOR CONTINUOUS DUTY AND THE LOAD IS PERIODIC. SOLVE FOR THE MINIMUM SIZES OF BRANCH CIRCUIT ELEMENTS.

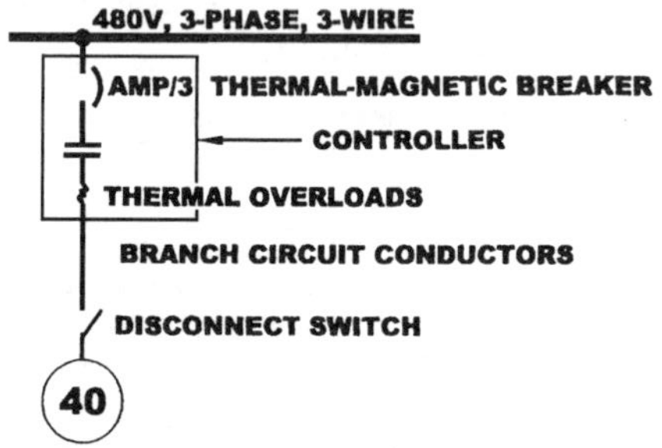

ELECTRICAL 1-LINE DRAWING OF PROBLEM

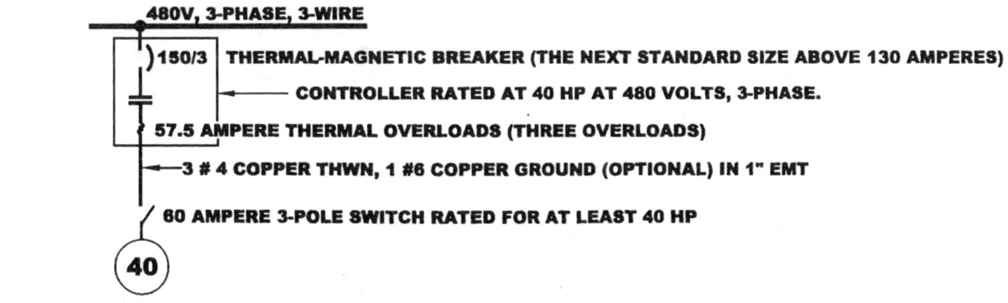

Figure 10-10 Solve for the wire ampere rating required for a continuous-duty ac motor driving a periodic-duty load.

PROBLEM

A 40 HORSEPOWER, 460 VOLT, 3-PHASE, CODE LETTER G, SERVICE FACTOR OF 1.0, IS PLANNED FOR OPERATION FROM A 480 VOLT, 3-PHASE SYSTEM. THE MOTOR HAS A NAMEPLATE FULL LOAD AMPERE RATING OF 50 AMPERES, AND THE TABLE AMPERE RATING FOR THIS MOTOR IS 52 AMPERES. THE MOTOR IS RATED FOR CONTINUOUS DUTY AND THE LOAD IS VARYING. SOLVE FOR THE MINIMUM SIZES OF BRANCH CIRCUIT ELEMENTS.

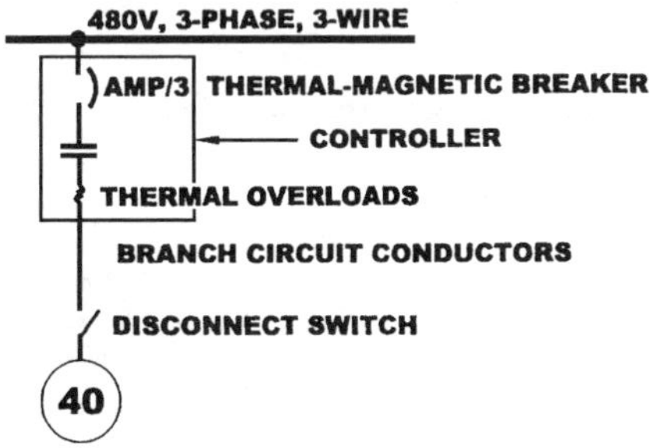

480V, 3-PHASE, 3-WIRE
AMP/3 THERMAL-MAGNETIC BREAKER
CONTROLLER
THERMAL OVERLOADS
BRANCH CIRCUIT CONDUCTORS
DISCONNECT SWITCH
40

ELECTRICAL 1-LINE DRAWING OF PROBLEM

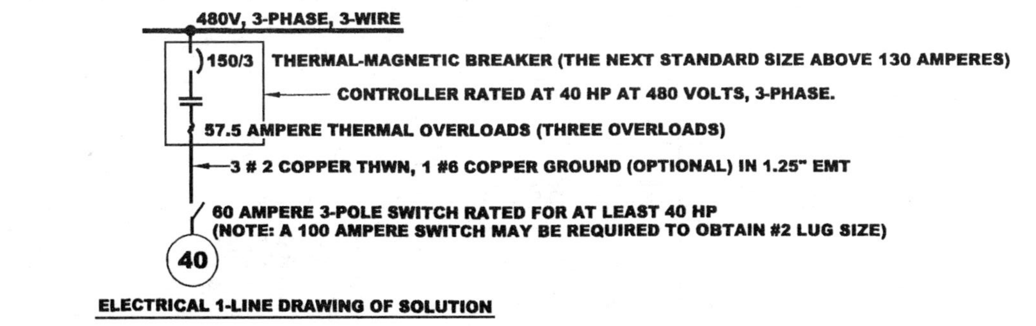

Figure 10-11 Solve for the wire ampere rating required for a continuous-duty ac motor driving a varying-duty load.

CHARACTERISTICS FOR THE PROPER SELECTION AND APPLICATION OF INDUCTION MOTORS

DESIGN LETTER	STARTING CURRENT IN % FULL-LOAD CURRENT	RELATIVE EFFICIENCY	SLIP IN % RPM	STARTING TORQUE IN % FULL-RATED TORQUE	STALLING TORQUE IN % FULL-RATED TORQUE
A	DEPENDS UPON NAMEPLATE CODE LETTER, NORMALLY 630% - 1000%	HIGH	3%	120% - 250%	200% - 275%
B	DEPENDS UPON NAMEPLATE CODE LETTER, NORMALLY 600% - 700%	HIGH	1.5% - 3%	120% - 250%	200% - 275%
C	DEPENDS UPON NAMEPLATE CODE LETTER, NORMALLY 600% - 700%	HIGH	1.5% - 3%	200% - 250%	190% - 225%
D	DEPENDS UPON NAMEPLATE CODE LETTER, NORMALLY 600% - 700%	MEDIUM	5% - 8%	275%	275%

Figure 10-12 Select the correct motor characteristics for the driven load from this chart.

NOTES

NOTES

Chapter 11

Raceways

Raceway Types and Their Characteristics

Although a very few types of conductors are rated for use without a raceway (such as for direct burial or for aerial installation), most conductors require protection in the form of a raceway. The most common raceways are conduit (both the metallic and nonmetallic) and sheet-metal wireways of various configurations. For the convenience of the engineer or designer selecting a raceway type for a particular installation, a brief summary of the types of raceways normally used, along with their trade names, is as follows:

Rigid metal conduit (RGS). Can be made of heavy-wall steel or aluminum.

Intermediate metal conduit (IMC). Normally made of steel conduit.

Electrometallic tubing (EMT). Thin-walled steel conduit.

Electric nonmetallic tubing (ENT). Corrugated plastic flexible raceway.

Nonmetallic underground conduit (PVC). Schedule 40 (heavy wall) or schedule 20 [called EB, for "encased burial" (in concrete)].

Flexible metallic tubing, Greenfield, spiral metal flexible conduit.

Liquidtite flexible metallic conduit, Sealtite, flexible metal conduit with an overall PVC waterproofing covering.

Surface metal and nonmetallic raceways, Wiremold.

Multioutlet assembly, Plugmold.

Cellular metal floor raceways, underfloor duct, Q-deck, Walker duct.

Cellular concrete floor raceways, Flexicore, Trenchduct.

Wireways.

Busways, bus duct.

Cablebus.

Boxes.

Auxiliary gutters.

Condulets and similar fittings.

The most common types of metallic raceways are electrometallic tubing (EMT) and rigid galvanized steel or rigid aluminum conduit. Rigid galvanized steel is available in both a standard wall thickness and a lighter-weight type known as *intermediate metal conduit* (IMC). Figure 11-1 shows the various common wiring methods in use today, along with the proper application for each to facilitate the correct selection of the wiring method type for a specific application.

The outside dimensions (OD) of these raceways are all approximately the same for each trade size, so the cross-sectional area available within each type of raceway for the placement of conductors varies with the type of raceway. Figure 11-2 on p. 315 shows the available cross-sectional areas within each type of raceway for the installation of conductors.

Pulling conductors into conduits requires considerations of pull-in friction and cable pinching at conduit bends, and operating conductors within raceways requires considerations of heat retention and temperature increase within the

WIRING METHOD	NORMAL USE
Aerial conductor	Outdoor distribution, overhead on poles
Direct burial cable	Outdoor distribution, underground in trench
Embedded PVC conduit	Underground or in concrete slab
Embedded Steel Conduit	Underground or in concrete slab
EMT	Indoors above ground, less commonly outdoors above ground
Rigid Steel Conduit	In all locations, including hazardous locations
IMC Conduit	In all locations, including hazardous locations
Aluminum Rigid Conduit	Above ground in all locations, but not embedded in concrete
ENT	Indoors above ground
Cables in Cable Tray	Outdoors in industrial plants, less commonly indoors
Nonmetallic Cable	Indoors above ground, most commonly for residential installations
Armored "AC" Cable ("BX")	Indoors above ground, normally for commercial installations
Armored "MC" Cable	Specific types of this cable are used in all locations

Figure 11-1 Common wiring methods and locations where each is normally implemented.

Heavy-wall rigid steel conduit in an industrial plant.

raceway. These considerations are both lessened in effect by limiting the raceway fill to 40 percent for three or more conductors, to 31 percent for two conductors, and to 53 percent for one single round conductor. Accordingly, Fig. 11-2 also provides the usable cross-sectional area within each type and size of raceway for 100 and 40 percent fill. Figure 11-3 on p. 316 provides a sample conduit fill calculation.

Many different types of enclosures are available, and they are selected to match the environment into which they will be located. Figure 11-4 on p. 317 is a listing of the National Electrical Manufacturers Association (NEMA) enclosures along with a description of the installation condition for which each is intended.

Trade Size (inches)	EMT 100% of C.S.A.*	EMT 40% OF C.S.A.	RIGID METAL 100% of C.S.A.	RIGID METAL 40% OF C.S.A.	IMC 100% of C.S.A.	IMC 40% OF C.S.A.	SCH 40 PVC 100% of C.S.A.	SCH 40 PVC 40% OF C.S.A.	FLEX METAL 100% of C.S.A.	FLEX METAL 40% OF C.S.A.
0.5	0.304	0.122	0.314	0.125	0.342	0.137	0.285	0.114	0.317	0.127
0.75	0.533	0.213	0.549	0.22	0.586	0.235	0.508	0.203	0.533	0.213
1	0.864	0.346	0.888	0.355	0.959	0.384	0.832	0.333	0.817	0.327
1.25	1.496	0.598	1.526	0.61	1.646	0.658	1.453	0.581	1.277	0.511
1.5	2.036	0.814	2.071	0.829	2.223	0.889	1.986	0.794	1.857	0.743
2	3.356	1.342	3.408	1.363	3.629	1.452	3.291	1.361	3.269	1.307
2.5	5.858	2.343	4.866	1.946	5.135	2.054	4.695	1.878	4.909	1.964
3	8.846	3.358	7.499	3	7.922	3.169	7.268	2.907	7.069	2.827
3.5	11.545	4.618	10.01	4.004	10.584	4.234	9.737	3.895	9.621	3.848
4	14.753	5.901	12.883	5.153	13.631	5.452	12.554	5.022	12.566	5.027
5	N.A.	N.A.	20.213	8,085	N.A.	N.A	19.761	7.904	N.A.	N.A.
6	N.A.	N.A.	29.158	11.663	N.A.	N.A.	28.567	11.427	N.A.	N.A.

* C.S.A. IS THE CROSS-SECTIONAL AREA OF THE INSIDE OF THE CONDUIT, MEASURED IN SQUARE INCHES

Figure 11-2 Cross-sectional areas of commonly used different types of raceways.

PROBLEM: A raceway must be selected for use in a grade-level concrete slab to contain the following conductors:

3 - 300 kCMIL THHN

1 - 0000 THHN

SOLUTION STEP 1 For installation in a concrete slab, select rigid steel conduit.

SOLUTION STEP 2 Calculate the required minimum size of conduit:

300 kCMIL THHN:	0.4608 sq. in.
300 kCMIL THHN:	0.4608 sq. in.
300 kCMIL THHN:	0.4608 sq. in.
0000 THHN:	0.3237 sq. in.
TOTAL WIRE C.S.A.	1.7061 sq. in.

For 4 or more conductors, select 40% conduit fill.

From the conduit fill tables, select the next conduit size having greater than 1.7061 sq. in. in the 40% fill column. **Therefore select 2.5 inch rigid steel conduit.**

Figure 11-3 Solve for minimum conduit size given wire insulation type, wire size, and wire quantity.

NEMA TYPE NUMBER	APPLICATION
1	INDOOR PROTECTION AGAINST ACCIDENTAL HUMAN CONTACT OF LIVE PARTS
2	INDOOR PROTECTION AGAINST FALLING DIRT AND ACCIDENTAL HUMAN CONTACT OF LIVE PARTS
3	OUTDOOR PROTECTION AGAINST WINDBLOWN RAIN, SNOW, SLEET, AND ACCIDENTAL HUMAN CONTACT OF LIVE PARTS
3R	OUTDOOR PROTECTION AGAINST FALLING RAIN AND ICE ON THE ENCLOSURE, BUT NO PROTECTION AGAINST DUST
3S	OUTDOOR PROTECTION AGAINST FALLING RAIN, ICE, WINDBLOWN DUST, AND HOSEDOWN WITH WATER
4	INDOOR AND OUTDOOR PROTECTION AGAINST WINDBLOWN RAIN AND DUST, SPLASHING WATER, HOSEDOWN WITH WATER, AND ICE ON THE ENCLOSURE
4X	INDOOR AND OUTDOOR PROTECTION AGAINST CORROSION, WINDBLOWN RAIN AND DUST, SPLASHING WATER, HOSEDOWN WITH WATER, AND ICE ON THE ENCLOSURE
5	INDOOR PROTECTION AGAINST FALLING DIRT, FALLING LIQUIDS, DUST, AND FIBERS
6	INDOOR AND OUTDOOR PROTECTION AGAINST WATER - TEMPORARY SUBMERSION AT SHALLOW DEPTH
6P	INDOOR AND OUTDOOR PROTECTION AGAINST WATER - PROLONGED SUBMERSION AT SHALLOW DEPTH
7	CLASS I DIVISION 1 AND 2 LOCATIONS TO CONTAIN AN EXPLOSION AND NOT IGNITE SURROUNDING EXTERNAL GASES
9	CLASS II LOCATIONS TO NOT IGNITE COMBUSTIBLE DUSTS
12	INDOOR PROTECTION AGAINST DUST, FALLING DIRT, AND NON-CORROSIVE LIQUIDS.
13	INDOOR PROTECTION AGAINST DUST, SPRAYING WATER, OIL AND NON-CORROSIVE LIQUIDS.

Figure 11-4 This is a listing of NEMA enclosures and the environments for which each is suitable.

NOTES

Chapter

12

Overcurrent Devices

Overcurrent Devices: Fuses and Circuit Breakers

A tremendous amount of electrical energy is available in almost every electrical power system, so every part of an electrical installation must be protected from excessive current flow. Excessive current flow can be considered in two distinct categories:

1. Instantaneous current from inrush on start-up or from a short circuit
2. Long-time overload current

Overcurrent devices are available in several forms. At low voltage, the most common forms are

1. Non-time-delay fuse
2. Time-delay dual-element fuse
3. Magnetic-only, instantaneous-trip circuit breaker
4. Thermal-magnetic-trip circuit breaker

Figure 12-1 shows the time-current characteristics of the most common of these types of overcurrent devices for a

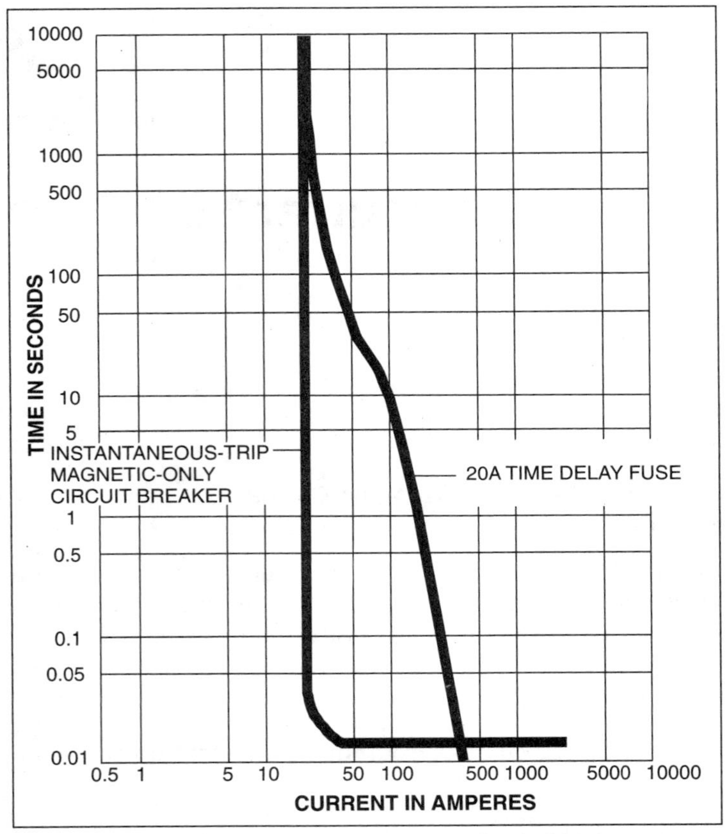

Figure 12-1 Time-current characteristic curves of typical 20-A overcurrent devices.

standard 20-ampere (A) device. Note that for instantaneous-only protection, a magnetic-only circuit breaker unlatches and trips (opens the power circuit) immediately on reaching the preset ampere value, as does the thermal-magnetic-trip circuit breaker. However, the instantaneous-trip setting on a thermal-magnetic-trip circuit breaker is normally set at a higher ampere rating than would be a magnetic-only breaker because the thermal element of the thermal-magnetic-trip circuit breaker adequately provides protection within the ampere range of maximum safe operating current. The thermal-magnetic-trip breaker curve and the curve of the time-delay fuse are very similar to each other because the thermal-magnetic-trip breaker curve is designed to mimic the curve of the time-delay fuse.

Fuses

A fuse heats internally due to I^2R heating, and after enough heat builds up, the thermal element in the fuse simply melts, opening the circuit. A time-delay dual-element fuse simply contains additional thermal mass that requires additional I^2R heating over time before reaching the melting temperature of the fuse element.

Although special fuses and circuit breakers are available in every ampere rating for special applications, their standard ampere ratings are 15, 20, 25, 30, 35, 40, 45, 50, 60, 70, 80, 90, 100, 110, 125, 150, 175, 200, 225, 250, 300, 350, 400, 450, 500, 600, 700, 800, 1000, 1200, 1600, 2000, 2500, 3000, 4000, 5000, and 6000 A. Additionally, the standard ampere rating for fuses includes 1, 3, 6, 10, and 601 A.

Short-time fuse operation

When a short circuit occurs in a phase conductor and a large value of current flows, the short-circuit current must flow through the fuse because the fuse is in series with the phase conductor. Inside of the fuse are (1) a conductor of copper or silver having a small cross-sectional area and therefore a high resistance compared with the resistance of the other circuit elements and (2) a spring that is continuously trying to pull apart the parts of the small cross-sectional conductor. Due to the electrical property that heat is equal to the square of the current multiplied by the resistance, the high fault current creates very high temperatures in the fuse element in a very short time (for large fault currents, less than $^1/_4$ cycle). As the fuse element melts, the spring pulls it apart, causing an arc and interrupting the circuit current.

Where low values of fault (less than 10 kA) are available to flow into a fault, the plain atmosphere in simple unfilled fuses is sufficient to extinguish the arc within the fuse barrel. However, where large amounts of fault current are available, fuses are filled with silica sand or a similar material that melts in the established arc, flowing into the arc and extinguishing it. After this type of *current-limiting* fuse "blows," microscopic bits of fuse element conductor are mixed with sand particles throughout the fuse barrel. A check of the continuity of this type of fuse after it has blown will reveal that it conducts current through an ohmmeter; however, when placed into a circuit where actual load current can be drawn, the resistance of the fuse is shown to be an open circuit.

A current-limiting fuse exhibits the ability to actually limit the available fault current downstream of the fuse as long as the available fault current upstream of the fuse is great enough. Current-limiting fuses cannot reduce fault current while they are interrupting current in the lower ranges of their current-limiting capability.

Current-limiting fuse characteristic curves are available from fuse manufacturers showing how much current each type and rating of fuse will let through compared with the amount of fault current available to flow immediately upstream of the fuse. Each fuse requires its own curve.

Long-time fuse operation

Keeping in mind that temperature is a function of heat flow or lack of heat flow, while heat energy is being released, a fuse that is conducting a small (10 to 25 percent) overload current through its high-resistance fuse element gains I^2R heat. Some of this heat continually flows away from the fuse element into the surroundings, but not all the heat flows away. Over time, this heat builds up until the fusible element melts and is pulled apart by the spring in the fuse. When the element is pulled apart, an arc occurs in much the same way as the arc occurs during a short circuit.

Fuse characteristic curves are available from fuse manufacturers showing how long each fuse can carry each value of current before melting. As with current-limiting fuses, each fuse requires its own curve.

Dual-element (time-delay) fuse operation

During fuse construction, placing a thermal mass, or heat sink, on the fuse element keeps the fuse element cooler for a few moments while the I^2R heat generated within the fuse during overload conditions flows to the thermal mass. This effectively delays the melting of the fuse element during overload time. However, the thermal mass does not affect fuse quick-blow operation during short-circuit time, nor does it affect the current-limiting capability of the fuse.

Circuit Breakers

The *National Electrical Code* defines a *circuit breaker* to be "a device designed to open and close a circuit automatically on a predetermined overcurrent without damage to itself when properly applied within its rating." Besides the current magnitude parameter, circuit breakers also operate within the time domain. Some circuit breakers trip to open the circuit instantly on reaching a predetermined setting, whereas others require overcurrent to flow for a certain time duration before tripping. In addition, a given circuit breaker is, by code definition, supposed to open a circuit while interrupting the current for which it is rated and to do so "without damage to itself." In actuality, circuit breakers that interrupt their rated current must be overhauled before being put back into service. This is due mainly to the pitted main contacts that result from the heat associated with interrupting a large current arc. That is, the apparent code definition of *damage* is not what many people normally would consider to be damage, and instead, the code definition of *damage* has to do with pieces of the breaker flying apart in an explosion or dismemberment of the mechanical links of the trip assembly. It is with this in mind that many engineers overspecify the current-interrupting duty of circuit breakers to minimize breaker maintenance and overhaul expense, but the expense of overspecification is considerable as well. In general, for low-voltage breakers, increasing the circuit breaker to the next standard fault duty increases the cost of the circuit breaker by approximately 25 percent.

The *National Electrical Code* lists several types of circuit breakers, including adjustable, instantaneous-trip, inverse-time, and nonadjustable.

Magnetic-only circuit breakers

Circuit breakers contain contacts that are held together under spring tension by an arm that, in normal "breaker closed" position, is "almost tripped." That is, with hardly any force on the trip arm at all, the springs within the breaker pull the contacts apart (instead of holding them closed) and keep them apart until the breaker is reset. The force to trip the breaker by moving the trip arm can come from a small solenoid driven by a ground-fault current monitor system, and this trip device is called a *ground-fault trip unit.* Or the force to trip the breaker by moving the trip arm can come from a magnetic force induced to trip the arm from a magnetic coil that can be in series with the main power circuit. When too much current flows, the magnetic forces set up by it, measured in ampere-turns, simply move the trip arm and cause the breaker to trip. This magnetic solenoid device also can be within an adjustable magnetic trip system, but the operation of all these breakers is still the same: The magnetic trip causes the breaker to "trip open" immediately when the line current reaches a predetermined ampere value.

Due to the operational requirement that a magnetic-only circuit breaker permit inrush current to flow in the start-up of a motor or other appliance, when used with loads having large inrush currents, magnetic-only circuit breakers must have trip ratings that are greater than the inrush current values, and this is a great drawback to their use.

Thermal-magnetic-trip circuit breakers

An improvement in the circuit breaker trip action is offered in the thermal-magnetic-trip circuit breaker in that it can permit large inrush currents for short time durations while still maintaining the ability to trip instantly on short-circuit current flow.

A thermal-magnetic-trip circuit breaker contains the same magnetic-only trip unit of the magnetic-only breaker, but the trip unit is placed on a moving platform that rests on a bimetal. The platform mechanically moves toward the tripping direction as the bimetal temperature gets hotter and away from the tripping direction as the bimetal cools. Given the laws of heat flow in classical thermodynamics, when current flows through the bimetal, the heat it builds up flows away from the bimetal less and less slowly as the difference in temperature between the bimetal and its surroundings lessens. The result of this is that a given ampere rating of current will cause the thermal-magnetic-trip circuit breaker to trip after that current flows for a given time, but the breaker will trip after a shorter current flow when the current magnitude is greater. In the electrical industry, this is called an *inverse-trip circuit breaker* because the greater the current flow, the shorter is the time before the breaker trips and opens the circuit. The operations of these circuit breakers are depicted by time-current curves that roughly follow the short-time inrush current and long-time low-amperage running current of motors, heaters, and many other appliances; thus they are normally the circuit breaker type chosen for most services of this type.

Medium-Voltage and Special-Purpose Circuit Breakers and Relay Controllers

Sometimes it is important for an overcurrent device to have exacting trip characteristics (depicted by a special shape of time-current trip curve), but no standard circuit breaker can be found with these characteristics. In such cases, as with the case of medium-voltage characteristics, either breakers with solid-state controllers or special-purpose relays are used to monitor the power circuit and trip the breaker at the proper preselected time. For certain applications, such as for generator protection, groups of relays are interconnected into special-purpose relay systems to perform the specified functions. In this way, trip curves having almost any desired characteristics can be achieved.

A relay can be found for almost any purpose, and multipurpose relays are available as well. Relays that perform as phase-overcurrent devices are available, as well as almost a hundred other types. Each type of relay is designated by modern standards with an alphanumeric nomenclature. The following is a list of the identifying symbol and function of the most frequently used relay types:

12	Overspeed
14	Underspeed
15	Speed matching
21	Distance relay that functions when the circuit admittance, reactance, or impedance changes beyond set values
25	Synch-check
27	Undervoltage
32	Directional power
37	Undercurrent or underpower
40	Field undercurrent
46	Reverse phase or phase balance
47	Phase sequence
49	Thermal
50	Instantaneous
51	ac time-overcurrent
52	ac circuit breaker control
59	Overvoltage
60	Voltage balance or current balance
62	Time delay
64	Ground detector
65	Governor
67	ac directional overcurrent
71	Level switch
74	Alarm
76	dc overcurrent
81	Frequency or change of frequency
86	Lockout
87	Differential protection

Templates showing examples of the uses of these relays are as follows:

Figure 12-2: Solve for relay selection and connections of a medium-voltage circuit breaker that incorporates both instantaneous and time-overcurrent relay protection for a feeder

Figure 12-3: Solve for relay selection and placement for the protection of a small generator

Figure 12-4: Solve for relay selection and placement for the protection of a large generator

Figure 12-5: Solve for relay selection and drawing of transformer protection that includes transformer differential protection

Figure 12-6: Solve for relay selection and placement of relay protection for a large induction motor

In Fig. 12-2, the 50 relays are instantaneous-trip devices that trip immediately on the flow of a set value of current, and one of these is required to protect each of the three phases. The 51 overcurrent relays are time-overcurrent relays whose time and current settings can be preprogrammed to settings that can protect the load circuit and that can be coordinated with upstream overcurrent devices. The 50N and 51N relays monitor phase-imbalance current and can be set to trip on a predetermined neutral current value.

In Fig. 12-3, the overcurrent relay (51V) is voltage restrained to correctly modify the time-current trip curve as the generator voltage changes. The 32 directional power relay prevents the generator from running as a motor instead of generating power into the bus, and the 46 relay performs a similar function while monitoring negative sequence currents and phase-imbalance currents. The 87 differential relay set guards against a fault within its protective zone, which extends from one set of three current transformers to the other set of current transformers, with the generator itself within the zone of protection.

As with the protection scheme for a small generator in Fig. 12-3, in Fig. 12-4 for a large generator, the overcurrent

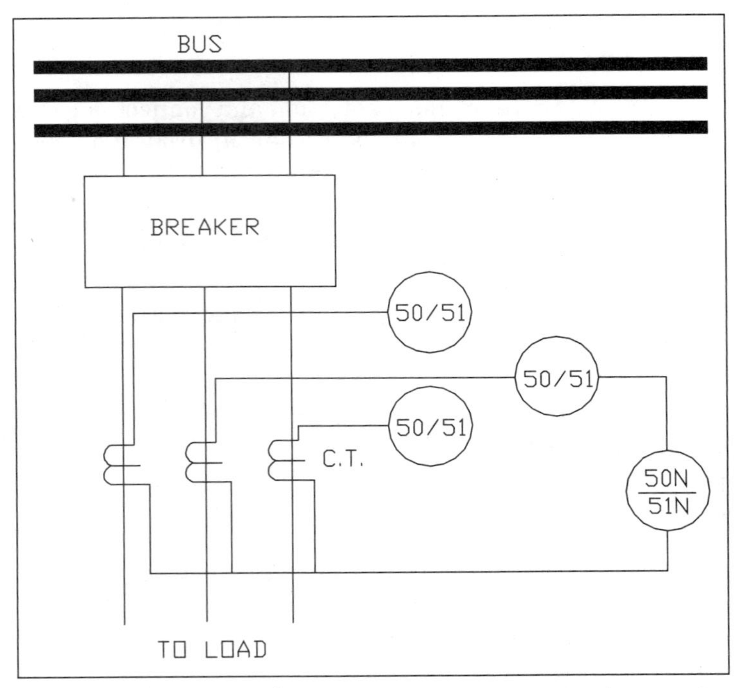

Figure 12-2 Solve for the relay selection and connections for the protection of a medium-voltage feeder breaker.

relay (51V) is voltage restrained to correctly modify the time-current trip curve as the generator voltage changes. The 40 directional field relay prevents problems of low field current, and the 46 relay monitors negative sequence currents and phase-imbalance currents. The 87 differential relay set around the controller guards against a fault within its protective zone, which extends from one set of three current transformers to the other set of current transformers, with the generator itself within the zone of protection. The other 87 relay set, the 87G, guards against phase-to-ground faults within the generator, and the 51G monitors for ground-fault current anywhere in the system.

In Fig. 12-5, the primary circuit breaker is used for transformer protection. The basic internal zone short-circuit protection is provided by the 87T differential relays, where all

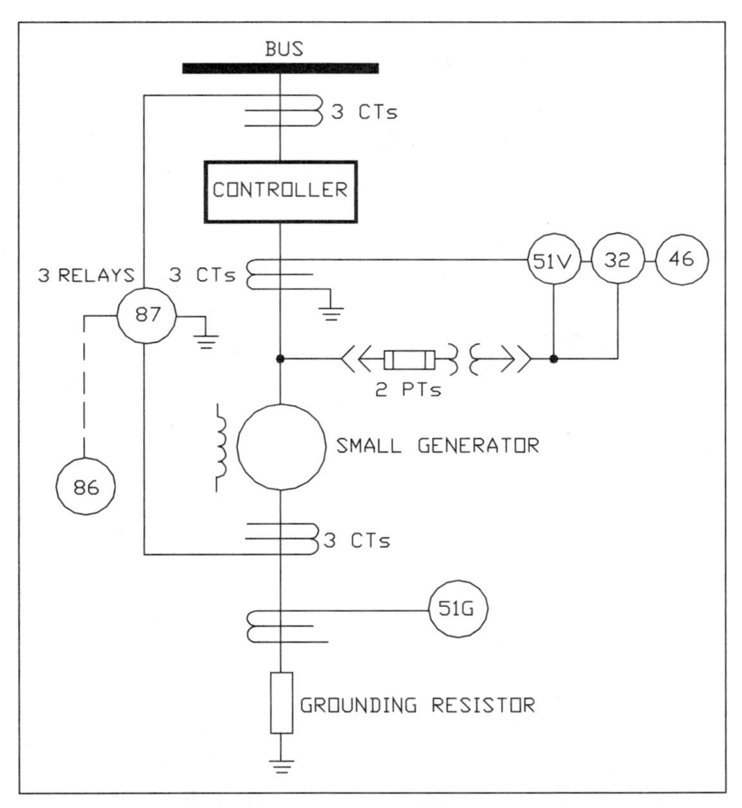

Figure 12-3 Solve for relay selection and connections for the protection of a small generator.

the power that enters this protective zone must exit this protective zone or else the circuit breaker is signaled to trip. The 50/51 provides backup fault protection through instantaneous and long-time overcurrent trips. The 50N/51N functions as backup ground-fault protection. Transformer overload and load-side conductor protection are provided by the 51 on the secondary side of the transformer. Since the low-voltage side is resistance grounded, the 51G-1 ground relay should be connected to trip breaker 52-1 for secondary side ground faults between the transformer and the secondary breaker and for resistor thermal protection. Device 51G-2 should be connected to trip breaker 52-2 to provide bus

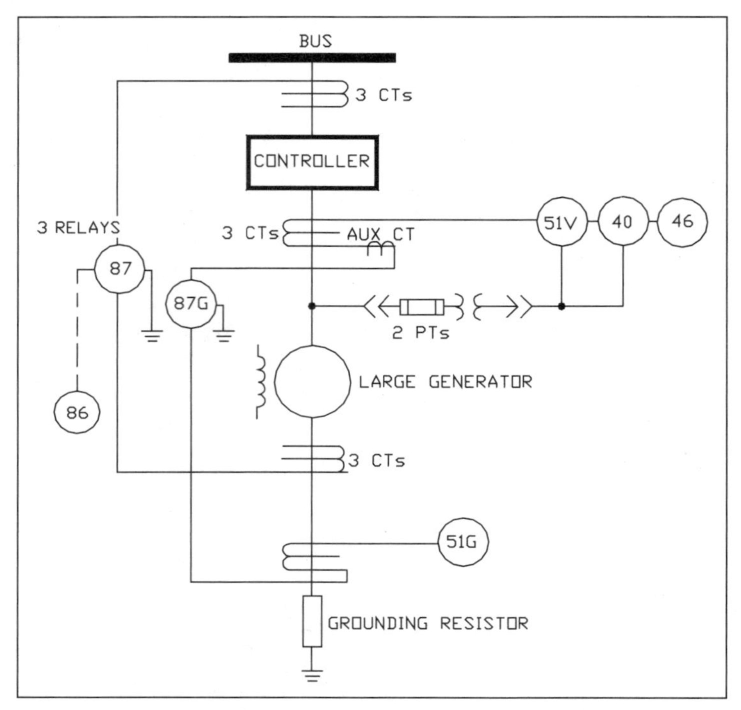

Figure 12-4 Solve for relay selection and connections for the protection of a large generator.

ground-fault protection and feeder ground backup. Device 63 is a sudden-pressure switch or Buckholtz relay that operates on a given value of pressure or on a given rate of change in pressure. It is highly sensitive to internal transformer faults. Note that the current transformers on the wye side of the transformer are connected in a delta and the current transformers on the delta side of the transformer are connected in a wye. In this way, the 30° phase shift from delta to wye in the power transformer is accommodated in the −30° phase shift in the current transformers from wye to delta. Some of the new electronic relays allow both sides to be connected in a wye and take care of the phase shift internally.

In Fig. 12-6, the motor controller can be either a contactor or an electrically operated circuit breaker. The principal

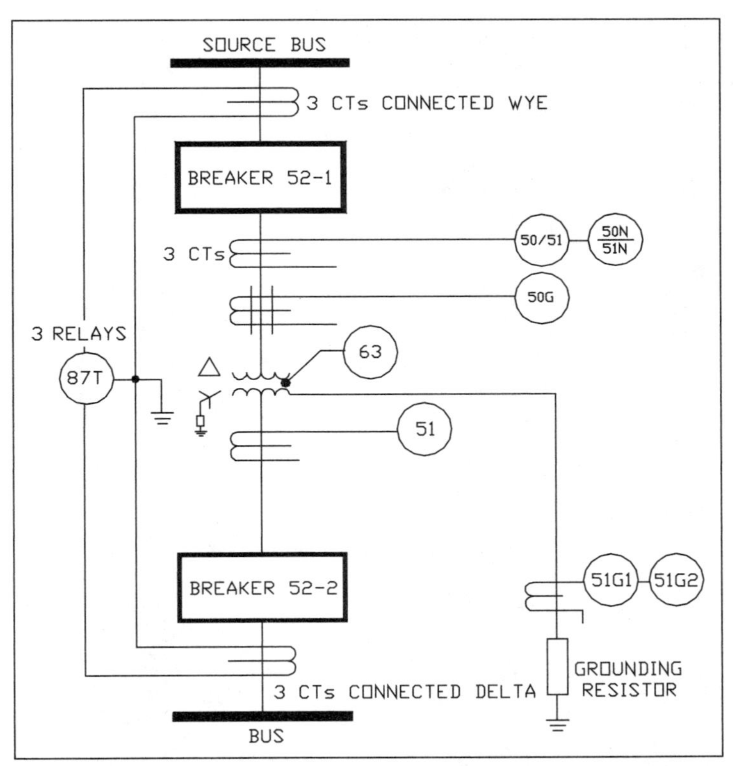

Figure 12-5 Solve for relay selection and connections for the protection of a large transformer with differential protection.

motor monitor is the 50 instantaneous overcurrent relay, and it is augmented by a 46 phase-balance relay to prevent negative sequence currents in the rotor of the motor. The 48 relay operates as phase failure protection and motor starting protection, tripping the motor off-line if its starting time exceeds the predetermined value set into the relay. This function is augmented by the 49S relay that mimics the thermal state of the motor, tripping it off just as a bimetal heater would trip off a low-voltage motor in a low-voltage motor starter, and this contact is thermally monitored by the 49 RTD in the motor stators (of which there are normally at least two and often six RTDs). The 50GS trips the

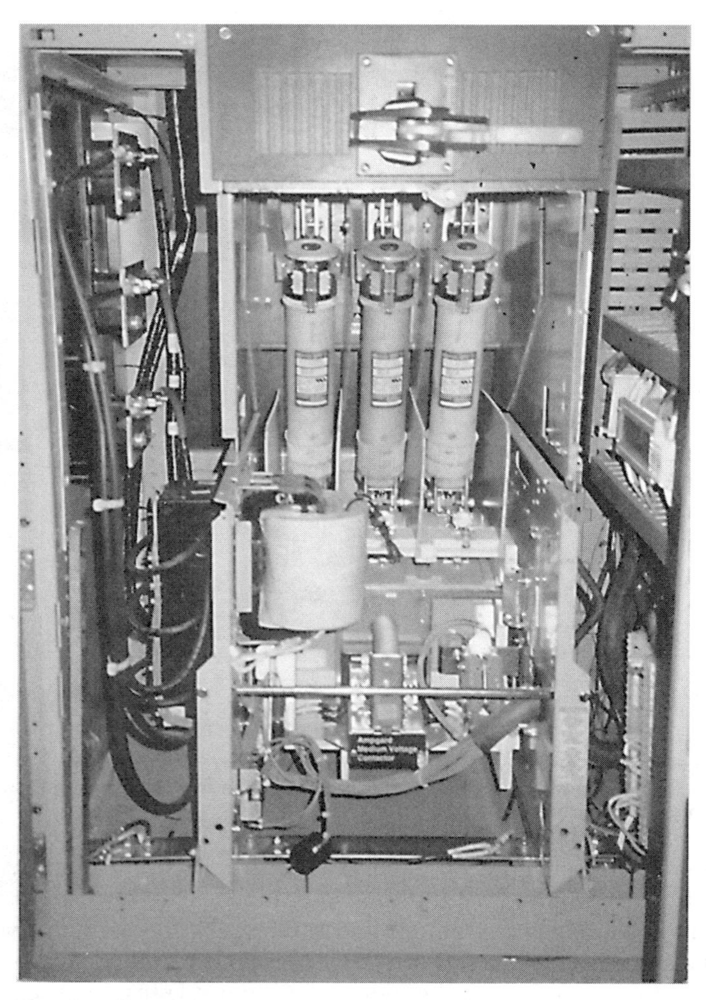

Fused medium-voltage contactors are normally fitted with relays as well as fuses for overcurrent protection.

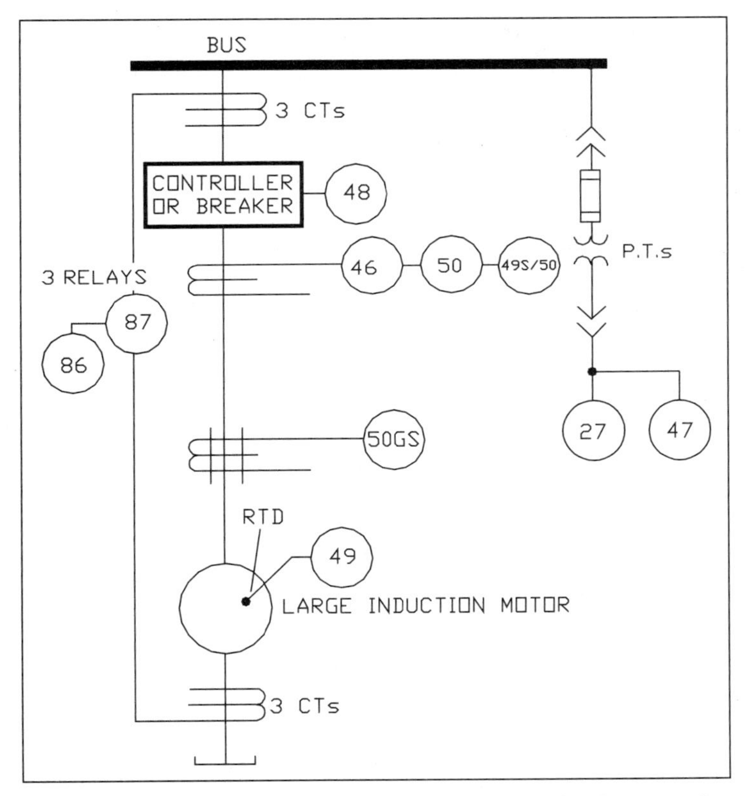

Figure 12-6 Solve for relay selection and connections for the protection of a large induction motor.

motor circuit in the event of a motor ground fault, and the 87 differential relay set trips the motor circuit in the event of a fault within the motor circuit zone. The 50, 49S, and 49 relays must be set individually for each type and size of motor, and these are often contained within one overall computer-based static relay package along with many other relays.

NOTES

Chapter

13

Circuits for Special Loads

Designing Circuits for Various Electrical Loads

The normal procedure used to determine circuit sizes and characteristics for typical loads is to determine the wiring method and conductor and insulation types to be used from the installation-site environmental data, solve for the current that will be drawn by the load, select the conductor size, and then determine the raceway size. After all this is completed, a calculation of the voltage drop in the system is normally done to determine that sufficient operating voltage is available at the terminals of the load for it to operate satisfactorily. A solution using this methodology is shown in Fig. 13-1. When the load can operate for 3 hours or longer, it is considered to be a continuous-duty load, and the circuit must be designed for continuous operation. Figure 13-2 is the solution method used for designing the electrical circuit to a continuous load.

There are many types of electrical loads that exhibit special operational characteristics, such as large inrush currents on initial energization. The electrical system must be designed to

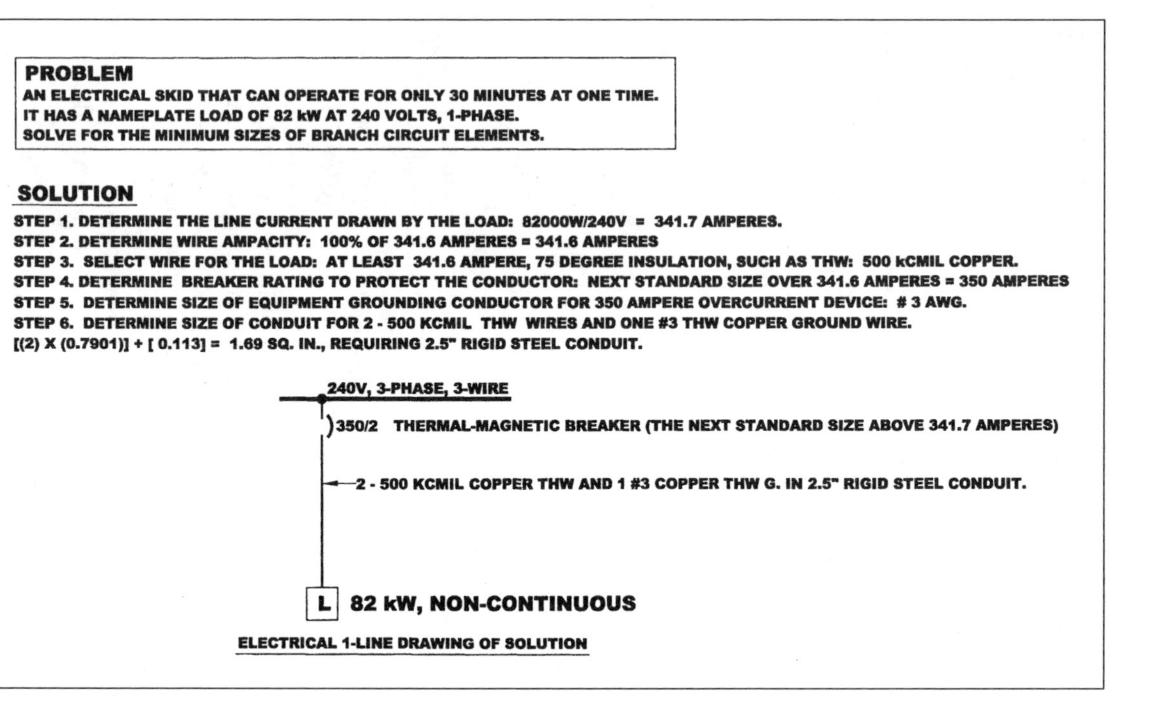

Figure 13-1 Solve for the conductor size and overcurrent device rating for a general load.

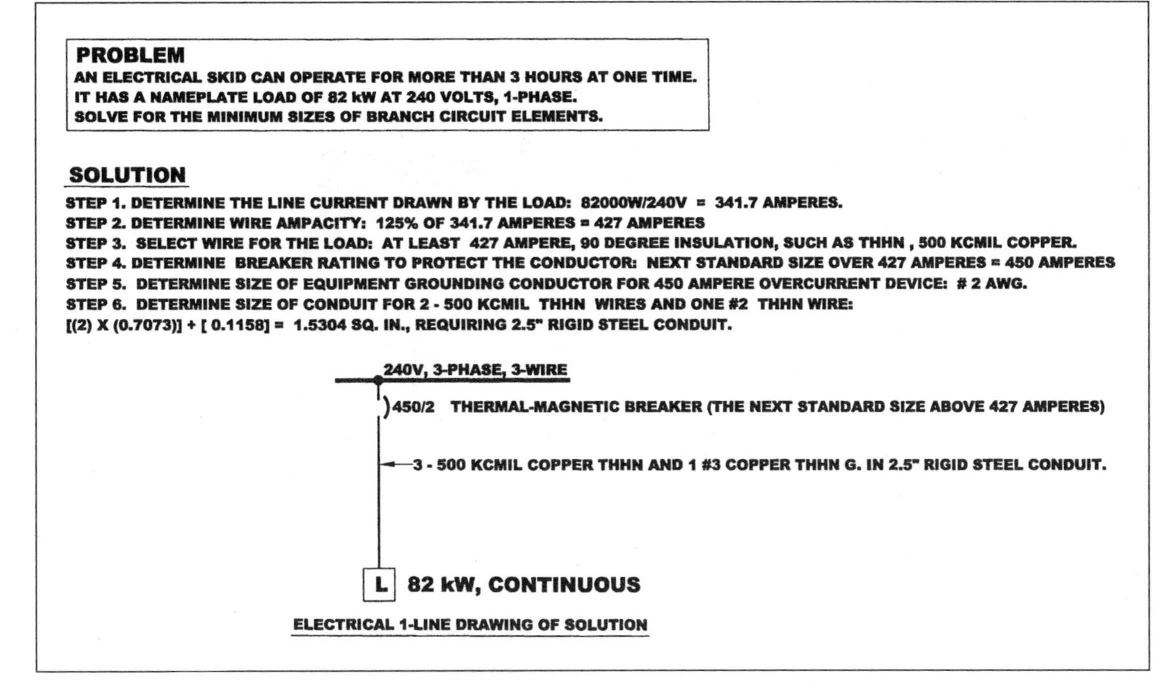

Figure 13-2 Solve for the conductor size and overcurrent device rating for a general continuous load.

Switchboard and transfer switches require working clearance in front, but not to the side.

permit these unique loads to start and operate successfully while still providing protection from abnormal current flow. For example, when starting a motor, a large inrush current flows until the motor can establish a "back emf" to limit line current. The electrical system must permit this inrush current to flow to start the motor, but it also must be able to interrupt it if the motor does not accelerate to speed in normal time, lest the motor be damaged or destroyed by excessive I^2R heat. Then the same electrical system must continuously monitor motor running current to also prevent long-time overload from damaging the motor. The following figures illustrate calculations required by specific types of appliances in the design of electric circuits to permit unique loads to start and to continue to protect them during operation:

Figure 13-3: Air-conditioning equipment

Figure 13-4: Household appliances

Figure 13-5: Heat tracing with self-regulating cable (see Note below)

Figure 13-6: Heat tracing with constant-wattage cable

Figure 13-7: Lighting fixtures

Figure 13-8: Electrical power receptacles with unidentified loads

Figure 13-9: Electrical power receptacle with specific load

Note: Some of these examples use actual field experience rather than code requirements for sizing. For example, field experience with self-regulating heat-tracing cable shows that initial energization inrush current of up to 4.9 times full-load rating current occurs for 3 to 5 minutes in dry locations and indefinitely in wet locations. Therefore, the circuit breaker and conductors must be sized to deliver this current, lest the electrical power system fail or trip "off" to this most important load. For this reason, the electrical engineer and designer should make every effort to obtain the appliance manufacturer's actual load characteristics for each electrical load.

Designing an Electrical System for a Commercial Building

The electrical system for a commercial building must be large enough to safely supply the facility electrical loads. While this can be done based on physics, over the last century, the best minds in the electrical industry have contributed information

PROBLEM

AN AIR CONDITIONER PACKAGE UNIT HAS A NAMEPLATE
BRANCH CIRCUIT SELECTION CURRENT OF 342 AMPERES, 3-PHASE, 240 VOLT.
SOLVE FOR THE MINIMUM SIZES OF BRANCH CIRCUIT ELEMENTS.

SOLUTION

STEP 1. DETERMINE THE LINE CURRENT TO BE USED IN EQUIPMENT SIZING: 342 AMPERES.
STEP 2. DETERMINE WIRE AMPACITY: 125% OF 342 AMPERES = 427.5 AMPERES
STEP 3. SELECT WIRE FOR THE LOAD: AT LEAST 427.5 AMPERE, 90 DEGREE INSULATION, SUCH AS THHN: 500 KCMIL COPPER.
STEP 4. DETERMINE BREAKER RATING: (175% OF 342 = 598.5 AMPERES); NEXT STANDARD SIZE OVER 598.5 AMPERES = 600 AMPERES
STEP 5. DETERMINE SIZE OF EQUIPMENT GROUNDING CONDUCTOR FOR 600 AMPERE OVERCURRENT DEVICE: # 1 AWG.
STEP 6. DETERMINE SIZE OF CONDUIT FOR 3 - 500 KCMIL THHN WIRES AND ONE #1 THHN COPPER GROUND WIRE.
[(3) X (0.7073)] + [0.1562] = 2.2781 SQ. IN., REQUIRING 3" RIGID STEEL CONDUIT.
STEP 7. DETERMINE DISCONNECT SWITCH AMPERE RATING: 115% OF 342 = 393.3 AMPERES. THE NEXT STANDARD SIZE IS 400 AMPERES.

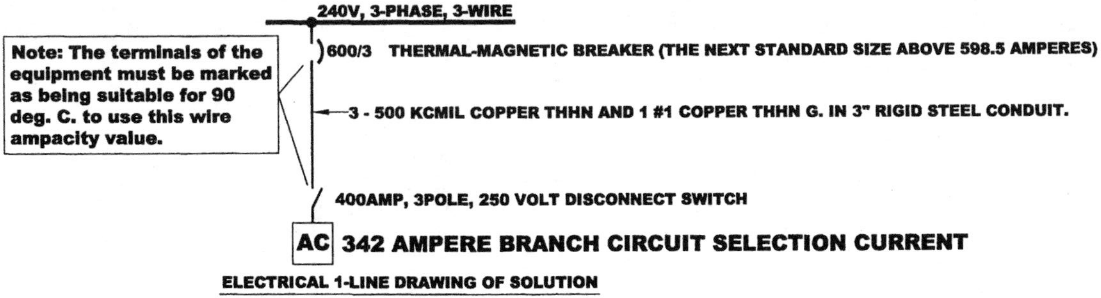

Figure 13-3 Solve for the conductor size, overcurrent device rating, and disconnect rating for an HVAC load.

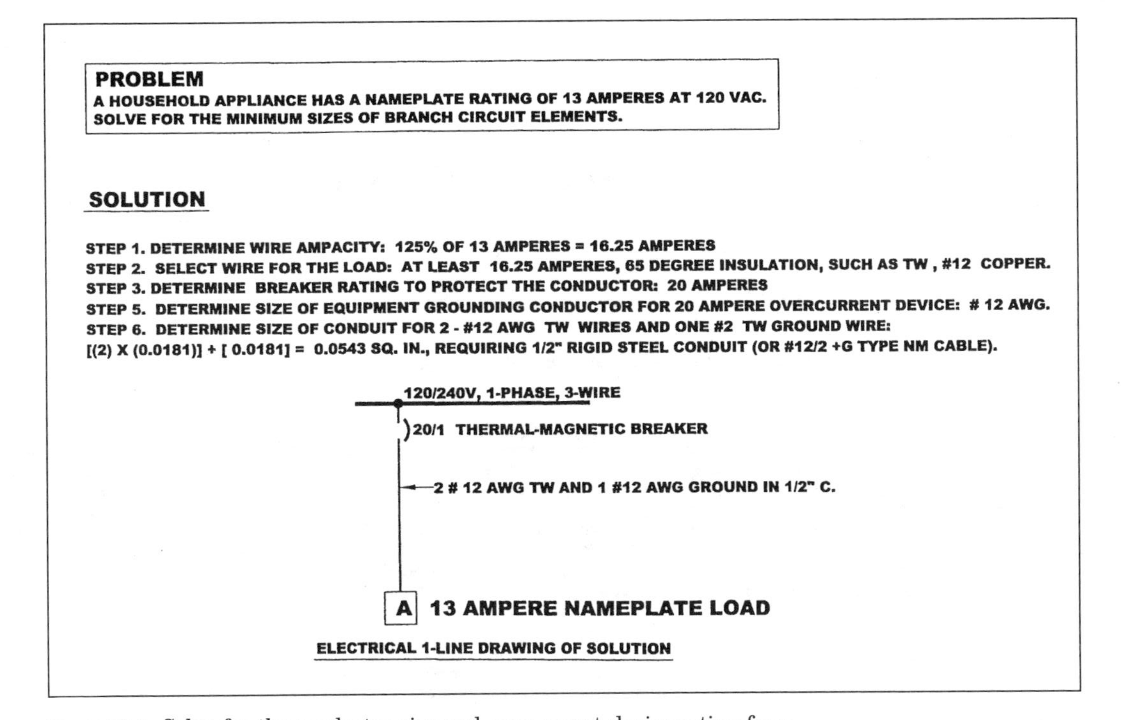

Figure 13-4 Solve for the conductor size and overcurrent device rating for a general household appliance.

PROBLEM

AN INDUSTRIAL PLANT HAS AN ELECTRICAL SYSTEM THAT INCLUDES A SECTION OF SELF-REGULATED HEAT TRACING CABLE. THE CABLE IS 255 FEET LONG AND IS RATED AT 3 WATTS PER LINEAL FOOT AT 120 VAC. DETERMINE THE CIRCUIT TO THE CABLE.

SOLUTION

THIS CABLE CAN DRAW UP TO 490% OF ITS NAMEPLATE RATING CONTINUOUSLY, SO THE ELECTRICAL SYSTEM MUST BE ABLE TO DELIVER (4.9) X (3 WATTS/FT) (255 FT) = 3748 WATTS. 3748 WATTS/120V = 31.2 AMPERES.

STEP 1. DETERMINE WIRE AMPACITY: 100% OF 31.2 AMPERES = 31.2 AMPERES. (490% EXCEEDS THE 125% CONTINUOUS DUTY REQUIREMENT).

STEP 2. SELECT WIRE FOR THE LOAD: AT LEAST 31.2 AMPERES, 90 DEGREE INSULATION, SUCH AS XHHW-2, #8 COPPER.
NOTE: #10 XHHW CAN NOT BE USED FOR 31.2 AMPERES DUE TO THE 30 AMPERE OVERCURRENT DEVICE LIMITATION FOR #10.

STEP 3. DETERMINE BREAKER RATING TO PROTECT THE CONDUCTOR AND DELIVER THE CURRENT 40 AMPERE GROUND FAULT.

STEP 5. DETERMINE SIZE OF EQUIPMENT GROUNDING CONDUCTOR FOR 40 AMPERE OVERCURRENT DEVICE: # 10 AWG.

STEP 6. DETERMINE SIZE OF CONDUIT FOR 2 - #8 AWG XHHW WIRES AND ONE #10 XHHW GROUND WIRE:
[(2) X (0.0437)] + [0.0243] = 0.1117 SQ. IN., REQUIRING 1/2" RIGID STEEL CONDUIT.

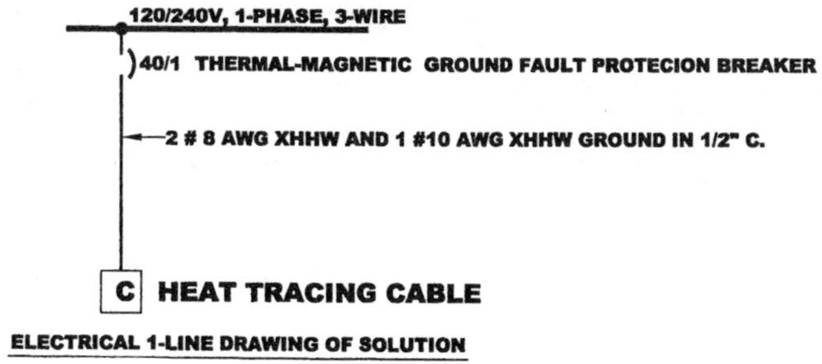

120/240V, 1-PHASE, 3-WIRE

40/1 THERMAL-MAGNETIC GROUND FAULT PROTECION BREAKER

2 # 8 AWG XHHW AND 1 #10 AWG XHHW GROUND IN 1/2" C.

C HEAT TRACING CABLE

ELECTRICAL 1-LINE DRAWING OF SOLUTION

Figure 13-5 Solve for the conductor size and overcurrent device rating for a self-regulated heat tracing cable.

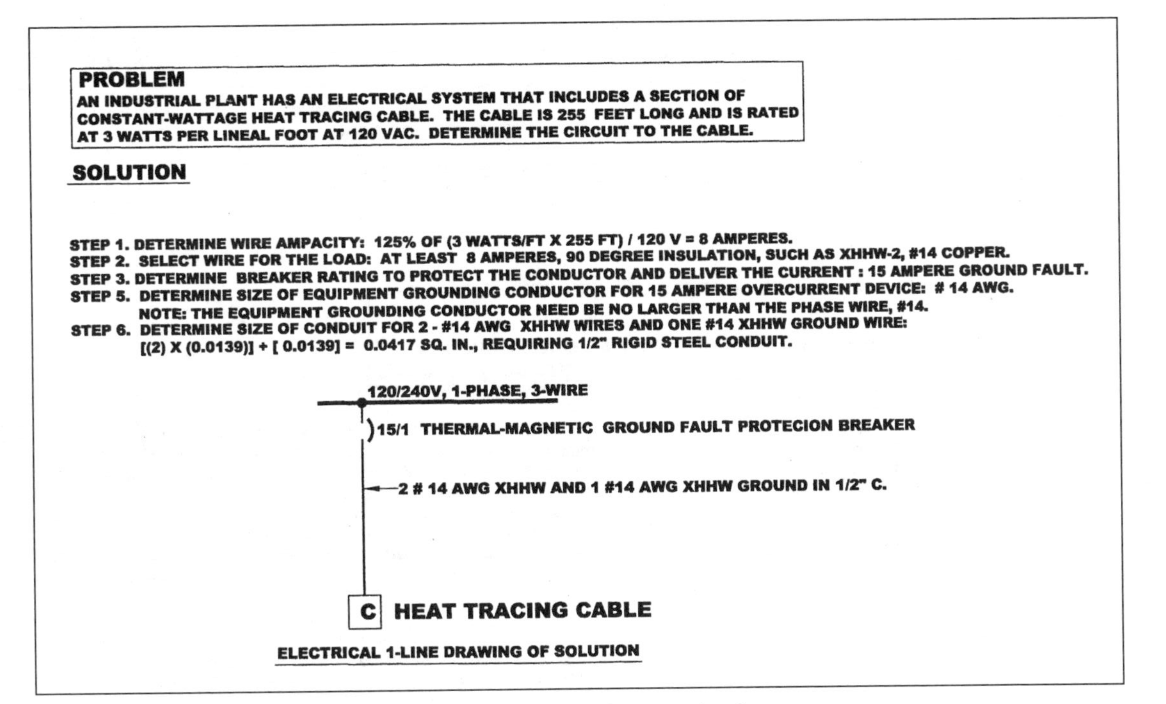

Figure 13-6 Solve for the conductor size and overcurrent device rating for a constant-wattage heat tracing cable.

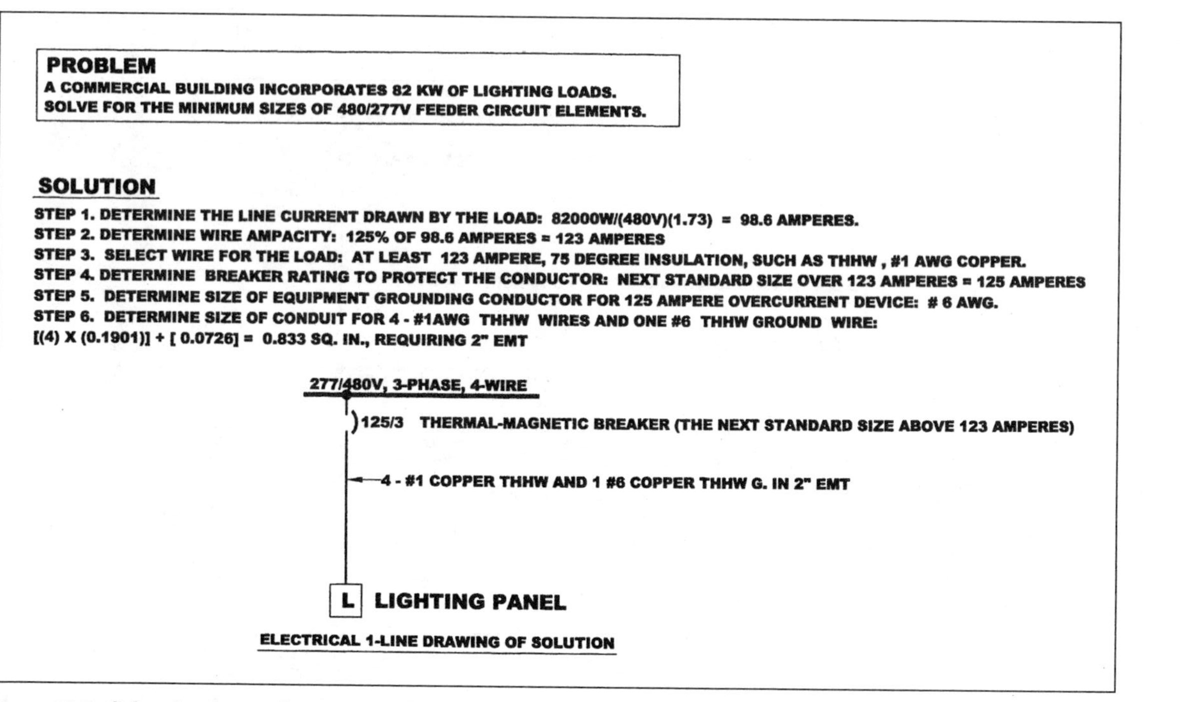

Figure 13-7 Solve for the conductor size and overcurrent device rating for a commercial lighting load.

PROBLEM

A COMMERCIAL BUILDING INCORPORATES 167 CONVENIENCE DUPLEX RECEPTACLES.
SOLVE FOR THE MINIMUM SIZES OF THE 120/208V FEEDER CIRCUIT ELEMENTS TO THE BRANCH CIRCUIT PANELBOARD.

SOLUTION

STEP 1. DETERMINE THE LINE CURRENT DRAWN BY THE LOAD: (180VA/REC) X (167 REC) / (208V)(1.73) = 83.5 AMPERES.
STEP 2. DETERMINE WIRE AMPACITY: 100% OF 83.5 AMPERES = 83.5 AMPERES (THE LOAD IS CONSIDERED NON-CONTINUOUS)
STEP 3. SELECT WIRE FOR THE LOAD: AT LEAST 83.5 AMPERE, 75 DEGREE INSULATION, SUCH AS THHW , #4 AWG COPPER.
STEP 4. DETERMINE BREAKER RATING TO PROTECT THE CONDUCTOR: NEXT STANDARD SIZE OVER 83.5 AMPERES = 90 AMPERES
STEP 5. DETERMINE SIZE OF EQUIPMENT GROUNDING CONDUCTOR FOR 90 AMPERE OVERCURRENT DEVICE: # 8 AWG.
STEP 6. DETERMINE SIZE OF CONDUIT FOR 4 - #4AWG THHW WIRES AND ONE #8 THHW GROUND WIRE:
[(4) X (0.0973)] + [0.0556] = 0.4448 SQ. IN., REQUIRING 1.25" EMT

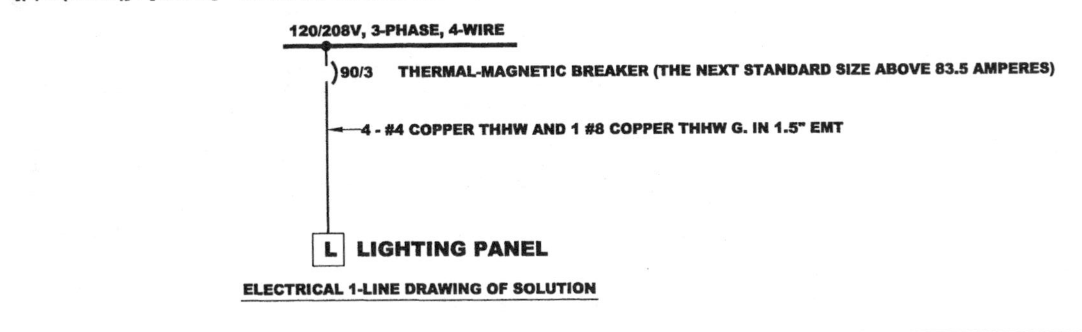

Figure 13-8 Solve for the conductor size and overcurrent device rating for *general* receptacle loads.

PROBLEM

A COMMERCIAL BUILDING CONTAINS 67 RECEPTACLES FOR SPECIAL PURPOSE CONTINUOUS LOADS OF 1400 VA PER RECEPTACLE. SOLVE FOR THE MINIMUM SIZES OF THE 120/208V FEEDER CIRCUIT ELEMENTS TO THE BRANCH CIRCUIT PANELBOARD THAT FEEDS ALL OF 67 OF THESE LOADS.

SOLUTION

STEP 1. DETERMINE THE LINE CURRENT DRAWN BY THE LOAD: (1400VA/REC) X (67 REC) / (208V)(1.73) = 260 AMPERES.
STEP 2. DETERMINE WIRE AMPACITY: 125% OF 260AMPERES = 325 AMPERES
STEP 3. SELECT WIRE FOR THE LOAD: FOUR CONDUCTORS, EACH AT LEAST 325 AMPERES, 75 DEGREE INSULATION, SUCH AS THHW , 500 KCMIL COPPER. (DO NOT SELECT 400 KCMIL, SINCE THIS SIZE IS NORMALLY UNAVAILABLE IN THE INDUSTRY).
STEP 4. DETERMINE BREAKER RATING TO PROTECT THE CONDUCTOR: NEXT STANDARD SIZE OVER 325 AMPERES = 350 AMPERES
STEP 5. DETERMINE SIZE OF EQUIPMENT GROUNDING CONDUCTOR FOR 350 AMPERE OVERCURRENT DEVICE: # 3 AWG.
STEP 6. DETERMINE SIZE OF CONDUIT FOR 4 - 500 KCMIL THHW WIRES AND ONE #3 AWG THHW GROUND WIRE:
[(4) X (0.7901] + [0.1134] = 3.2738 SQ. IN., REQUIRING 3" EMT

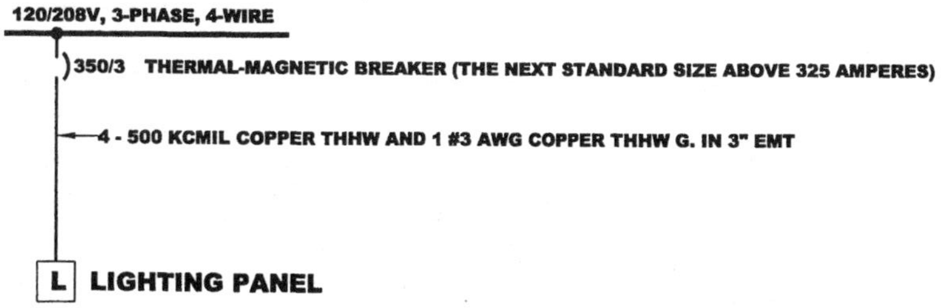

Figure 13-9 Solve for the conductor size and overcurrent device rating for *specific* receptacle loads.

to the *National Electrical Code* (NEC), which now sets the standards for the characteristics of the required electrical system. Accordingly, this book both points out the requirements using calculation methodology and provides NEC reference information where the engineer and designer can obtain further information.

For every feeder and switchgear bus, panelboard bus, or motor control center bus, a separate calculation must be made; however, these calculations are all very similar, with only the connected loads changing. The first of the calculations that must be made is for the service feeder and service equipment.

The following six general groups of loads must be considered within commercial buildings:

1. Lighting
2. Receptacle loads
3. Special appliance loads
4. Motor loads other than heating, ventilation, and air-conditioning (HVAC) loads
5. The greater of
   a. HVAC compressor loads and hermetically sealed motor loads, or
   b. Heating loads

Lighting loads consist of

1. The greater of 125 percent (for continuous operation) of the quantity of voltamperes per square foot shown in NEC Table 220-3(a) or 125 percent of the actual lighting fixture load, including low-voltage lighting (Article 411), outdoor lighting, and 1200-voltampere (VA) sign circuit [600-4(b)(3)].
2. 125 percent of show window lighting [220-12(a)].
3. Track lighting at 125 percent of 150 VA per lineal foot [220-12(b)].

Receptacle loads consist of

1. 100 percent of the quantity of 1 VA/ft^2 shown in Table 220-3(a) or 100 percent of 180 VA per receptacle [220-3(b)(9)].

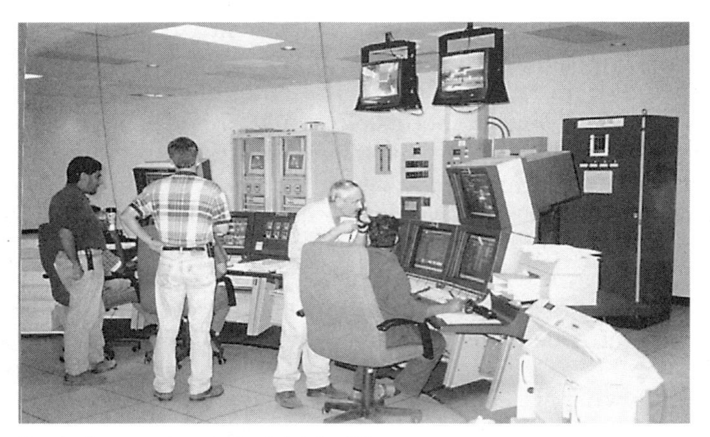

Control room of an industrial facility.

Induction motors driving pumps in industrial facility.

2. 100 percent of specific appliance loads that "plug into" receptacles or 125 percent of specific appliance loads that can operate continuously (for 3 hours or longer) that "plug into" receptacles.
3. 100 percent of 180 VA per 5 ft of multioutlet assembly or 100 percent of 180 VA per 1 ft of multioutlet assembly, when every receptacle will be used.

Special appliance loads consist of 100 percent of the voltampere rating of each appliance that will not run continuously and 125 percent of each appliance that will run continuously.

Motor loads other than HVAC loads, in accordance with NEC Article 430, consist of 125 percent of the largest table ampere size motor plus 100 percent of all other motors, except that any motor that is noncoincidental with another need not be considered. For example, if a building contains two chilled water pumps, P-1A and P-1B, where P-1B serves as a backup to P-1A, then only one of the two pumps can be expected to be running. In this case, the load flow analysis should include P-1A but not P-1B.

HVAC loads consist of heating and cooling loads. Frequently, these heating and cooling loads are noncoincidental, with heating loads only running when cooling is not required, and vice versa. The feeder that serves both noncoincidental loads only needs to be sized for the larger of the two. Note, however, that some loads, such as air-handling loads, operate with either heating or cooling and must be added to the load listing for either case.

Demand factors can be applied to receptacle loads (NEC Table 220-13 lists specific factors), to commercial cooking units (NEC Table 220-20 lists specific factors), to welders (630-11), to cranes (610-14), and to motors that will not be running simultaneously (430-26). The demand factor for a group of motor loads must be determined by the engineer or designer.

As an example, Fig. 13-10 shows a calculation of the minimum service feeder ampacity required for a commercial building that consists of 100,000 ft^2 of office space with one 50-horsepower (hp) air-handling unit, one 250-hp chiller with a branch circuit selection current of 250 A at 480 V three-phase, one 10-hp domestic water pump, and one 30-kilowatt (kW) heating coil that provides reheat dehumidification.

Designing an Electrical System for an Industrial Facility

Engineering and design work in an industrial facility is similar to that in commercial buildings, except that the loads are

PROBLEM

COMMERCIAL BUILDING WITH 100,000 SQ. FT. OF OFFICE SPACE AND THE FOLLOWING ELECTRICAL LOADS:

1 - 50HP AIR HANDLING UNIT MOTOR
1 - 250HP CHILLER MOTOR
1 - 10HP DOMESTIC WATER PUMP
1 - 30kW HEATING COIL FOR REHEAT

LOAD ANALYSIS FOR SERVICE FEEDER:

	THREE-PHASE CONDUCTORS	NEUTRAL CONDUCTORS
LIGHTING LOAD:		
100,000 SQ. FT. X 3.5 VOLT-AMPERES PER SQ. FT. =	350,000 VOLT-AMPERES	350,000 VOLT-AMPERES
25% LIGHTING LOAD FOR CONTINUOUS DUTY=	87,500 VOLT-AMPERES	87,500 VOLT-AMPERES
SIGN CIRCUIT, 1200 VOLT-AMPERES	1,200 VOLT-AMPERES	1,200 VOLT-AMPERES
GENERAL PURPOSE RECEPTACLE LOAD:		
100,000 SQ. FT. X 1 VOLT-AMPERE PER SQ. FT. = 100,000 VOLT AMPERES. PER TABLE 220-13,		
THE FIRST 10kVA @ 100% =	10,000 VOLT-AMPERES	10,000 VOLT-AMPERES
THE REMAINING 90,000 VA @ 50%	45,000 VOLT-AMPERES	45,000 VOLT-AMPERES
50 HP AIR HANDLING UNIT MOTOR PER TABLE 430-150:		
65 AMPERES x 460V x 1.732 =	51,788 VOLT-AMPERES	0
10 HP DOMESTIC WATER PUMP PER TABLE 430-150"		
14 AMPERES x 460V x 1.732 =	11,154 VOLT-AMPERES	0
30 kW HEATING COIL @ 125% FOR CONTINUOUS DUTY=	37,500 VOLT-AMPERES	0
250 HP CHILLER: 125% OF BRANCH CIRCUIT SELECTION		
CURRENT = 125% OF 250 AMP =		
250 X 1.25 X 460V X 1.732 =	248,975 VOLT-AMPERES	0
SUMMARY OF MINIMUM RATING OF SERVICE FEEDER:	843,117 VOLT-AMPERES	493,700 VOLT-AMPERES

AMPACITY OF SERVICE FEEDER CONDUCTORS:

PHASE CONDUCTORS: 843,117 / (460 X 1.732) = **1058 AMPERES**

NEUTRAL CONDUCTORS: 493,700/ (460 X 1.732) = **620 AMPERES**

Figure 13-10 Solve for service for a commercial building.

Motor control center.

quite different in type and rating. The electrical load of the buildings in a typical industrial plant is normally much smaller than the large-process load, and the design reliability considerations are different. In commercial buildings, the emphasis is on personnel safety first and equipment functionality second. In many industrial plants, the process equipment is unmanned, and process interruptions can cause

PROBLEM

INDUSTRIAL PLANT INCORPORATES TWO COMMERCIAL BUILDINGS.
COMMERCIAL BUILDINGS TOTAL 5,000 SQ. FT. OF FLOOR SPACE
PLUS THE FOLLOWING PROCESS EQUIPMENT:

- 1 - 10 kVA UPS
- 1 - Pump set P-1101A and B, 10HP, 460 volts
- 1 - Pump set P-1601A and B, 50HP, 460 volts
- 10 - 30HP Finfans, 460 volt
- 1 - 20 kW Cathodic Protection Unit
- 50 kW of process area lighting
- 1 - 6000 kW compressor motor that operates at a power factor of 0.9 at 13.8 kV

LOAD ANALYSIS FOR SERVICE FEEDER:

BUILDING LIGHTING LOAD:	13.8 kV CONDUCTORS	460V CONDUCTORS	NEUTRAL CONDUCTORS
5,000 SQ. FT. X 2 VOLT-AMPERES PER SQ. FT. =	10,000 VOLT-AMPERES	10,000 VOLT-AMPERES	10,000 VOLT-AMPERES
25% LIGHTING LOAD FOR CONTINUOUS DUTY=	2,500 VOLT-AMPERES	2,500 VOLT-AMPERES	2,500 VOLT-AMPERES
PROCESS LIGHTING: 50kW X 125% =	62,500 VOLT-AMPERES	62,500 VOLT-AMPERES	62,500 VOLT-AMPERES
10 kVA UPS at 125%	12500 VOLT-AMPERES	12500 VOLT-AMPERES	0
P-1101A AND B (NON-COINCIDENTAL LOADS, SO ONLY ONE IS RUNNING)			
10 HP - 14 AMPERES x 460V x 1.732 =	11,154 VOLT-AMPERES	11,154 VOLT-AMPERES	0
P-1601A AND B (NON-COINCIDENTAL LOADS, SO ONLY ONE IS RUNNING)			
50 HP - 65 AMPERES x 460V x 1.732 =	51,788 VOLT-AMPERES	51,788 VOLT-AMPERES	0
TEN FINFAN MOTORS, EACH IS 30 HP, 460V			
30 HP - 40 AMPERES x 460V x 1.732 X 10 MOTORS =	318,697 VOLT-AMPERES	318,697 VOLT-AMPERES	0
20 kW CATHODIC PROTECTION @ 125% FOR CONTINUOUS DUTY=	25,000 VOLT-AMPERES	25,000 VOLT-AMPERES	0
6000 kW COMPRESSOR MOTOR @ 13.8 kV' @ pf = 0.9	6,666,667 VOLT-AMPERES		
CURRENT = 125% OF 6000 kW (LARGEST MOTOR)	1,667,667 VOLT-AMPERES		
		12,947 VOLT-AMPERES	0
		(25% LARGEST MOTOR)	0
SUMMARY OF MINIMUM RATING OF SERVICE FEEDER:	8,827,139 VOLT-AMPERES	507,086 VOLT-AMPERES	75,000 VOLT-AMPERES

AMPACITY OF SERVICE FEEDER CONDUCTORS:
13.8 kV PHASE CONDUCTORS: 8,849,973 / (13800 X 1.732) = **370 AMPERES**
460 VOLT PHASE CONDUCTORS: 507,086 / (460 X 1.713) = **636 AMPERES**
NEUTRAL CONDUCTORS: 75,000/ (460 X 1.732) = **94 AMPERES**

Figure 13-11 Solve for service for an industrial plant.

Industrial plants commonly incorporate prefabricated switchgear buildings.

millions of dollars in damage losses. Therefore, the continuity of electrical power to the process equipment is frequently treated with as much design care as the design for continuity of electrical power to the life safety electrical system branch of a hospital electrical system. These two changes, the size of the loads and system redundancy requirements, are shown clearly in the following example of sizing of an electrical system in an industrial plant, as shown in Fig. 13-11. This system consists of the following loads:

1. A 3000-ft^2 control building
2. A 2000-ft^2 switchgear building
3. A 10-kilovoltampere (kVA) redundant uninterruptible power system (UPS) to energize the distributed control system (DCS)
4. The following series of pumps and fans:
   a. P-1101A and P-1101B, 10 horsepower (hp), 460 volts (V)
   b. P-1601A, P1601B, 50 hp, 460 V
   c. Ten 30-hp finfan motors, 460 V
   d. One 20-kW cathodic protection system
   e. 50 kW of lighting in the process facility
5. One 10-ton air-conditioner package unit with a 50-hp motor

6. One 6000-kW compressor motor that has an operating power factor of 0.90 at 13.8 kV

The electrical power system to the industrial plant is redundant from the electrical utility, serving the industrial plant at 13.8 kV; therefore, each service is capable of operating the entire plant and is sized to support the load of the entire plant.

NOTES

NOTES

Chapter

14

Electrical Design and Layout Calculations

Many design and layout issues arise daily in the work of electrical engineers and designers, such as: What are the minimum dimensions of a straight-through pull box, an angle pull box, or a junction box? How close to a wall can a 480/277-volt (V) panel be placed? How close together can two opposing 13.8-kilovolt (kV) switchgear layouts be placed? How close together can knockouts be punched without causing locknuts to physically overlap and interfere with one another? What minimum phase-to-phase and phase-to-ground dimensions must be maintained when constructing an auxiliary wireway? This chapter provides convenient answers to these questions by providing the rules for each, along with completed "go by" calculations that engineers and designers can use as templates for their specific calculations simply by changing certain values.

Straight-Through Pull Box in a Conduit System

When wires are drawn through a conduit, the friction between the conductor insulation and the conduit can become too great, damaging the conductors. Therefore, sometimes intermediate pulling points are required within what would otherwise be a continuous conduit run.

When wires are pulled in a conduit run that contains pull boxes, the wires are completely drawn through and out of the open cover of the pull box and then are refed into the next part of the conduit run. As most of the wire length is pulled into the next part of the conduit run, the wire loop at the cover of the pull box forms an increasingly smaller radius until the wire is all in the box. The minimum bending radius of the conductor, the size of conduit that each conductor circuit will normally fit within, and the possibility of physically damaging the conductor on the sharp edges of the pull box flanges have been considered in specifying the minimum dimensions of a straight-through pull box, which is eight times the trade size of the largest conduit entering the box. See Fig. 14-1 for an example of this calculation.

The other box dimensions are determined by the physical space required to place a locknut on the conduit fitting, as shown in Fig. 14-6 and replicated for convenience in Fig. 14-1.

Angle Pull Box in a Conduit System

When conduits enter adjacent walls of a pull box, the loop that the wire makes as it is pulled in is less of an issue, and so the box can be smaller (only six times the trade size of the largest conduit, plus other considerations). When more than one conduit enters the wall of a box, then the additional conduit trade sizes must be added to the "six times the trade size of the largest conduit" to determine the dimension to the opposite wall of the box. In addition, raceways enclosing the same conductors are required to have a minimum separation between them of six times the trade size of the conduit to provide adequate space for the conductor to make the bend. Figure 14-2 illustrates these calculations.

Working Space Surrounding Electrical Equipment

Sufficient working space must be provided for workers to operate safely around electrical equipment, and *National*

Electrical Code Table 110-26(a) defines the minimum dimensions of this working space. The height of the working space must be the greater of (1) 6.5 feet (ft) or (2) the actual height of the electrical equipment, and the width of the working space must be the greater of (1) 30 inches (in), or (2) the width of the electrical equipment.

There are three distinct possibilities, or conditions, that determine the depth of the working space:

Condition 1. Exposed live parts on one side and no live or grounded parts on the other side of the working space or exposed live parts on both sides effectively guarded by suitable wood or other insulating materials. Insulated wire or insulated busbars are considered to be live parts only if their potential exceeds 300 V to ground.

Condition 2. Exposed live parts on one side and grounded parts on the other side. Concrete, brick, and tile walls are all considered to be grounded, just as is sheetrock screwed to metal studs where the metal screws are accessible.

Condition 3. Exposed live parts on both sides of the work space with the operator between them.

There is one more consideration regarding working clearance depth. The doors of the electrical equipment must be able to open a full 90°.

Note that practically everywhere in the code and in the electrical industry, voltage is given as phase-to-phase voltage, but in this case voltage is stated in phase-to-ground units, and with ungrounded systems, the voltage to ground is taken as the phase-to-phase voltage.

These requirements apply to equipment that is likely to require examination, adjustment, servicing, or maintenance. Some examples of such equipment include circuit breaker panelboards, fused switch panelboards, disconnect switches, individual circuit breaker enclosures, motor starters, and switchboards.

Generally, the working clearance requirements for the rear of rear-accessible electrical equipment are the same as for the front of the equipment, except that if the equipment can only be worked on in a deenergized state, then the 36-in

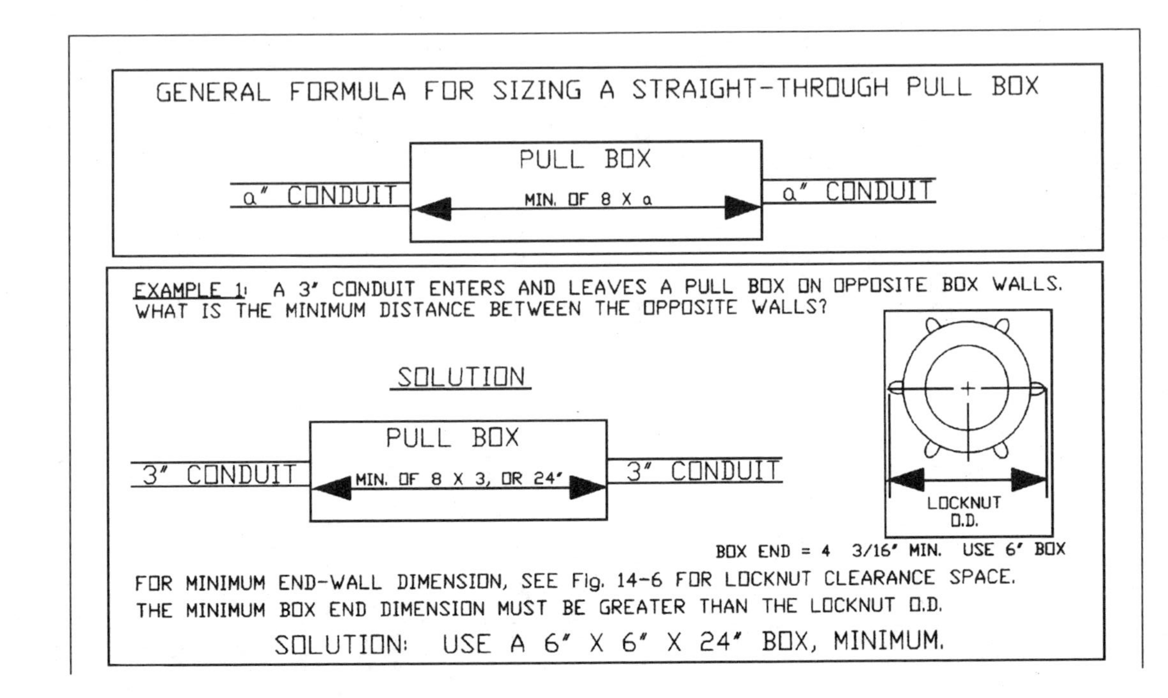
GENERAL FORMULA FOR SIZING A STRAIGHT-THROUGH PULL BOX
PULL BOX
a" CONDUIT
MIN. OF 8 X a
a" CONDUIT
EXAMPLE 1: A 3" CONDUIT ENTERS AND LEAVES A PULL BOX ON OPPOSITE BOX WALLS.
WHAT IS THE MINIMUM DISTANCE BETWEEN THE OPPOSITE WALLS?
SOLUTION
PULL BOX
3" CONDUIT
MIN. OF 8 X 3, OR 24"
3" CONDUIT
LOCKNUT O.D.
BOX END = 4 3/16" MIN. USE 6" BOX
FOR MINIMUM END-WALL DIMENSION, SEE Fig. 14-6 FOR LOCKNUT CLEARANCE SPACE.
THE MINIMUM BOX END DIMENSION MUST BE GREATER THAN THE LOCKNUT O.D.
SOLUTION: USE A 6" X 6" X 24" BOX, MINIMUM.

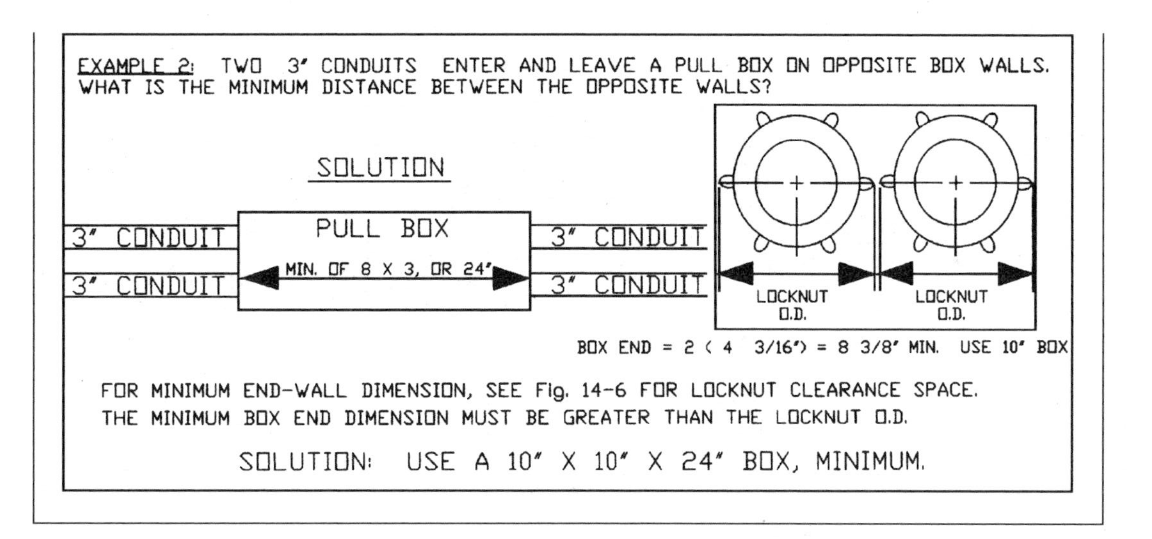

Figure 14-1 Solve for dimensions of a straight-through pull box in a conduit system.

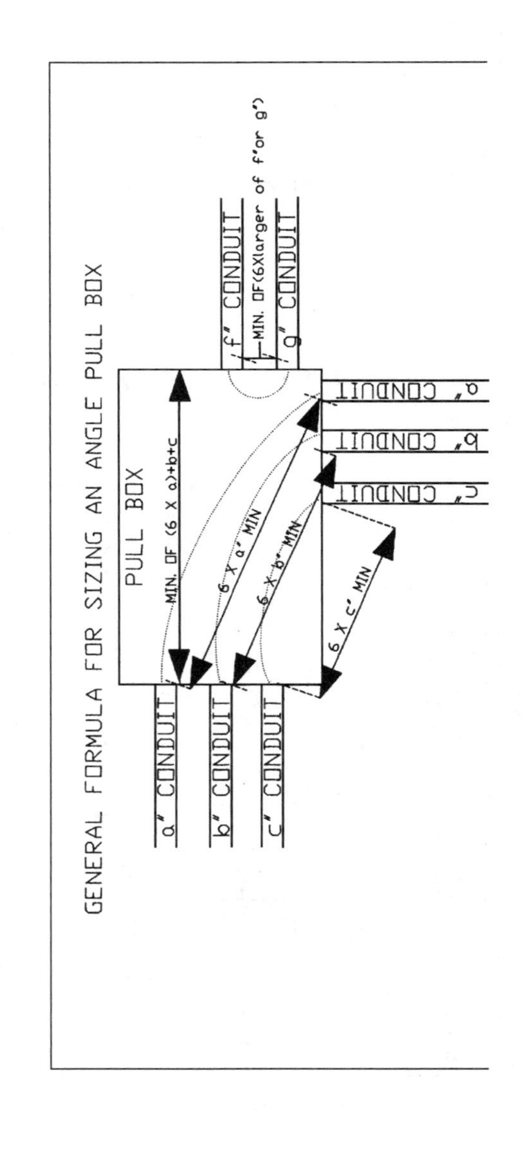
GENERAL FORMULA FOR SIZING AN ANGLE PULL BOX
PULL BOX
MIN. OF (6 X a)+b+c
a" CONDUIT
b" CONDUIT
c" CONDUIT
6 X a" MIN
6 X b" MIN
6 X c" MIN
f" CONDUIT
MIN. OF(6Xlarger of f"or g")
g" CONDUIT
c" CONDUIT
b" CONDUIT
a" CONDUIT

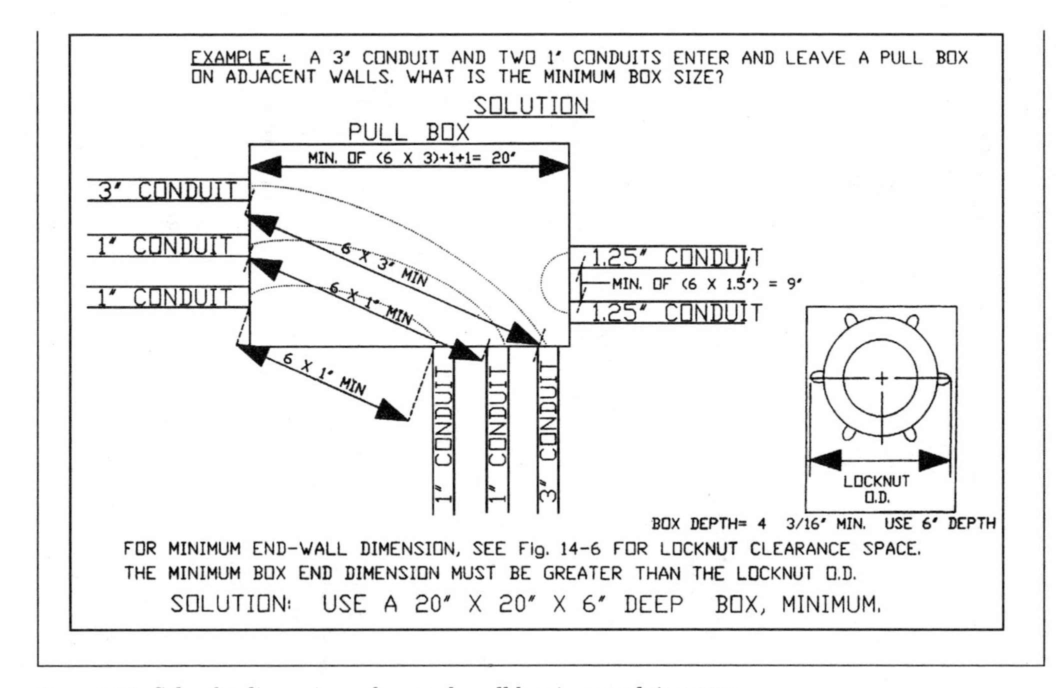

Figure 14-2 Solve for dimensions of an angle pull box in a conduit system.

minimum condition 1 clearance for up to 150 V to ground may be reduced to 30 in. No rear clearance space is required for equipment that requires no rear access.

The requirements for working space increase as the voltage to ground increases. Figure 14-3 shows the required space for electrical equipment operating at from 0 to 150 V to ground, Fig. 14-4 on p. 366 shows the required space for electrical equipment operating at from 151 to 600 V to ground, and Fig. 14-5 on p. 367 shows the required space for electrical equipment operating at voltages above 600 V to ground.

Minimum Centerline-to-Centerline Dimensions of Knockouts to Provide for Locknut Clearance

While trade sizes of conduit are well known, the dimensions of the locknuts that secure conduit connectors are less established. When planning for conduits to enter a wall of a junction box or wireway, it is necessary to provide physical space for the locknuts in addition to providing for the conduit opening space. When determining the exact centerlines of knockouts, it is essential to provide for the locknut space, or else the connectors will not fit beside one another in the box wall. Figure 14-6 on pp. 368 and 369 provides exact layout dimensions that can be used to determine the minimum dimensions from one conduit knockout centerline to the next, regardless of the sizes of the conduits involved. For example, reading directly from Fig. 14-6, a knockout centerline for a 2-in trade size conduit can be no closer than 3.375 in to the sidewall of a box or else the locknut will not fit between the connector and the box wall. Also reading directly from Fig. 14-6, the minimum centerline-to-centerline dimension from the knockout for a 2.5-in conduit to a 1.25-in conduit would be 3 in, but 3.25 in is the recommended minimum distance to allow some clearance between the locknuts.

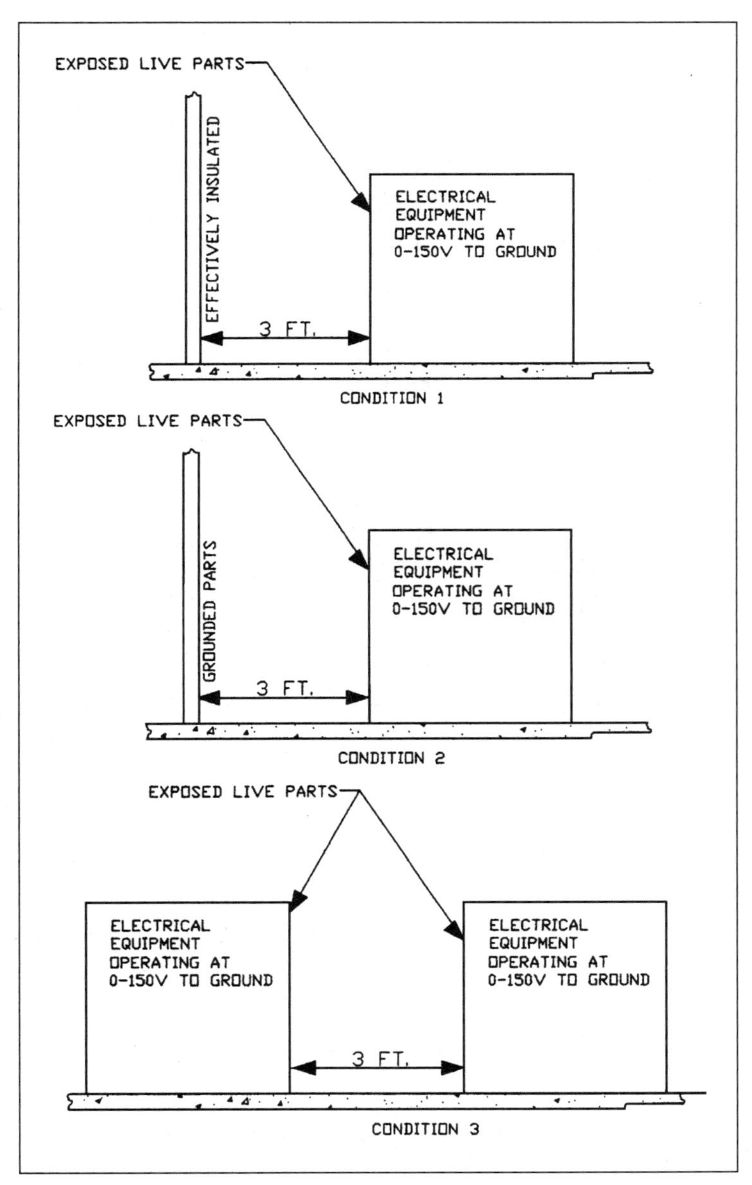

Figure 14-3 Solve for working space in front of equipment operating at 0–150 V to ground.

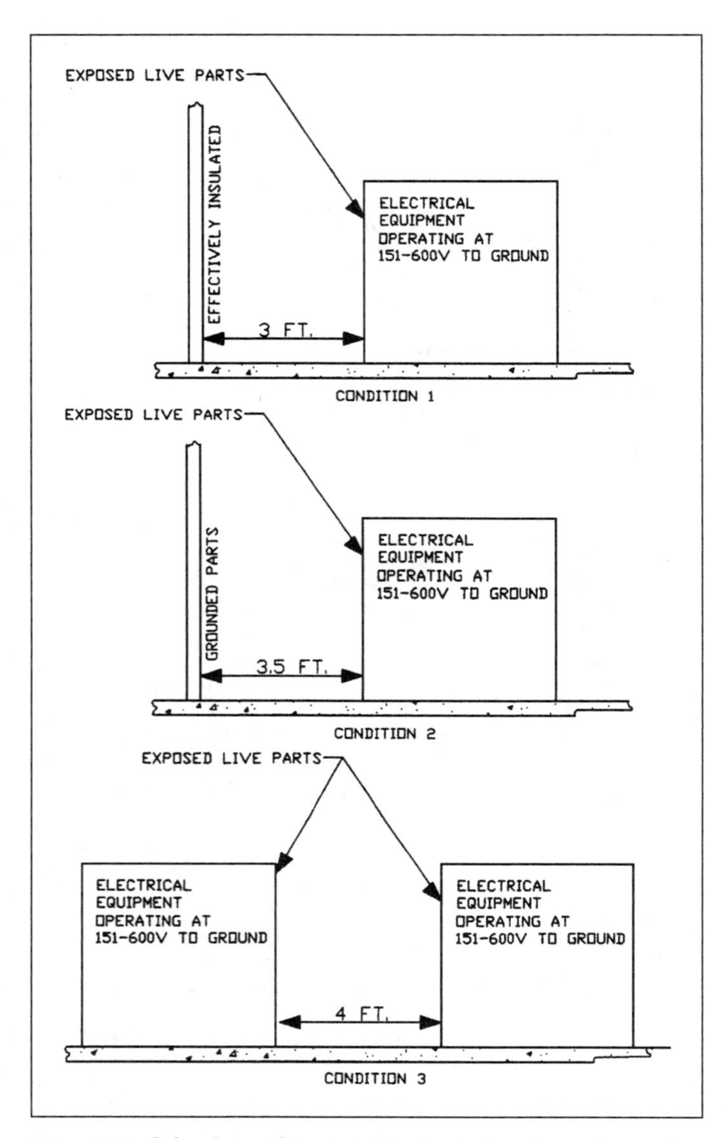

Figure 14-4 Solve for working space in front of equipment operating at 151–600 V to ground.

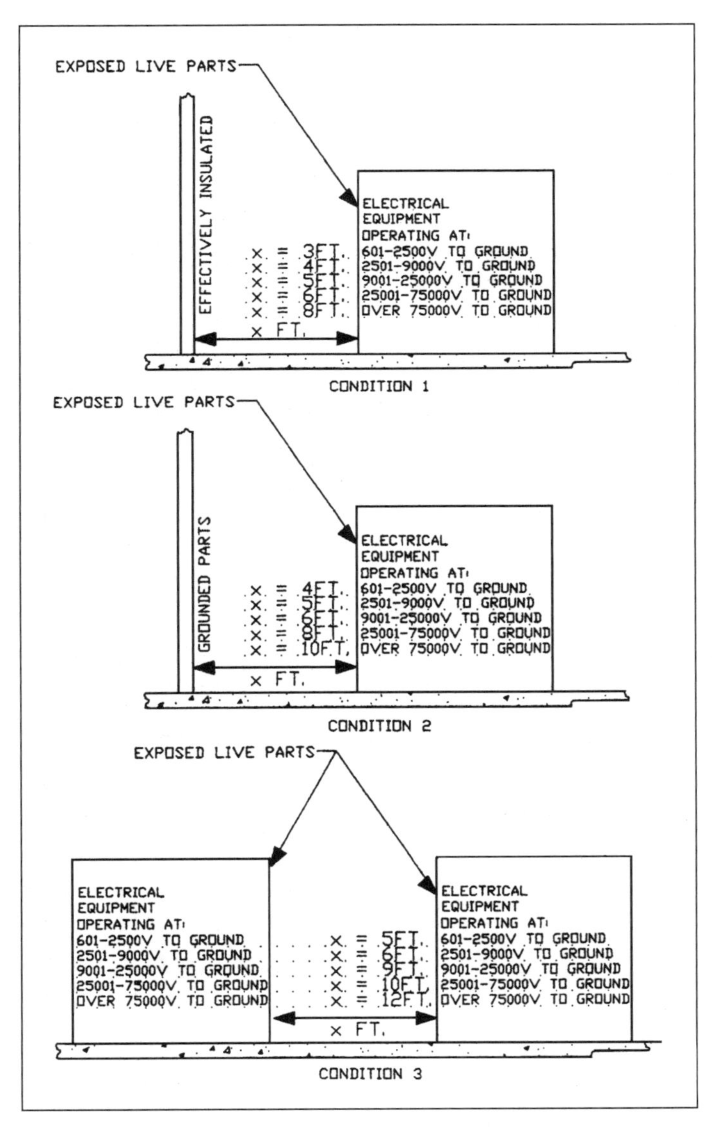

Figure 14-5 Solve for working space in front of equipment operating at over 600 V to ground.

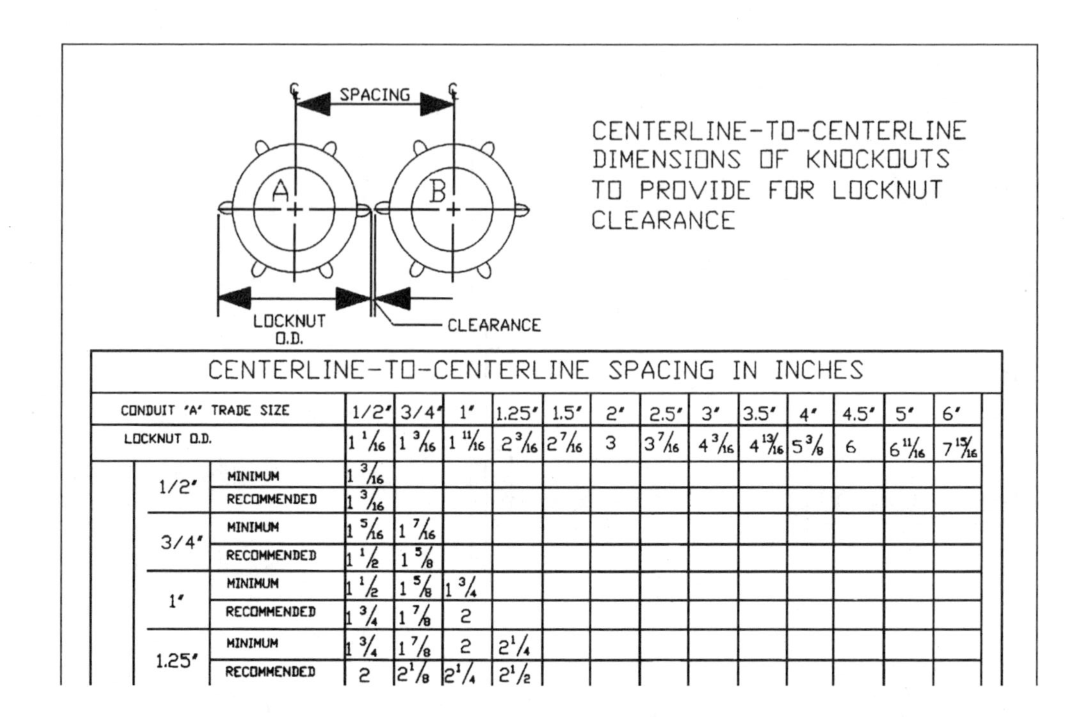

CENTERLINE-TO-CENTERLINE SPACING IN INCHES

CONDUIT 'A' TRADE SIZE		1/2″	3/4″	1″	1.25″	1.5″	2″	2.5″	3″	3.5″	4″	4.5″	5″	6″
LOCKNUT O.D.		1 1/16	1 3/16	1 11/16	2 3/16	2 7/16	3	3 7/16	4 3/16	4 13/16	5 3/8	6	6 11/16	7 15/16
1/2″	MINIMUM	1 3/16												
	RECOMMENDED	1 3/16												
3/4″	MINIMUM	1 5/16	1 7/16											
	RECOMMENDED	1 1/2	1 5/8											
1″	MINIMUM	1 1/2	1 5/8	1 3/4										
	RECOMMENDED	1 3/4	1 7/8	2										
1.25″	MINIMUM	1 3/4	1 7/8	2	2 1/4									
	RECOMMENDED	2	2 1/8	2 1/4	2 1/2									

CONDUIT 'B' TRADE SIZE														
1.5'	MINIMUM	1 15/16	2 1/16	2 3/16	2 7/16	2 9/16								
	RECOMMENDED	2 1/8	2 1/4	2 3/8	2 5/8	2 3/4								
2'	MINIMUM	2 3/16	2 5/16	2 1/2	2 3/4	2 7/8	3 1/8							
	RECOMMENDED	2 3/8	2 1/2	2 3/4	3	3 1/8	3 3/8							
2.5'	MINIMUM	2 7/16	2 9/16	2 3/4	3	3 1/8	3 3/8							
	RECOMMENDED	2 5/8	2 3/4	3	3 1/4	3 3/8	3 5/8	4						
3'	MINIMUM	2 13/16	2 15/16	3 1/16	3 5/16	3 7/16	3 3/4	4	4 5/16					
	RECOMMENDED	3	3 1/8	3 3/8	3 5/8	3 3/4	4	4 3/8	4 3/4					
3.5'	MINIMUM	3 1/8	3 1/4	3 3/8	3 5/8	3 3/4	4 1/16	4 5/16	4 5/8	4 5/16				
	RECOMMENDED	3 3/8	3 1/2	3 5/8	3 7/8	4	4 3/8	4 5/8	5	5 3/8				
4'	MINIMUM	3 7/16	3 9/16	3 11/16	3 15/16	4 1/16	4 3/8	4 5/8	4 5/16	5 1/4	5 9/16			
	RECOMMENDED	3 3/4	3 7/8	4	4 1/4	4 3/8	4 3/4	5	5 3/8	5 3/8	6			
4.5'	MINIMUM	3 3/4	3 7/8	4	4 1/4	4 3/8	4 5/8	4 7/8	5 1/4	5 9/16	5 7/8			
	RECOMMENDED	4	4 1/8	4 1/4	4 1/2	4 3/4	5	5 1/4	5 5/8	6	6 1/4	6 1/2		
5'	MINIMUM	4 1/8	4 1/4	4 3/8	4 5/8	4 3/4	5	5 1/4	5 9/16	5 7/8	6 3/16	6 1/2	6 13/16	
	RECOMMENDED	4 3/8	4 1/2	4 5/8	4 7/8	5	5 3/8	5 5/8	6	6 1/4	6 5/8	7	7 1/4	
6'	MINIMUM	4 3/4	4 7/8	5	5 1/4	5 3/8	5 5/8	5 7/8	6 3/16	6 1/2	6 13/16	7 1/8	7 7/16	8 1/8
	RECOMMENDED	5	5 1/8	5 1/4	5 1/2	5 5/8	6	6 1/4	6 5/8	7	7 1/4	7 5/8	8	8 5/8

Figure 14-6 Solve for minimum centerline-to-centerline dimensions of knockouts to provide for locknut clearance.

NOTES

Chapter

15

Electrical Cost Estimating

Determining the cost of an electrical installation is done professionally using a set mechanical technique that terminates in a complete listing of electrical equipment, raceways, luminaires, and devices to "build into" the installation and a detailed listing of installation personnel hours that the construction work will require. This chapter provides insights and templates for use in calculating electrical estimates.

Electrical Takeoff and Personnel-Hour Cost Estimating

Calculating the value of a typical electrical construction project consists of counting the luminaires and devices, measuring the cables and raceways, listing the electrical equipment, and determining the necessary fittings and hardware required to complete the installation. The first step in this work is the takeoff sequence, which includes the following:

- List switchgear, including switchboards, panels, transformers, bus duct, motor starters and motor control centers, and similar equipment items

- Count lighting fixtures, lamps, and hanging material
- Count switch and receptacle devices
- Count devices in special systems, such as fire alarm systems, sound systems, security systems, telephone systems, and other special systems
- Determine the amount of feeder conduit and wire, listing fittings and elbows as well
- Count junction and specialty boxes
- Count fuses
- Count connections to large pieces of equipment, such as air-handling units, chillers, pumps, kitchen equipment, and similar items
- Determine the amount of branch-circuit conduit and wiring
- Sketch and take off control wiring for HVAC and other systems
- Count and list all items on drawings or in specifications that are not colored in at this stage of the takeoff procedure.

After listing the items to be installed, they must be entered into an estimate sheet such as the one shown in Fig. 15-17, onto which materials prices and labor unit pricing can be entered. The labor unit insertion is to determine how many personnel hours of installation time will be required to complete the work. This estimate is most easily and exactly done from personnel-hour tables made from recently completed similar work. When such tables are not available, the following can be used to determine approximate installation personnel-hour budgets:

Lighting fixtures (Fig. 15-1)

Switches and receptacle devices (Fig. 15-2)

Outlet boxes (Fig. 15-3)

Conduit elbows and fittings—EMT (Fig. 15-4)

Conduit and elbows—heavy-wall rigid steel (Fig. 15-5)

Conduit and elbows—IMC steel (Fig. 15-6)
Conduit and elbows—aluminum (Fig. 15-7)
Conduit and elbows—plastic underground (Fig. 15-8)
Cables (Fig. 15-9)
Wires and connectors (Fig. 15-10)
Transformers (Fig. 15-11)
Switches (Fig. 15-12)
Panelboards (Fig. 15-13)
Cable tray (Fig. 15-14)
Motor connections, by horsepower (Fig. 15-15)
Motor controller, individual (Fig. 15-16)

The blank pricing sheet shown in Fig. 15-17 is useful in preparing electrical estimates. An example of its use is shown in Fig. 15-18 for the following project: A single-floor commercial building will contain ten 2/96 fluorescent luminaires, five 20-ampere (A) duplex receptacles, and one light switch. It will be energized through a 100-A, 480-volt (V), two-fuse switch, a 15-kilovoltampere (kVA) 480-120/240-V transformer, a 6-circuit 480/277-V lighting panel, and a 12-circuit 120/240-V panel. The service feeder is four no. 6 copper THHN conductors in 1-inch (in) EMT that is 100 feet (ft) in length. Determine the installation cost of this electrical system.

The electrical estimate must be done in stages, none of which should be omitted:

1. Perform an electrical equipment listing. This is done most easily from the electrical one-line drawing.
2. Perform a feeder takeoff using the feeder takeoff sheet found in Fig. 15-19. The completed feeder takeoff for the present project is shown in Fig. 15-20.
3. On a blank sheet, make a branch-circuit takeoff sheet by listing all the symbols shown on the electrical drawing. One symbol is for the fluorescent luminaire, another is for the light switch, and another is for the duplex

INSTALLATION MANHOUR RATES FOR

LIGHTING FIXTURE INSTALLATION LABOR

(MANHOURS PER EACH)

LUMINAIRE DESCRIPTION	EASIEST (BELOW 10 FT) DENSE LAYOUT REPETITIVE	NORMAL (BELOW 10') DENSE LAYOUT NON-REPETITIVE	MOST DIFFICULT (BELOW 20 FT) SCATTERED PRECISE
INCANDESCENT			
SURFACE-MTD DRUM	0.3	0.4	0.6
RECESSED CAN	0.8	1	1.5
WALL BRACKET	0.35	0.45	0.67
FLUORESCENT			
2-LAMP 4' STRIP	0.8	0.9	1.35
2-LAMP 8' STRIP	1.1	1.2	1.8

2-LAMP 4' INDUSTRIAL	1.4	1.3	1.9
2-LAMP 8' INDUSTRIAL	2	2.5	3.5
4' WRAP-AROUND	0.9	1	1.45
8' WRAP-AROUND	1.2	1.3	1.9
2' X 2' LAY-IN	0.7	0.9	1.25
2' X 4' LAY-IN	0.75	1	1.3
HID			
70W HPS WALL BRACKET	0.5	0.7	1
400 WATT WALL PACK FLOOD	4	4.5	6
PENDANT MTD INDUSTRIAL	1.6	1.8	2.2
400 WATT FLOODLIGHT	4	4.5	7

LABOR UNITS INCLUDE UNLOADING, JOBSITE STORAGE AND RETRIEVAL, LAYOUT, FIXTURE INSTALLATION, WIRE CONNECTIONS, LAMP INSTALLATION, REMOVAL OF SHIPPING CARTONS.

Figure 15-1 Lighting fixture installation personnel hours.

INSTALLATION MANHOUR RATES FOR

DEVICE INSTALLATION LABOR

(MANHOURS PER EACH)

DEVICE DESCRIPTION	EASIEST (BELOW 5 ') SURFACE REPETITIVE	NORMAL (BELOW 5') IN HOLLOW WALL NON-REPETITIVE	MOST DIFFICULT (BELOW 12 ') IN MASONRY WALL 5 OR FEWER
WALL-MOUNTED LIGHT SWITCH			
15 AMPERE DUPLEX RECEPTACLE	0.15	0.2	0.3
20 AMPERE DUPLEX RECEPTACLE	0.2	0.25	0.3
30 AMPERE SINGLE RECEPTACLE	0.25	0.35	0.5
50 AMPERE SINGLE RECEPTACLE	0.45	0.55	0.8

20 AMPERE 1-P LIGHTING SWITCH	0.2	0.25	0.4
20 AMPERE 3-WAY LIGHTING SWITCH	0.3	0.35	0.5
20 AMPERE 4-WAY LIGHTING SWITCH	0.35	0.45	0.7
600W INCANDESCENT DIMMER	0.3	0.4	0.8
1000 WATT INCANCDESCENT DIMMER	0.4	0.5	0.75
1500 WATT INCANDESCENT DIMMER	0.5	0.6	0.8
2000 WATT INCANDESCENT DIMMER	0.8	1.2	1.5
PHOTOCELL	0.3	0.5	2
1-GANG DEVICE PLATE	0.08	0.1	0.15
2-GANG DEVICE PLATE	0.1	0.12	0.15
3-GANG DEVICE PLATE	0.11	0.14	0.21
WEATHERPROOF DEVICE COVER PLATE	0.2	0.25	0.3

LABOR UNITS INCLUDE UNLOADING, JOBSITE STORAGE AND RETRIEVAL, LAYOUT, INSTALLATION, WIRE CONNECTIONS, AND REMOVAL OF CARTONS

Figure 15-2 Switch and receptacle device installation personnel hours.

INSTALLATION MANHOUR RATES FOR

OUTLET BOX INSTALLATION LABOR

(MANHOURS PER EACH)

OUTLET BOX DESCRIPTION	EASIEST BELOW 4 FT WITHIN IN WOOD STUD WALLS REPETITIVE	NORMAL BELOW 8' OR ABOVE A SUSPENDED CEILING NON-REPETITIVE	MOST DIFFICULT (BELOW 16 FT) IN CONCRETE WALLS PRECISE
4" SQUARE BOX	0.3	0.4	0.6
4" OCTAGONAL BOX	0.3	0.4	0.6
4 11/16 SQUARE BOX IN CEILING	0.3	0.4	0.6
HANDY SWITCH BOX	0.25	0.3	0.5
2-GANG BOX	0.25	0.3	0.5
3-GANG BOX	0.27	0.35	0.55
4-GANG BOX	0.3	0.4	0.6
5-GANG BOX	0.37	0.5	0.7
6-GANG BOX	0.45	0.6	0.8
7-GANG BOX	0.5	0.7	0.9
8-GANG BOX	0.6	0.8	1
THROUGH-THE-WALL BOX	0.3	0.4	0.6
BOX EXTENSION	0.15	0.2	0.3
PLASTER RING	0.13	0.15	0.25
FLAT BLANK BOX COVER	0.08	0.1	0.15

LABOR UNITS INCLUDE UNLOADING, JOBSITE STORAGE AND RETRIEVAL, LAYOUT, BOX INSTALLATION, WIRE CONNECTIONS, AND REMOVAL OF SHIPPING CARTONS.

Figure 15-3 Outlet box installation personnel hours.

receptacle. Then count the symbols from the drawing, marking them out fully on the drawing with a colored pencil as their quantities are placed onto the takeoff sheet.

4. Under the quantity of each symbol on the branch-circuit takeoff sheet, write down every piece of material required to install that symbol, along with how many pieces of that material will be required for each symbol installation. For example, mounting a chain-hung 8-ft fluorescent luminaire would require two pieces of allthread rod, two quick bolts, two allthread rod couplings, four hex nuts, four fender washers, two red/yellow wire nuts, 6 ft of $^1/_2$-in flexible conduit, 18 ft of no. 14 THHN wire, two $^1/_2$-in flexible conduit connectors, and two lamps.
5. List each piece of electrical equipment on the pricing sheet found on the branch-circuit takeoff sheet and on the feeder takeoff sheet, marking it off in colored pencil from the takeoff sheet as it is transferred to the pricing sheet. When all items are listed on the pricing sheet, then all items on the drawing will be colored and all items on the takeoff sheets will be colored as well.
6. Price and assign personnel hours to the pricing sheet, referring to the personnel-hour tables found earlier in this chapter. Current material pricing is normally procured from an electrical supply house due to the fact that it changes frequently, and therefore, material pricing is not shown here.
7. Multiply the quantity of each item by its price and then by its personnel-hour installation requirement, and then sum the columns. The totals are the project totals for "raw cost," to which labor rate ($/hour), expendable tools, miscellaneous expenses, and the "labor burden" of fringe benefits, taxes, and insurance must be added to obtain actual cost. A project expense sheet, as described below, should be completed for every project to determine the cost of all project-related items that are not actual materials and installation labor items. A complete pricing sheet for the subject project is shown in Fig. 15-18.

(Text continues on p. 397.)

INSTALLATION MANHOUR RATES FOR

EMT CONDUIT INSTALLATION LABOR

(MANHOURS)

DESCRIPTION	EASIEST BELOW 4 FT WITHIN IN WOOD STUD WALLS REPETITIVE	NORMAL BELOW 8' OR ABOVE A SUSPENDED CEILING NON-REPETITIVE	MOST DIFFICULT (BELOW 16 FT) IN CONCRETE WALLS PRECISE	PER
1/2" EMT CONDUIT	3.5	4.5	7	C
3/4" EMT CONDUIT	4	5.5	8	C
1" EMT CONDUIT	5	7	10	C
1.25" EMT CONDUIT	6	8.5	12	C
1.5" EMT CONDUIT	7	9	13	C
2" EMT CONDUIT	8	11	16	C
2.5" EMT CONDUIT	9	12	14	C
3" EMT CONDUIT	10	13	17	C
4" EMT CONDUIT	12	16	24	C

1" EMT CONDUIT ELBOW	0.15	0.25	0.37	EA
1.25" EMT CONDUIT ELBOW	0.22	0.3	0.45	EA
1.5" EMT CONDUIT ELBOW	0.26	0.35	0.52	EA
2" EMT CONDUIT ELBOW	0.3	0.4	0.6	EA
2.5" EMT CONDUIT ELBOW	0.45	0.6	0.9	EA
3" EMT CONDUIT ELBOW	0.5	0.7	1	EA
4" EMT CONDUIT ELBOW	0.75	1	1.5	EA
1/2" EMT CONNECTOR	0.06	0.08	0.12	EA
3/4" EMT CONNECTOR	0.07	0.1	0.15	EA
1" EMT CONNECTOR	0.09	0.12	0.18	EA
1.25" EMT CONNECTOR	0.11	0.15	0.22	EA
1.5" EMT CONNECTOR	0.13	0.17	0.25	EA
2" EMT CONNECTOR	0.2	0.24	0.4	EA
2.5" EMT CONNECTOR	0.25	0.3	0.5	EA
3" EMT CONNECTOR	0.3	0.4	0.6	EA
4" EMT CONNECTOR	0.45	0.6	0.9	EA

NOTE: LABOR FOR CONDUIT COUPLINGS IS IN THE CONDUIT LABOR
HANGER LABOR IS NOT INCLUDED IN THE CONDUIT LABOR

Figure 15-4 EMT conduit, elbow, and fitting installation personnel hours.

INSTALLATION MANHOUR RATES FOR

HEAVY WALL RIGID STEEL CONDUIT INSTALLATION LABOR

(MANHOURS)

DESCRIPTION	EASIEST GROUND LEVEL OR UNDERSLAB NO BENDS PER RUN	NORMAL BELOW 8' OR ABOVE A SUSPENDED CEILING 2 BENDS PER RUN	MOST DIFFICULT (BELOW 16 FT) IN CONCRETE WALLS 4 BENDS PER RUN	PER
1/2" RIGID STEEL CONDUIT	4	5.5	8	C
3/4" RIGID STEEL CONDUIT	5	6.5	10	C
1" RIGID STEEL CONDUIT	6	8	12	C
1.25" RIGID STEEL CONDUIT	7.5	10	15	C
1.5" RIGID STEEL CONDUIT	9	12	18	C
2" RIGID STEEL CONDUIT	11	14	22	C
2.5" RIGID STEEL CONDUIT	12	16	24	C
3" RIGID STEEL CONDUIT	15	20	30	C
4" RIGID STEEL CONDUIT	21	28	42	C

1/2" RIGID STEEL CONDUIT ELBOW	0.25	0.35	0.5	
3/4" RIGID STEEL CONDUIT ELBOW	0.25	0.35	0.5	
1" RIGID STEEL CONDUIT ELBOW	0.25	0.35	0.5	EA
1.25" RIGID STEEL CONDUIT ELBOW	0.3	0.3	0.5	EA
1.5" RIGID STEEL CONDUIT ELBOW	0.26	0.35	0.6	EA
2" RIGID STEEL CONDUIT ELBOW	0.3	0.4	0.7	EA
2.5" RIGID STEEL CONDUIT ELBOW	1.1	1.5	2.25	EA
3" RIGID STEEL CONDUIT ELBOW	1.5	2	3	EA
4" RIGID STEEL CONDUIT ELBOW	2	2.5	3.75	EA
1/2" INSULATED BUSHING	0.06	0.08	0.12	EA
3/4" INSULATED BUSHING	0.07	0.1	0.15	EA
1" INSULATED BUSHING	0.09	0.12	0.18	EA
1.25" INSULATED BUSHING	0.11	0.15	0.22	EA
1.5" INSULATED BUSHING	0.13	0.17	0.25	EA
2" INSULATED BUSHING	0.2	0.24	0.4	EA
2.5" INSULATED BUSHING	0.25	0.3	0.5	EA
3" INSULATED BUSHING	0.3	0.4	0.6	EA
4" INSULATED BUSHING	0.45	0.6	0.9	EA

NOTE: LABOR FOR CONDUIT COUPLINGS IS IN THE CONDUIT LABOR
HANGER LABOR IS NOT INCLUDED IN THE CONDUIT LABOR
CONDUIT TERMINATION LABOR IS IN THE INSULATED BUSHING LABOR

Figure 15-5 Heavy-wall rigid-steel installation personnel hours.

INSTALLATION MANHOUR RATES FOR

INTERMEDIATE METAL STEEL CONDUIT (IMC) INSTALLATION LABOR

(MANHOURS)

DESCRIPTION	EASIEST GROUND LEVEL OR UNDERSLAB NO BENDS PER RUN	NORMAL BELOW 8' OR ABOVE A SUSPENDED CEILING 2 BENDS PER RUN	MOST DIFFICULT (BELOW 16 FT) IN CONCRETE WALLS 4 BENDS PER RUN	PER
1/2" IMC CONDUIT	4	5	8	C
3/4" IMC CONDUIT	4.5	6	9	C
1" IMC CONDUIT	5	7	10	C
1.25" IMC CONDUIT	6	8	12	C
1.5" IMC CONDUIT	7.5	10	15	C
2" IMC CONDUIT	9	12	18	C
2.5" IMC CONDUIT	10	13	20	C
3" IMC CONDUIT	12	16	24	C
4" IMC CONDUIT	15	20	30	C

1/2" IMC ELBOW	0.25	0.35	0.5	
.75" IMC ELBOW	0.3	0.4	0.6	
1" IMC ELBOW	0.35	0.5	0.75	EA
1.25" IMC ELBOW	0.45	0.6	0.9	EA
1.5" IMC ELBOW	0.5	0.75	1.1	EA
2" IMC ELBOW	0.7	0.9	1.35	EA
2.5" IMC ELBOW	1.2	1.5	3	EA
3" IMC ELBOW	1.5	2	3	EA
4" IMC ELBOW	2.25	3	4.5	EA
1/2" INSULATED BUSHING	0.06	0.08	0.12	EA
3/4" INSULATED BUSHING	0.07	0.1	0.15	EA
1" INSULATED BUSHING	0.09	0.12	0.18	EA
1.25" INSULATED BUSHING	0.11	0.15	0.22	EA
1.5" INSULATED BUSHING	0.13	0.17	0.25	EA
2" INSULATED BUSHING	0.2	0.24	0.4	EA
2.5" INSULATED BUSHING	0.25	0.3	0.5	EA
3" INSULATED BUSHING	0.3	0.4	0.6	EA
4" INSULATED BUSHING	0.45	0.6	0.9	EA

NOTE: LABOR FOR CONDUIT COUPLINGS IS IN THE CONDUIT LABOR
HANGER LABOR IS NOT INCLUDED IN THE CONDUIT LABOR
CONDUIT TERMINATION LABOR IS IN THE INSULATED BUSHING LABOR

Figure 15-6 IMC installation personnel hours.

INSTALLATION MANHOUR RATES FOR

RIGID ALUMINUM CONDUIT INSTALLATION LABOR

(MANHOURS)

DESCRIPTION	EASIEST GROUND LEVEL OR UNDERSLAB NO BENDS PER RUN	NORMAL BELOW 8' OR ABOVE A SUSPENDED CEILING 2 BENDS PER RUN	MOST DIFFICULT (BELOW 16 FT) IN CONCRETE WALLS 4 BENDS PER RUN	PER
1/2" ALUMINUM CONDUIT	4	5	8	C
3/4" ALUMINUM CONDUIT	4.5	6	9	C
1" ALUMINUM CONDUIT	5.5	7.5	11	C
1.25" ALUMINUM CONDUIT	7	9	14	C
1.5" ALUMINUM CONDUIT	8.5	11	17	C
2" ALUMINUM CONDUIT	9	12	18	C
2.5" ALUMINUM CONDUIT	10	13	19	C
3" ALUMINUM CONDUIT	11	14	21	C
4" ALUMINUM CONDUIT	14	19	30	C

1/2" ALUMINUM ELBOW	0.18	0.25	0.4	
.75" ALUMINUM ELBOW	0.22	0.3	0.45	
1" ALUMINUM ELBOW	0.26	0.35	0.5	EA
1.25" ALUMINUM ELBOW	0.3	0.4	0.6	EA
1.5" ALUMINUM ELBOW	0.37	0.5	0.75	EA
2" ALUMINUM ELBOW	0.48	0.65	1	EA
2.5" ALUMINUM ELBOW	0.75	1	1.5	EA
3" ALUMINUM ELBOW	1	1.35	2	EA
4" ALUMINUM ELBOW	1.5	2	3	EA
1/2" INSULATED BUSHING	0.06	0.08	0.12	EA
3/4" INSULATED BUSHING	0.07	0.1	0.15	EA
1" INSULATED BUSHING	0.09	0.12	0.18	EA
1.25" INSULATED BUSHING	0.11	0.15	0.22	EA
1.5" INSULATED BUSHING	0.13	0.17	0.25	EA
2" INSULATED BUSHING	0.2	0.24	0.4	EA
2.5" INSULATED BUSHING	0.25	0.3	0.5	EA
3" INSULATED BUSHING	0.3	0.4	0.6	EA
4" INSULATED BUSHING	0.45	0.6	0.9	EA

NOTE: LABOR FOR CONDUIT COUPLINGS IS IN THE CONDUIT LABOR
HANGER LABOR IS NOT INCLUDED IN THE CONDUIT LABOR
CONDUIT TERMINATION LABOR IS IN THE INSULATED BUSHING LABOR

Figure 15-7 Aluminum rigid conduit installation personnel hours.

INSTALLATION MANHOUR RATES FOR

SCHEDULE 40 PLASTIC RIGID CONDUIT INSTALLATION LABOR

(MANHOURS)

DESCRIPTION	EASIEST GROUND LEVEL OR UNDERSLAB NO BENDS PER RUN	NORMAL BELOW 8' OR ABOVE A SUSPENDED CEILING 2 BENDS PER RUN	MOST DIFFICULT (BELOW 16 FT) IN CONCRETE WALLS 4 BENDS PER RUN	PER
1/2" PVC CONDUIT	4	5	8	C
3/4" PVC CONDUIT	5	6	9	C
1" PVC CONDUIT	6	7	11	C
1.25" PVC CONDUIT	7	9	14	C
1.5" PVC CONDUIT	8.5	11	17	C
2" PVC CONDUIT	10	12	18	C
2.5" PVC CONDUIT	11	13	20	C
3" PVC CONDUIT	12	16	24	C
4" PVC CONDUIT	16	22	33	C
1/2" PVC ELBOW	0.19	0.25	0.4	EA
.75" PVC ELBOW	0.22	0.3	0.45	EA

1" PVC ELBOW	0.26	0.35	0.5	EA
1.25" PVC ELBOW	0.3	0.4	0.6	EA
1.5" PVC ELBOW	0.37	0.5	0.75	EA
2" PVC ELBOW	0.45	0.6	1.2	EA
2.5" PVC ELBOW	0.6	0.8	1.65	EA
3" PVC ELBOW	0.8	1.4	2	EA
4" PVC ELBOW	1.35	1.8	2.7	EA
1/2" MALE ADAPTER	0.1	0.14	0.21	EA
3/4" MALE ADAPTER	0.12	0.16	0.15	EA
1" MALE ADAPTER	0.14	0.18	0.24	EA
1.25" MALE ADAPTER	0.16	0.2	0.3	EA
1.5" MALE ADAPTER	0.2	0.17	0.37	EA
2" MALE ADAPTER	0.2	0.3	0.4	EA
2.5" MALE ADAPTER	0.3	0.4	0.6	EA
3" MALE ADAPTER	0.4	0.55	0.82	EA
4" MALE ADAPTER	0.7	0.9	1.35	EA

NOTE: LABOR FOR CONDUIT COUPLINGS IS IN THE CONDUIT LABOR
HANGER LABOR IS NOT INCLUDED IN THE CONDUIT LABOR
CONDUIT TERMINATION LABOR IS IN THE MALE ADAPTER LABOR

Figure 15-8 Schedule 40 plastic rigid conduit installation personnel hours.

INSTALLATION MANHOUR RATES FOR

CABLE INSTALLATION LABOR

[MANHOURS PER THOUSAND (M) FT]

DESCRIPTION 600 VOLT, THERMOPLASTIC, COPPER 3/C + G	EASIEST CONDUIT IN TRENCH DIRECTLY BURIED 1 BEND PER PULL	NORMAL PULLED INTO CONDUIT OR TRAY BELOW 10 FT. 2 - 3 BENDS PER PULL	MOST DIFFICULT PULLED INTO TRAY BELOW 20 FT. 4 BENDS PER PULL	PER
#14 TC CABLE - BRANCH CIRCUIT	27	35	52	M
#12 TC CABLE - BRANCH CIRCUIT	30	40	60	M
#10 TC CABLE - BRANCH CIRCUIT	35	45	67	M
#8 TC, NM, or UF CABLE - BRANCH CIRCUIT	40	50	75	M
#14 Type AC (BX) CABLE, BRANCH CIRCUIT	32	40	60	M
#12 Type AC (BX) CABLE, BRANCH CIRCUIT	36	45	67	M
#10 Type AC (BX) CABLE, BRANCH CIRCUIT	40	50	75	M

#6 TC, NM, or UF CABLE - BRANCH CIRCUIT	50	60	90	M
#4 TC, NM, or UF CABLE - FEEDER	15	22	27	M
#3 TC, NM, or UF CABLE - FEEDER	16	23	28	M
#2 TC, NM, or UF CABLE - FEEDER	19	26	32	M
#1 TC, NM, or UF CABLE - FEEDER	22	30	37	M
0 TC, NM, or UF CABLE - FEEDER	24	34	42	M
00 TC, NM, or UF CABLE - FEEDER	26	38	47	M
000 TC, NM, or UF CABLE - FEEDER	30	42	52	M
0000 TC, NM, or UF CABLE - FEEDER	35	46	57	M
250 KCMILTC, NM, or UF CABLE - FEEDER	42	50	62	M
300 KCMILTC, NM, or UF CABLE - FEEDER	45	54	67	M
350 KCMILTC, NM, or UF CABLE - FEEDER	49	58	72	M
500 KCMILTC, NM, or UF CABLE - FEEDER	56	68	85	M
750 KCMILTC, NM, or UF CABLE - FEEDER	68	80	100	M

Figure 15-9 Installation personnel hours for cables for branch circuits and feeders.

INSTALLATION MANHOUR RATES FOR

WIRE INSTALLATION LABOR

[MANHOURS PER THOUSAND (M) FT]

DESCRIPTION 600 VOLT, THERMOPLASTIC, COPPER	EASIEST CONDUIT IN TRENCH OR UNDERSLAB 4 - 8 WIRES PER RUN	NORMAL GENERAL WIRING SUSPENDED CEILING 2 - 3 WIRES PER RUN	MOST DIFFICULT (BELOW 16 FT) IN CONCRETE WALLS 1 - 2 WIRES PER RUN	PER
#16	4.5	5.5	8.3	M
#14	4.8	6	9	M
#12	5.6	7	10.5	M
#10	6.8	8.5	12.75	M
#8	9	11	16.5	M
#6	9.5	12	18	M
#4	11.25	14	21	M
#3	13	16	24	M
#2	14.5	18	27	M
#1	16	20	30	M
0	18	22.5	34	M
00	20	25	38	M
000	22.5	28	42	M
0000	25	31	47	M
250 KCMIL	27	34	51	M

300 KCMIL	30	37	56	M
350 KCMIL	32	40	60	M
500 KCMIL	40	49	72	M
750 KCMIL	48	60	90	M
#12 COMPRESSION LUG		0.25	0.3	EA
#10 COMPRESSION LUG		0.25	0.3	EA
#8 COMPRESSION LUG		0.27	0.35	EA
#6 COMPRESSION LUG		0.3	0.35	EA
#4 COMPRESSION LUG		.35	0.4	EA
#3 COMPRESSION LUG		0.35	0.45	EA
#2 COMPRESSION LUG		0.4	0.45	EA
#1 COMPRESSION LUG		0.45	0.5	EA
0 COMPRESSION LUG		0.5	0.55	EA
00 COMPRESSION LUG		0.6	0.7	EA
000 COMPRESSION LUG		0.7	0.8	EA
0000 COMPRESSION LUG		0.85	1	EA
250 KCMIL COMPRESSION LUG		1	1.1	EA
300 KCMIL COMPRESSION LUG		1.25	1.4	EA
350 KCMIL COMPRESSION LUG		1.25	1.4	EA
500 KCMIL COMPRESSION LUG		1.5	1.7	EA
750 KCMIL COMPRESSION LUG		1.75	2.25	EA

Figure 15-10 Wire and wire connector installation personnel hours.

INSTALLATION MANHOUR RATES FOR

TRANSFORMER INSTALLATION LABOR

MANHOURS PER EACH

DESCRIPTION	TRANSFORMER WEIGHT (POUNDS)	EASIEST ON FLOOR OR PREFAB SHEET GOOD ACCESS	NORMAL ON WALL BELOW 5 FT. GOOD ACCESS	MOST DIFFICULT SUSPENDED BELOW 20 FT. POOR ACCESS
DRY TYPE TRANSFORMER				
UP TO 600 VOLT PRIMARY - 1-PHASE				
1 kVA	20	2	3.5	5.5
2 kVA	35	3	4	6
3 kVA	45	4	4.25	6.5
5 kVA	70	4.5	5.25	8
7.5 kVA	140	5	8	18
10 kVA	160	6	9	19.5
15 kVA	300	7	12	25
25 kVA	300	12	N.A.	27
37.5 kVA	400	14	N.A.	31
50 kVA	500	16	N.A.	33
75 kVA	600	18	N.A.	45
100 kVA	700	20	N.A.	N.A.
167 kVA	1200	25	N.A.	N.A.
250 kVA	1800	30	N.A.	N.A.

UP TO 600 VOLT PRIMARY - 3-PHASE				
3 kVA	50	7	8	17
6 kVA	110	9	10	21
9 kVA	150	11	12	24
15 kVA	250	13	14	27
30 kVA	350	15	16	30
45 kVA	500	16	17	34
75 kVA	700	18	N.A.	N.A.
112.5 kVA	800	20	N.A.	N.A.
150 kVA	1000	22	N.A.	N.A.
225 kVA	1600	28	N.A.	N.A.
300 kVA	2000	32	N.A.	N.A.
500 kVA	3000	40	N.A.	N.A.
600 kVA	3500	45	N.A.	N.A.
750 kVA	4000	50	N.A.	N.A.
LIQUID-FILLED TRANSFORMER, ALL VOLTAGES				
150 kVA (including connections & crane operator)	1200	40	N.A.	N.A.
225 kVA (including connections & crane operator)	2000	48	N.A.	N.A.
300 kVA (including connections & crane operator)	3000	52	N.A.	N.A.
500 kVA (including connections & crane operator)	5000	60	N.A.	N.A.
750 kVA (including connections & crane operator)	8000	70	N.A.	N.A.

Figure 15-11 Transformer installation labor personnel hours.

INSTALLATION MANHOUR RATES FOR

SAFETY SWITCH INSTALLATION LABOR

MANHOURS PER EACH

DESCRIPTION	WEIGHT (POUNDS)	2-WIRE	3-WIRE	4-WIRE
NON-FUSIBLE 250 VOLT SAFETY SWITCH				
30 AMPERE	7	2	2.75	3
60 AMPERE	12	2.5	3	3.5
100 AMPERE	34	3	4	4.25
200 AMPERE	54	5.5	6.25	6.5
400 AMPERE	150	8.5	10	10.25
600 AMPERE	180	12	15.25	15.5
800 AMPERE	450	16.5	18.75	19
1200 AMPERE	500	22.5	22.75	23
FUSIBLE 250 VOLT SAFETY SWITCH				
30 AMPERE	9	2.5	3	3.25
60 AMPERE	14	3	3.25	3.5
100 AMPERE	36	3.5	4.25	4.5
200 AMPERE	57	6	6.5	6.75
400 AMPERE	155	9	10.25	10.5
600 AMPERE	186	12.5	15.5	15.75
800 AMPERE	460	17	19	19.25
1200 AMPERE	520	22	23	23.25
NON-FUSIBLE 600 VOLT SAFETY SWITCH				
30 AMPERE	7	2.75	3.5	3.75
60 AMPERE	12	3.25	3.75	5.25
100 AMPERE	34	3.75	4.75	5
200 AMPERE	54	6.25	7	7.25
400 AMPERE	150	9.25	10.25	11
600 AMPERE	180	12.25	16	16.25
800 AMPERE	450	17.25	19.5	19.75
1200 AMPERE	500	23.25	23.5	23.75
FUSIBLE 600 VOLT SAFETY SWITCH				
30 AMPERE	9	3	3.75	4
60 AMPERE	14	3.5	4	4.5
100 AMPERE	36	4	5	5.25
200 AMPERE	57	6.5	7.25	7.5
400 AMPERE	155	9.5	11	11.25
600 AMPERE	186	12.5	16.25	16.5
800 AMPERE	460	17.5	19.75	20
1200 AMPERE	520	23.5	23.75	24

Figure 15-12 Safety switch installation labor personnel hours.

INSTALLATION MANHOUR RATES FOR

CIRCUIT BREAKER PANELBOARD INSTALLATION LABOR

DESCRIPTION	MANHOURS PER EACH
FOR TOTAL LABOR, ADD PANEL CAN AND AND MCB LABOR TO BRANCH CIRCUIT LABOR	
100 A MCB & PANEL CAN	12
225 A MCB & PANEL CAN	16
400 A MCB & PANEL CAN	30
800 A MCB & (FREE-STANDING) PANEL CAN	45
1200 A MCB & (FREE-STANDING) PANEL CAN	60

QUANTITY OF ACTIVE CIRCUITS	20 AMP	50 AMP	100 AMP	200 AMP	400 AMP
2	0.3	0.4	0.6	0.9	1.6
4	0.6	0.8	1.2	1.8	3.2
6	0.9	1.2	1.8	2.7	4.8
8	1.2	1.6	2.4	3.6	6.4
10	1.5	2	3	4.5	8
12	1.8	2.4	3.6	5.4	9.6
14	2.1	2.8	4.2	6.3	11.2
16	2.4	3.2	4.8	7.2	12.8
18	2.7	3.6	5.4	8.1	14.4
20	3	4	6	9	16
22	3.3	4.4	6.6	9.9	17.6
24	3.6	4.8	7.2	10.8	19.2
26	3.9	5.2	7.8	11.7	20.8
28	4.2	5.6	8.4	12.6	22.4
30	4.5	6	9	13.5	24
32	4.8	6.4	9.6	14.4	25.6
34	5.1	6.8	10.2	15.3	27.2
36	5.4	7.2	10.8	16.2	28.8
38	5.7	7.6	11.4	17.1	30.4
40	6	8	12	18	32
42	6.6	8.8	13.2	19.8	35.2

Figure 15-13 Circuit breaker panelboard installation labor personnel hours.

Factoring of Labor Units

The labor units provided in the preceding figures include the following approximate amounts of work:

1. Actual installation: 69 percent
2. Nonproductive labor: 2.5 percent
3. Studying plans: 2.5 percent
4. Material procurement: 2.5 percent
5. Receiving and storage: 2.0 percent

INSTALLATION MANHOUR RATES FOR

STEEL CABLE TRAY INSTALLATION LABOR

(MANHOURS)

DESCRIPTION	EASIEST UP TO 5 FT REPETITIVE	NORMAL UP TO 15 FT REPETITIVE	MOST DIFFICULT UP TO 25 FT NON-REPETITIVE	PER
6" SIDE RAIL LADDER CABLE TRAY				
6" WIDE	15	18	27	C
9" WIDE	16	19	30	C
12" WIDE	17	20	30	C
18" WIDE	18	21	32	C
24" WIDE	19	22	33	C
30" WIDE	20	24	36	C
36" WIDE	21	26	39	C

6" ELBOW	1.2	1.6	2.25	EA
9" ELBOW	1.3	1.65	2.5	EA
12" ELBOW	1.45	1.8	2.75	EA
18" ELBOW	1.6	2	3	EA
24" ELBOW	1.85	2.35	3.5	EA
30" ELBOW	2.1	2.65	4	EA
36" ELBOW	2.4	3	4.5	EA
6" TEE	3.2	4	6	EA
9" TEE	3.4	4.25	6.35	EA
12" TEE	3.6	4.5	6.75	EA
18" TEE	4	5	7.5	EA
24" TEE	4.5	5.5	8	EA
30" TEE	5	6	9	EA
36" TEE	5.25	6.5	11	EA

Figure 15-14 Cable tray installation personnel hours.

INSTALLATION MANHOUR RATES FOR

MOTOR CONNECTION INSTALLATION LABOR

(MANHOURS)

DESCRIPTION	EASIEST FLOOR LEVEL	NORMAL UP TO 5 FT	MOST DIFFICULT UP TO 10 FT	PER
0 - 7.5 HP MOTOR	2	2.6	3.5	EA
10 HP MOTOR	2	2.6	3.5	EA
15 HP MOTOR	2.25	3	4	EA
20 HP MOTOR	2.5	3.25	4.35	EA
25 HP MOTOR	2.75	3.6	4.75	EA
30 HP MOTOR	3	4	5	EA
40 HP MOTOR	3.5	4.5	6	EA
50 HP MOTOR	3.75	5	6.5	EA
60 HP MOTOR	4	5.2	6.75	EA
75 HP MOTOR	5	6.5	8.5	EA
100 HP MOTOR	6	7.8	10	EA
125 HP MOTOR	7	9.1	12	EA
150 HP MOTOR	8	10.4	14	EA
200 HP MOTOR	10	13	17	EA

WIRE LUG AND CABLE CONNECTOR LABOR MUST BE LABORED SEPARATELY FROM ABOVE

Figure 15-15 Motor connection personnel hours.

INSTALLATION MANHOUR RATES FOR MOTOR CONTROLLER INSTALLATION LABOR		
DESCRIPTION	(MANHOURS) NORMAL UP TO 5 FT	PER
0 - 10 HP MOTOR CONTROLLER	6	EA
15 - 25 HP MOTOR CONTROLLER	8	EA
30 - 50 HP MOTOR CONTROLLER	12	EA
60 - 100 HP MOTOR CONTROLLER	16	EA
125-150 HP MOTOR CONTROLLER	40	EA
200 HP MOTOR CONTROLLER	60	EA

Figure 15-16 Motor controller personnel hours.

6. Mobilization: 5.0 percent
7. Layout: 5.0 percent
8. Cleanup: 2.0 percent
9. Punch-list work: 5.0 percent
10. Coffee breaks: 3.5 percent

If one or more of the preceding categories of work will not be a part of the work scope of a given installation, then the total personnel-hour budget can be adjusted accordingly.

When a building is more than one floor in height, then the overall personnel hours for the project should be increased by 1 percent per floor.

When installing large conduit in parallel runs, less labor is required than for installing individual runs. Besides the labor column differences in the personnel-hour tables, the following further adjustments should be made to the conduit labor:

- For two parallel conduits, deduct 10 percent of the conduit labor.

PRICING SHEET

PROJECT NAME			DATE					
DESCRIPTION OF WORK								
ITEM DESCRIPTION	**QUANTITY**	**PRICE**	**PER**	**MULT**	**MAT'L EXT**	**LABOR**	**PER**	**LAB.EXT.**

Totals for sheet			material			manhours		

Figure 15-17 Blank pricing sheet for electrical estimating.

PRICING SHEET

PROJECT NAME	Example		DATE					
DESCRIPTION OF WORK	lighting, branch wiring, & feeders							
ITEM DESCRIPTION	**QUANTITY**	**PRICE**	**PER**	**MULT**	**MAT'L EXT**	**LABOR**	**PER**	**LAB.EXT.**
100A, 2F, 600V, N-1 Switch	1							
Meter Socket	1							
15 kVA, 480:120/240V xfmr	1							
Circuit Breaker Panelboard:								
80/2 MCB	2							
20/1 branch circuit breakers	12							
#2 Cu THHN Wire	30							
#6 Cu THHN Wire	390							
#12 THHN Wire	1000							
#14 THHN Wire	160							
2/96 fluorescent luminaire	10							
F96T8 Lamp	20							
1" EMT Conduit	100							
1" EMT Elbow	3							
1/2" EMT Conduit	300							
1" Emt Coupling	15							
1/2" EMT Coupling	33							
1" EMT Connector	3							
1/2" EMT Connector	28							
1" EMT Conduit Strap	15							

1/2" EMT Conduit Strap	35							
1" x 12" nipple	1							
4S 1/2 Outlet Box	14							
4S Switch Cover	1							
4S Duplex Receptacle Cover	5							
20 Ampere Switch	1							
20 Ampere Duplex Receptacle	5							
1" Scrutite Hub	1							
1" Weatherhead	1							
7-Point Service Rack	1							
60 Ampere Fuses	2							
Water Seal at Conduit entrance	1							
5/8" x 10ft ground rod	1							
#6/1 Armored Ground Cable	50							
Cold Water Pipe Ground Clamp	1							
1/4" Allthread Rod	80							
1/4" Allthread Rod Couplings	16							
1/4" Hex nuts	28							
Fender Washers	32							
R/Y Wire Nuts	16							
1/2" Flex Conduit	48							
1/2" Flex Connector	16							
4S Box flat blank cover	8							
1/4" quickbolts	28							
Ground Rod Clamp	1							
Totals for sheet			material			manhours		

Figure 15-18 Completed pricing sheet.

Staging of material at an electrical construction site is part of the manhour requirement for the project.

- For three parallel conduits, deduct 15 percent of the conduit labor.
- For four parallel conduits, deduct 20 percent of the conduit labor.
- For five parallel conduits, deduct 25 percent of the conduit labor.
- For over five parallel conduits, deduct 30 percent of the conduit labor.

When installing multiple conductors in a common conduit or in parallel conduit runs, less labor is required than for installing individual conductor pulls. Therefore, in addition to the labor column differences in the personnel-hour tables, the following further adjustments should be made to the wire labor:

- For three conductors pulled during the same pull, deduct 20 percent of the wire labor.
- For five conductors pulled during the same pull, deduct 35 percent of the wire labor.

FEEDER TAKE-OFF SHEET

PROJECT:__________ DATE:______ T.O. SHEET # ____ OF ____ T.O. SHEETS

FEEDER NUMBER	FROM	TO	CONDUIT LENGTH	CONDUIT SIZE	CONDUIT TYPE	ELBOWS	CONDUIT TERM.	WIRE QUANTITY	WIRE SIZE & TYPE	TOTAL LENGTH	MISCELLANEOUS	

Figure 15-19 Sample blank feeder takeoff sheet.

FEEDER TAKE-OFF SHEET

	PROJECT: EXAMPLE 1-FLOOR COMMERCIAL BUILDING					DATE:		T.O. SHEET # 1 OF 1 T.O. SHEETS			
FEEDER NUMBER	FROM	TO	CONDUIT LENGTH	CONDUIT SIZE	CONDUIT TYPE	ELBOWS	CONDUIT TERM.	WIRE QUANTITY	WIRE SIZE & TYPE	TOTAL LENGTH	MISCELLANEOUS
1	Weatherhead	Meter	20 ft	1"	EMT	0	1 - hub	3	#6 cu THHN	90 ft	1 - 1" Weatherhead
2	Meter	Service Switch	1" x 12" nipple			0	2	3	#6 cu THHN	30 ft	1 - 7-pt rack
3	Service Switch	Transformer	80 ft	1"	EMT	3	2	3	#6 cu THHN	270 ft	2 - 60 amp fuses
			4 ft	1"	Flex		2		incl above		1- 100A 2F 600V switch
4	Transformer	Panelboard	4 ft.	1.25"	Flex		2	3	#2 cu THHN	30 ft	1- Seal at conduit entrance
											1 - ground rod
											50 ft - 6/1 armored grnd cable
											1 - cold water clamp
											1 - 15 kVA transformer

Figure 15-20 Completed feeder takeoff sheet.

- For seven conductors pulled during the same pull, deduct 45 percent of the wire labor.
- For over seven conductors pulled simultaneously, deduct 50 percent of the wire labor.

Estimate of Project Expense

In addition to the material and labor required for an electrical installation, all projects require other work that costs money, and some projects require more than others. To accommodate these project expenses, the following checklist should be used on each project after determining the total materials and labor for the job:

1. Service truck time should be determined, as well as the number of service trucks multiplied by their hourly rental rate.
2. Premium labor cost must be added. This is the price of labor per hour for time spent working in excess of the normal work week. The normal work week is 40 hours in most of the country but is as little as 32 hours per week in some locations. The premium cost is normally 50 percent of the labor rate of straight-time work, but in some locations it can be as much as 200 to 300 percent of the base rate for straight-time work.
3. Excessive guarantee-warranty costs must be determined and added to the estimate.
4. Rental rates for lifting equipment such as scaffolds, sissors lifts, JLG lifts, etc., must be determined and added to the estimate.
5. Hoisting costs for crane rental, forklift trucks, etc., must be determined and added to the estimate. Often a general contractor will provide the hoisting equipment, but sometimes the general contractor will "backcharge" the electrical contractor for its use and for the cost of the equipment operator.
6. Travel costs must be determined and added. These costs may consist of travel to the job site, if the job site is in

another city, or these costs may be required travel to factories that are manufacturing equipment to be installed at the job site.

7. Room and board must be determined and added for everyone who must be away from home because of the project. Sometimes these costs and travel expenses can be as large as the actual costs of the electrical work itself.
8. Storage costs must be determined, even if they are necessary as part of the staging of equipment for the project. Also, if the equipment must be handled by your forces at the storage site, then the cost for the personnel hours to be spent performing this extra handling must be included within the estimate.
9. The cost of job-site telephone service and the cost of long-distance telephone expenses must be included in the estimate.
10. The cost of sanitary facilities should be included, unless they will be provided by the general contractor and there is an agreement that the general contractor will not backcharge the subcontractors for their cost.
11. The cost of ice water should be included, unless it will be provided by the general contractor and there is an agreement that the general contractor will not backcharge the subcontractors for its cost.
12. The cost of freight must be included. It either can be included as a part of the materials price that is normally quoted by an electrical supply house or can be included as a separate expense sheet item.
13. The purchase or rental cost of special tools must be included in the estimate by entering into the expense sheet.
14. The cost of manufacturer's engineers must be included if they will be required for start-up or to ensure warranty continuance.
15. The cost of temporary wiring, temporary lighting, temporary elevator, and temporary electrical energy must

be included either by the general contractor or by the electrical contractor.

16. The cost of "as built" drawings must be determined and included as an expense sheet item.
17. The cost of testing and relay setting must be determined and included as an expense sheet item.
18. The cost of purchasing or renting job-site trailers and offices must be determined and included within the estimate, along with the determination of where they can be located.
19. The cost of ladders and small tools must be determined and included within the estimate. Normally, these are estimated to be 3 percent of labor cost.
20. The cost of the electrical project manager and project supervision must be forecast and entered into the estimate sheet as a job cost.
21. The cost of parking for the workers must be determined and included within the estimate.

After all these project expense values have been determined, they must be added to the cost of material and installation labor to determine the final job cost. The final selling price is then determined by adding an "overhead and profit markup" to the final job cost.

Engineering Economics Calculations Considering the Time Value of Money

Once the cost of a piece of equipment or installation has been determined in terms of today's dollars, it is often of value to determine its cost over its entire life span, or sometimes it is necessary to compare the costs of one item or system to the costs of another both in terms of today's dollars and in terms of long-term operating and maintenance costs, including interest on the money. The remainder of this chapter provides information needed to make these cost comparisons that include the time value of money.

There are almost always many ways to accomplish the same result, and the choice between them is most often determined by costs. Among cost considerations, the initial (or capital) cost, the interest rate (also called the *time value of money*), and the operating and maintenance cost are the three most notable costs.

Although the value of a capital expenditure made today often can be obtained by a simple verbal inquiry, if the money to buy that item had to be borrowed at interest and repaid over the life of the item, the total money that would be spent to obtain the item would be considerably greater. And due to the time value of money, the dollar number would be different today than the dollar number at the end of the life of the item.

Since money can earn interest over time, it is necessary to recognize that a dollar received at some future date F is not worth as much as a dollar received in the present P. The reverse is true as well, since a hundred dollar bill received today, at P, is worth much more than a hundred dollar bill received in 10 years, at F. This is due to the fact that money has earning power over those 10 years n (where n is the number of years).

Economic analysis in engineering deals with the evaluation of economic alternatives. Most engineering cost alternatives can be determined approximately by simplifying real-world realities, such as ignoring inflation, taxes, and depreciation, and considering that interest is simple interest rather than compounded interest.

In most engineering economic calculations, only small parts of entire schemes are evaluated at one time, and then the pieces are added together as necessary to formulate the overall conclusions.

Simple interest is the money to be paid for money borrowed for a certain time, normally given on a per-annum basis (i.e., for 1 year). In its simplest form, the interest I that must be paid on such a loan is equal to the value V of the loan multiplied by the interest rate i per year multiplied by the quantity of years n. The general formula for this is

$$I = Vin$$

Solve for interest given loan value, interest rate, and time

As an example, if \$10,000 is loaned for 1 year at the rate of 10 percent per year, the interest cost for this loan will be

$$I = Vin$$

$$= (\$10{,}000)\,(0.10)\,(1) = \$1000$$

Repayment of the loan at the end of 1 year would require a one-time payment of \$11,000, the sum of the V (the initial dollar loan amount) plus I (the interest cost).

Solve for the future value given the present value, the compounded interest rate, and time over several years

If an amount P is disbursed today at the annual interest rate of i, one can also determine the total value of the loan F (interest + the original amount) over a span of n years by the following method. The formula used is

$$F = P\,(1 + i)^n$$

For example, if \$1000 is invested at 10 percent interest compounded annually for 4 years, it will have a future value F of

$$F = P\,(1 + i)^n$$

$$= \$1000\,(1 + 0.1)^4$$

$$= \$1000\,(1.1)^4$$

$$= (\$1000)\,(1.464)$$

$$= \$1464.10$$

This formula is useful in determining the total future-day investment of pieces of equipment that would each be purchased at different times in the life of the plant. Simply calculate the future value of each piece for the quantity of years that it will exist before the end of the plant life, turning the purchase investment of each into equivalent future dollars that simply can be summed together after each has been calculated individually. An example of this is shown in Fig. 15-21, where the owner is planning to purchase a truck and a trailer separately, the truck this year and the trailer when it becomes available after 3 years, and it is necessary to determine the total cost F of the completed truck-trailer combination at the end of its life in 5 years.

Solve for the present value given the future value, the compounded interest rate, and time over several years

The question might be asked as to how much must be invested now at 10 percent interest compounded annually so that $15,000 can be received in 5 years.

The single-payment compound-amount relationship may be solved for the present value P as follows:

$$P = F\left[\frac{1}{(1 + i)^n}\right] = \$15{,}000\left[\frac{1}{(1 + 0.1)^5}\right] = \$9314$$

Note that this is simply the original formula for F but solved instead for P. This formula is useful in determining the present-day value of pieces of equipment that would each be purchased at different times in the life of the plant. Simply perform the calculation for each piece to transform the various values into today's dollars and then sum the values of today's dollars together to obtain the total value in present-day money. An example of this is shown in Fig. 15-22, where a proposal for a maintenance contract has been received in which payment of $3,153,705.12 would be made each year for 20 years, and the buyer needs to calculate the total value of this maintenance cost in present-day dollars to compare against a similar competing proposal.

Graphic representation of cash flows

Since diagrams are common in engineering work, cash-flow diagrams are normal tools to engineers performing engineering economic studies. A cash-flow diagram shows any money received as an upward arrow, which represents an increase in cash, and it shows any money disbursed as a downward arrow, which represents a decrease in cash. These arrows are placed on an x-axis time scale that represents the duration of the transactions. Figure 15-23 shows a cash-flow diagram for the following transaction: Company A borrows the sum of $2000 for 3 years at an interest rate of 10 percent per annum. At the end of the first year, Company A must pay the interest on the loan of $200. At the end of the second and third years, Company A also must pay the interest on the loan. At the end of the fourth year, Company A not only must pay the $200 interest on the loan but also must repay the initial amount of the loan, $2000.

Solve for the single future value that would accumulate from a series of equal payments occurring at the end of succeeding annual interest periods

The compounding of interest paid on interest earned and saved can amount to a significant monetary sum. In many engineering economics studies, it is necessary to determine the single future value that would accumulate from a series of equal payments each occurring at the end of continual annual interest periods. The cash-flow diagram for the following transaction is shown in Fig. 15-24: An accessory to a machine saves $100 in electrical energy costs each year, and this money is set aside in a savings account having an interest-bearing rate of 8 percent per year. Solve for the total future amount F for which the accessory will have been directly responsible by the end of the seventh year.

Note that the compound amount shown at the end of 7 years is greatest for year 1, since its earned interest will have had the longest duration to earn interest on interest, and the total amount of F is equal to the sums of the compound amounts measured at the end of each year. However, the simplest way to calculate this final value for F is by simply applying this formula:

PROBLEM: CALCULATE THE TOTAL COST OF THE BRAND "A" TRUCK AND TRAILER IN TERMS OF DOLLARS AT THE END OF 5 YEARS, WHERE THE TRUCK IS PURCHASED TODAY FOR $ 100,000 AND THE TRAILER IS PURCHASED AFTER 3 YEARS FOR $ 50,000. THE INTEREST RATE (THE TIME VALUE OF MONEY) IS 10% PER YEAR.

TOTAL COST CALCULATION OF TRUCK

THE BRAND "A" TRUCK CAN BE PURCHASED THIS YEAR FOR A ONE-TIME CASH PRICE OF $ 100,000.
THE LIFE OF THE TRUCK IS 5 YEARS.

YEAR		ORIGINAL PRICE PAID FOR TRUCK	n YEARS FORWARD TO YEAR 5	FUTURE WORTH IN FUTURE DOLLARS CALCULATED BY FORMULA BELOW TABLE
		P	n	F
1	$	100,000.00	5	
2	$	-	4	
3	$	-	3	
4	$	-	2	
5	$	-	1	
TOTAL COST OF TRUCK AND INTEREST IN FUTURE DOLLARS				$ 161,051.00

THESE CALCULATIONS ARE MADE USING THE FOLLOWING FORMULA, AS ILLUSTRATED FOR n=5 YEARS FOR THE TRUCK.

F = P X (1 + i) EEX n F = $ 100,000 X (1 + 0.1) EEX 5 F = $ 100,000 X 1.61051 = $ 161,051

TOTAL COST CALCULATION OF TRAILER

THE BRAND "A" TRAILER CAN BE PURCHASED
AS SOON AS THE MODEL BECOMES AVAILABLE
IN THREE YEARS, FOR THE SUM OF $ 50,000.
THESE ARE FIRM QUOTED PRICES THAT WILL NOT CHANGE.
THE TRAILER WILL BE OF NO USE AFTER THE TRUCK
IS AT THE END OF ITS LIFE, SO THE TRAILER WILL
ONLY BE USED FOR 2 YEARS AND THEN BE SCRAPPED.

YEAR	ORIGINAL PRICE PAID FOR TRAILER THIS YEAR P	n YEARS FORWARD TO YEAR 5 n	FUTURE WORTH IN FUTURE DOLLARS CALCULATED BY FORMULA BELOW TABLE F
1	$ -	5	
2	$ -	4	
3	$ -	3	
4	$ 50,000.00	2	
5	$ -	1	
TOTAL COST OF TRAILER AND INTEREST IN FUTURE DOLLARS			$ 60,500.00

THESE CALCULATIONS ARE MADE USING THE FOLLOWING FORMULA, AS ILLUSTRATED FOR n=3 YEARS FOR THE TRAILER.

F = P X (1 + i) EEX n F = $ 50,000 X (1 + 0.1) EEX 2 F = $ 50,000 X 1.331 = $ 60,500

TOTAL COST CALCULATION OF TRUCK AND TRAILER

TOTAL COST OF THE TRUCK AND TRAILER IN FUTURE DOLLARS = $ 161,051 + $ 60,500 = **$ 221,551**

Figure 15-21 Solve for the future value F of a truck and trailer at the end of a 5-year life.

GAS TURBINE GENERATOR SET ENGINE MAINTENANCE COSTS (SET BY CONTRACT)

YEAR	QUANTITY OF GENSETS	COST PER FIRED HOUR		OPERATION DURING EACH YEAR		
				HOURS PER DAY	DAYS PER YEAR	OPERATING %AGE
1999	6	$	63.16	24	365	0.95
2000	6	$	63.16	24	365	0.95
2001	6	$	63.16	24	365	0.95
2002	6	$	63.16	24	365	0.95
2003	6	$	63.16	24	365	0.95
2004	6	$	63.16	24	365	0.95
2005	6	$	63.16	24	365	0.95
2006	6	$	63.16	24	365	0.95
2007	6	$	63.16	24	365	0.95
2008	6	$	63.16	24	365	0.95
2009	6	$	63.16	24	365	0.95
2010	6	$	63.16	24	365	0.95
2011	6	$	63.16	24	365	0.95
2012	6	$	63.16	24	365	0.95
2013	6	$	63.16	24	365	0.95
2014	6	$	63.16	24	365	0.95
2015	6	$	63.16	24	365	0.95
2016	6	$	63.16	24	365	0.95
2017	6	$	63.16	24	365	0.95
2018	6	$	63.16	24	365	0.95

THESE CALCULATIONS ARE MADE USING THE FOLLOWING FORMULA, AS ILLUSTRATED FOR n=2 YEARS.

$$P = F \text{ X } \frac{1}{(1 + i) \text{ EEX } n} = \$\,3{,}153{,}705.12 \text{ X } \frac{1}{(1 + 0.1) \text{ EEX } 2} =$$

$$(\$\,3{,}153{,}705.12) \text{ X } \frac{1}{1.21} = \$\,2{,}606{,}367.87$$

Figure 15-22 Solve for the present value P of maintenance costs over the 20-year life of the generators.

$$F = A\left[\frac{(1 + i)^n - 1}{i}\right]$$

Inserting the values for A, the annual payments, and i, the annual interest rate, into this formula yields

$$F = \$100\left[\frac{(1 + 0.08)^7 - 1}{0.08}\right] = \$892.28$$

CONTRACT MAINTENANCE PRICE PAID EACH YEAR	n YEARS BACK TO PRESENT YEAR 1999	TIME VALUE OF MONEY, i	PRESENT WORTH IN 1999 DOLLARS CALCULATED BY FORMULA BELOW TABLE
$ 3,153,705.12	0	N.A.	$ 3,153,705.12
$ 3,153,705.12	1	0.1	$ 2,867,004.65
$ 3,153,705.12	2	0.1	$ 2,606,367.87
$ 3,153,705.12	3	0.1	$ 2,369,425.33
$ 3,153,705.12	4	0.1	$ 2,154,023.03
$ 3,153,705.12	5	0.1	$ 1,958,202.76
$ 3,153,705.12	6	0.1	$ 1,780,184.32
$ 3,153,705.12	7	0.1	$ 1,618,349.38
$ 3,153,705.12	8	0.1	$ 1,471,226.71
$ 3,153,705.12	9	0.1	$ 1,337,478.83
$ 3,153,705.12	10	0.1	$ 1,215,889.85
$ 3,153,705.12	11	0.1	$ 1,105,354.41
$ 3,153,705.12	12	0.1	$ 1,004,867.64
$ 3,153,705.12	13	0.1	$ 913,516.04
$ 3,153,705.12	14	0.1	$ 830,469.13
$ 3,153,705.12	15	0.1	$ 754,971.93
$ 3,153,705.12	16	0.1	$ 686,338.12
$ 3,153,705.12	17	0.1	$ 623,943.75
$ 3,153,705.12	18	0.1	$ 567,221.59
$ 3,153,705.12	19	0.1	$ 515,655.99
TOTAL MAINTENANCE COST IN 1999 DOLLARS (PRESENT DOLLARS)			$ 29,534,196.45

Therefore, the accessory is responsible for a savings of \$892.28 over its 7 years of operation.

Solve for the value of equal year-end payments that would be required to accumulate a given future amount at an annual interest rate

Frequently, it is necessary to save and accumulate funds for a future expenditure goal, and it is necessary to determine

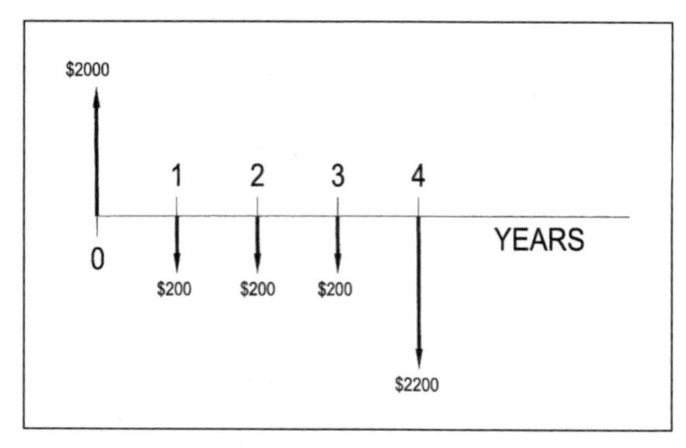

Figure 15-23 Solve for the cash-flow diagram of a loan transaction at interest.

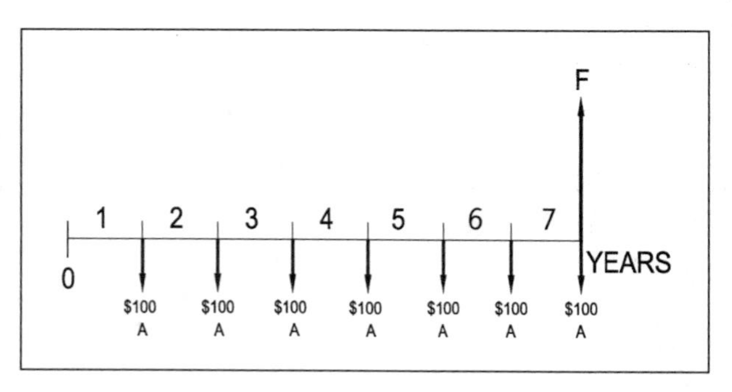

Figure 15-24 Solve for the cash-flow diagram of a recurrent savings set aside at interest.

how much money A must be saved each year to accumulate the desired future amount F if the savings are set at interest i. The cash-flow diagram of Fig. 15-25 shows this overall transaction, and the dollar value A of each annual payment can be calculated using this formula:

$$A = F\left[\frac{i}{(1 + i)^n - 1}\right]$$

For example, a future amount of $6000 is desired at the end of 5 years, and the savings account into which annual payments will be placed pays an interest rate of 6 percent per annum. Determine how much money must be placed

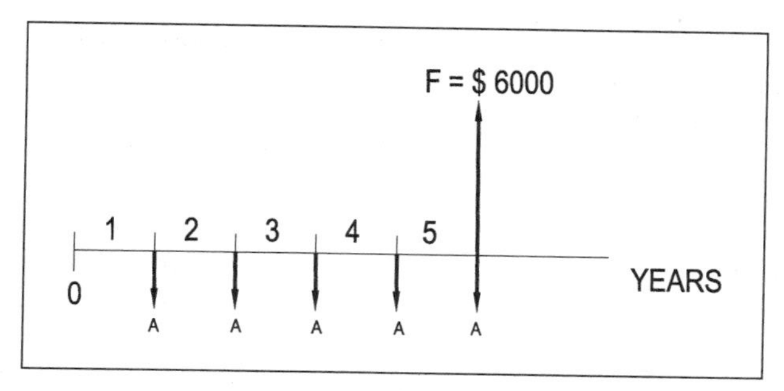

Figure 15-25 Solve for required payments to accumulate a given sum at a future time at interest.

PVC conduit stored for future installation to take advantage of volume discounts that exceed the time value of money costs.

into the savings account at the end of each year to have \$6000 at the end of the fifth year.

Inserting the values for F, the future goal, and i, the annual interest rate, into this formula yields

$$A = \$6000 \left[\frac{0.06}{(1 + 0.06)^5 - 1} \right]$$

$$= \$6000\ (0.177396)$$

$$= \$1064.38 \text{ per year}$$

Therefore, the sum of $1064.38 must be deposited into the savings account at the end of each of the 5 years to have $6000 at the end of the fifth year.

In all engineering economics studies, outgoing payments made to begin a transaction are made at the beginning of the transaction period, and payments that occur during the transaction are assumed to occur at the end of the interest period during which they occur; normally, this is at the end of the year. In addition to this, in all engineering cost studies, the following also should be kept in mind:

- The end of one year is the beginning of the next year.
- P is at the beginning of a year at a point in time that is designated the *present* time.
- F occurs at the end of the nth year from P.
- A occurs at the end of each year in the transaction period.
- When both P and A are involved in a calculation, the A occurs 1 year later than P. That is, A occurs at the end of the year, while P occurred at the beginning of that year.
- When both F and A are involved in a calculation, during the last year of the transaction, F and A occur at the same time.

NOTES

NOTES

Chapter

16

Conversion Calculations

Temperature Conversion Calculations

While almost all ambient temperatures in electrical engineering are given in degrees Fahrenheit, almost all temperatures of equipment operation are stated in degrees Celsius. Therefore, it is important that a quick and easy conversion method be available. Figure 16-1 shows how to make these conversions and provides a quick cross-reference tool from which the conversion can be derived instantly in either direction.

Frequently Used Conversion Calculations

For work in the electrical industry, many other conversion calculations are encountered repeatedly. Figure 16-2 provides the conversion formulas for these common conversions, and Fig. 16-3 is a set of typical sample calculations showing how to use the conversion formulas correctly. Each of these conversion calculations has been changed into a simple one-step calculation for ease of use.

Multiple Conversion Calculations

Sometimes it is necessary to multiply an item in one set of units by another item to determine an answer in a completely

TEMPERATURE CONVERSION TABLE

(DEGREES C TO DEGREES F)

$C = 5/9\ (F - 32)$

$F = [(9/5)C] + 32$

C	F
-80	-112
-70	-94
-60	-76
-50	-58
-40	-40
-30	-22
-20	-4
-10	14
0	32
10	50
20	68
30	86
40	104
50	122
60	140
70	158
80	176
90	194
100	212
110	230
120	248
130	266
140	284
150	302
160	320
170	338
180	356
190	374
200	392
210	410
220	428
230	446
240	464
250	482
260	500
270	518

Figure 16-1 Use these formulas and values to solve for degrees Celsius given degrees Fahrenheit or for degrees Fahrenheit given degrees Celsius.

CONVERSION FORMULAS			
FROM		X	= TO
AMPERE-TURNS	x	1.256637	= GILBERTS
AMPERES/SQ.CM.	x	6.4516	= AMPERES/SQ.IN.
ATMOSPHERES	x	29.9213	= INCHES OF MERCURY
ATMOSPHERES	x	14.696	= POUNDS/SQ.IN.
BARRELS (PETRO)	x	42	= GALLONS
BARS	x	1,000,000	= DYNES/SQ.CM.
BARS	x	14.5038	= POUNDS/SQ.IN.
BOARD FEET	x	144	= CUBIC INCHES
BTU	x	1.054E+010	= ERGS
BTU	x	777.649	= FOOT-POUNDS
BTU	x	3.928E-004	= HORSEPOWER-HOURS
BTU	x	1054.35	= JOULES
BTU	x	2.929E-004	= KILOWATT-HOURS
BTUs/HOUR	x	0.292875	= WATTS
BTUs/MINUTE	x	2.356E-002	= HORSEPOWER
BTUs/MINUTE	x	1.757E-002	= KILOWATTS
CALORIES	x	4.184	= JOULES
CANDELAS	x	12.56637	= LUMENS
CENTIMETERS	x	1.0E+008	= ANGSTROM UNITS
CENTIMETERS	x	0.3937008	= INCHES
CENTIMETERS	x	1.0E-005	= KILOMETERS
CENTIMETERS	x	393.7008	= MILS
CENTIMETERS/SEC	x	1.9685	= FEET/MINUTE
CIRCULAR MILS	x	5.067E-006	= SQUARE CENTIMETERS
CIRCULAR MILS	x	7.854E-007	= SQUARE INCHES
CIRCULAR MILS	x	5.067E-004	= SQUARE MILLIMETERS
CIRCULAR MILS	x	0.7853982	= SQUARE MILS
COULOMBS	x	1.036E-005	= FARADAY
CUBIC CENTIMETERS	x	3.531E-005	= CUBIC FEET
CUBIC CENTIMETERS	x	6.102E-002	= CUBIC INCHES
CUBIC CENTIMETERS	x	1.308E-006	= CUBIC YARDS
CUBIC CENTIMETERS	x	2.642E-004	= GALLONS
CUBIC CENTIMETERS	x	1.0E-003	= LITERS
CUBIC FEET	x	1728	= CUBIC INCHES
CUBIC FEET	x	2.832E-002	= CUBIC METERS
CUBIC FEET	x	7.4805	= GALLONS
CUBIC FEET	x	28.31605	= LITERS
CUBIC FEET	x	62.42618	= POUNDS OF WATER
CUBIC METERS	x	1.307951	= CUBIC YARDS
CUBIC METERS	x	264.1721	= GALLONS
DEGREES (ANGLE)	x	1.745E-002	= RADIANS
DEGREES PER SECOND	x	0.1666667	= REVOLUTIONS PER MINUTE
FATHOMS	x	6	= FEET
FEET	x	0.1666667	= FATHOMS
FEET	x	3.048E-004	= KILOMETERS
FEET	x	0.3048	= METERS
FEET	x	1.894E-004	= MILES
FEET	x	6.061E-002	= RODS
FEET PER MINUTE	x	1.667E-002	= FEET PER SECOND
FEET PER MINUTE	x	5.08E-003	= METERS PER SECOND
FEET PER MINUTE	x	1.136E-002	= MILES PER HOUR
FOOT-CANDLES	x	1	= LUMENS PER SQUARE FOOT
FOOT-CANDLES	x	10.7639	= LUX
FOOT-POUNDS	x	1.286E-003	= BRITISH THERMAL UNITS
FOOT-POUNDS PER MINUTE	x	3.03E-005	= HORSEPOWER
FOOT-POUNDS PER MINUTE	x	2.26E-005	= KILOWATTS
GALLONS	x	3.069E-006	= ACRE-FEET
GALLONS	x	2.381E-002	= BARRELS OF PETROLEUM
GALLONS	x	0.1336806	= CUBIC FEET
GALLONS	x	3.785E-003	= CUBIC METERS
GALLONS	x	4.95E-003	= CUBIC YARDS

Figure 16-2 Commonly used conversion formulas.

FROM		X	= TO
GALLONS	x	3.785306	= LITERS
GALLONS (BRITISH)	x	1.200949	= GALLONS (U.S.)
GALLONS (U.S.)	x	0.8326747	= GALLONG (BRITISH)
GALLONS PER MINUTE	x	2.228E-003	= CUBIC FEET PER SECOND
GAUSS	x	1	= LINES PER SQUARE CM.
GAUSS	x	6.4516	= LINES PER SQUARE INCH
GRAINS	x	2.286E-003	= OUNCES
GRAINS	x	2.083E-003	= OUNCES (TROY)
GRAINS	x	1.429E-004	= POUNDS
HECTARES	x	2.471058	= ACRES
HEMISPHERES	x	6.283185	= STERADIANS
HORSEPOWER	x	42.45282	= BTU PER MINUTE
HORSEPOWER	x	33000	= FOOT-POUNDS PER MINUTE
HORSEPOWER	x	0.7457	= KILOWATTS
HORSEPOWER	x	745.7	= WATTS
HORSEPOWER-HOURS	x	2546.136	= BRITISH THERMAL UNITS
INCHES	x	2.54	= CENTIMETERS
INCHES	x	2.540E-002	= METERS
INCHES	x	25400	= MICRONS
INCHES	x	25.4	= MILLIMETERS
INCHES	x	1000	= MILS
INCHES OF MERCURY	x	3.342E-002	= ATMOSPHERES
INCHES OF MERCURY	x	0.491154	= POUNDS PER SQUARE INCH
INCHES OF WATER	x	7.355E-002	= INCHES OF MERCURY
KILOGRAMS	x	2.204623	= POUNDS
KILOMETERS	x	3280.84	= FEET
KILOMETERS	x	0.6213712	= MILES
KILOMETERS PER HOUR	x	0.9113444	= FEET PER SECOND
KILOMETERS PER HOUR	x	0.5399568	= KNOTS
KILOMETERS PER HOUR	x	0.6213712	= MILES PER HOUR
KILOMETERS PER HOUR	x	54.68066	= FEET PER MINUTE
KILOWATT-HOURS	x	3414.426	= BTU
LIGHT YEARS	x	5.879E+012	= MILES
LINES PER SQUARE CM	x	1	= GAUSS
LINES PER SQUARE INCH	x	0.1550003	= GAUSS
LITERS	x	3.532E-002	= CUBIC FEET
LITERS	x	61.02543	= CUBIC INCHES
LITERS	x	0.2641794	= GALLONS
LITERS	x	33.81497	= FLUID OUNCES
MAXWELLS	x	1.0E-003	= KILOLINES
MAXWELLS	x	1.0E-008	= WEBERS
METERS	x	3.28084	= FEET
METERS	x	39.37008	= INCHES
METERS	x	6.214E-004	= MILES
METERS	x	1.093613	= YARDS
MILES	x	5280	= FEET
MILES	x	8	= FURLONGS
MILES	x	1.609344	= KILOMETERS
MILES	x	1609.344	= METERS
MILES	x	0.8689762	= NAUTICAL MILES
MILES PER HOUR	x	1.466667	= FEET PER SECOND
MILES PER HOUR	x	1.609344	= KILOMETERS PER HOUR
MILES PER HOUR	x	0.8689763	= KNOTS
MILLIBARS	x	0.750062	= TORR
MILS	x	2.54E-003	= CENTIMETERS
MILS	x	1.0E-003	= INCHES
MILS	x	2.54E-002	= MILLIMETERS
MOLECULES	x	1.66E-024	= MOLES
MOLES	x	6.023E+023	= MOLECULES
NAUTICAL MILES	x	1.852	= KILOMETERS

Figure 16-2 *(Continued)*

FROM	X	= TO
NAUTICAL MILES	x 1.150779	= MILES
NEWTON-METERS	x 0.737561	= FOOT-POUNDS
NEWTONS	x 0.2248089	= POUNDS
OERSTEDS	x 79.57747	= AMPERS PER METER
OUNCES	x 437.5	= GRAINS
OUNCES	x 28.34592	= GRAMS
OUNCES	x 6.2E-002	= POUNDS
OUNCES (FLUID)	x 1.804688	= CUBIC INCHES
OUNCES (FLUID)	x 2.957E-002	= LITERS
OUNCES (TROY)	x 480	= GRAINS
OUNCES (TROY)	x 31.10349	= GRAMS
PASCALS	x 4.015E-003	= INCHES OF WATER
PERCENT GRADE	x 1	= FEET PER 100 FEET
POUNDS	x 0.4535924	= KILOGRAMS
POUNDS	x 4.448222	= NEWTONS
POUNDS	x 16	= OUNCES
POUNDS	x 4.464E-004	= TONS (LONG)
POUNDS	x 4.4536E-004	= TONS (METRIC)
POUNDS	x 5.0E-004	= TONS (SHORT)
POUNDS PER SQUARE INCH	x 6.895E-002	= BARS
POUNDS PER SQUARE INCH	x 2.036021	= INCHES OF MERCURY
POUNDS PER SQUARE INCH	x 27.68068	= INCHES OF WATER
POUNDS PER SQUARE INCH	x 6.894745	= KILOPASCALS
POUNDS PER SQUARE INCH	x 6.805E-002	= ATMOSPHERES
RADIANS	x 57.29578	= DEGREES (ANGLE)
RADIANS	x 3437.747	= MINUTES (ANGLE)
RADIANS	x 0.1591549	= REVOLUTIONS
RADIANS PER SECOND	x 9.549297	= REVOLUTIONS PER MINUTE
REVOLUTIONS	x 6.283185	= RADIANS
REVOLUTIONS PER MINUTE	x 6	= DEGREES PER SECOND
RODS	x 16.5	= FEET
SQUARE CENTIMETERS	x 197352.5	= CIRCULAR MILS
SQUARE CENTIMETERS	x 1.076E-003	= SQUARE FEET
SQUARE CENTIMETERS	x 0.1550003	= SQUARE INCHES
SQUARE CENTIMETERS	x 155000.3	= SQUARE MILS
SQUARE FEET	x 2.296E-005	= ACRES
SQUARE FEET	x 9.29E-006	= HECTARES
SQUARE FEET	x 9.29E-002	= SQUARE METERS
SQUARE INCHES	x 1273240	= CIRCULAR MILS
SQUARE INCHES	x 6.4516	= SQUARE CENTIMETERS
SQUARE INCHES	x 645.16	= SQUARE MILLIMETERS
SQUARE INCHES	x 1,000,000	= SQUARE MILS
SQUARE MILS	x 1.27324	= CIRCULAR MILS
SQUARE MILS	x 1.0E-006	= SQUARE INCHES
TONS (LONG)	x 1016.047	= KILOGRAMS
TONS (LONG)	x 2240	= POUNDS
TONS (METRIC)	x 1000	= KILOGRAMS
TONS (METRIC)	x 2204.623	= POUNDS
TONS (SHORT)	x 907.1847	= KILOGRAMS
TONS (SHORT)	x 2000	= POUNDS
TORR	x 1.0E-0001	= CM OF MERCURY
TORR	x 3.937E-002	= INCHES OF MERCURY
TORR	x 1.333223	= MILLIBARS
WATT-HOURS	x 3.414426	= BTU
WATT	x 3.414426	= BTU PER HOUR
WATT	x 0.737562	= FOOT-POUNDS PER SECOND
WEBERS	x 1.0E+008	= MAXWELLS
YARDS	x 9.144E-004	= KILOMETERS
YARDS	x 0.9144	= METERS
YARDS	x 5.682E-004	= MILES

Figure 16-2 *(Continued)*

EXAMPLE 1. CONVERSIONS WITH NO POWER OF TEN

FROM	X	= TO
ATMOSPHERES	x 14.696 =	POUNDS/SQ.IN.

PROBLEM: HOW MANY POUNDS PER SQUARE INCH MUST A CONTAINER RESIST TO HOLD 15 ATMOSPHERES OF PRESSURE?

ANSWER:

15 ATMOSPHERES	x 14.696 =	**220.44 POUNDS/SQ.IN.**

EXAMPLE 2. CONVERSIONS WITH A POSITIVE POWER OF TEN

FROM	X	= TO
BTU	x 1.05E+10 =	ERGS

PROBLEM: HOW MANY ERGS ARE CONTAINED IN 33,000 BTU?

ANSWER

33,000BTU	x 1.05E+10 =	ERGS
3.30E+04	x 1.05E+10 =	[(3.3) X (1.05)] E (4 + 10) ERGS
3.30E+04	x 1.05E+10 =	**3.465E+14 ERGS.**

EXAMPLE 3. CONVERSIONS WITH A NEGATIVE POWER OF TEN

FROM	X	= TO
BTU	x 3.93E-04 =	HORSEPOWER-HOURS

PROBLEM: HOW MANY HORSEPOWER-HOURS ARE REPRESENTED BY A TOTAL OF 100,000 BTU?

ANSWER:

100,000BTU	x 3.93E-04 =	HORSEPOWER-HOURS
1.0E+5 BTU	x 3.93E-04 =	HORSEPOWER-HOURS
1.0E+5 BTU	x 3.93E-04 =	[(1.0) X (3.93)]E (+5 - 4)]HORSEPOWER-HOURS
1.0E+5 BTU	x 3.93E-04 =	[3.93]E (+1)]HORSEPOWER-HOURS
1.0E+5 BTU	x 3.93E-04 =	**39.3 HORSEPOWER-HOURS**

Figure 16-3 Use these calculation methods with the conversion formulas of Fig. 16-2.

different set of units. When this is necessary, it is frequently expedient to multiply by "one" in the calculation to make the units come out as desired. For example, the rental cost of a diesel generator is quoted at \$10 per gallon of diesel fuel used, and it is stated that the diesel engine consumes one 55-gallon (gal) barrel of fuel per 1-hour (h) running period. Determine the rental cost of the diesel generator for a 72-h running period.

The cost equals the dollar rate per hour times the quantity of hours in operation:

$$\text{Cost} = \$\text{ rate per hour} \times \text{quantity of hours in operation}$$

ABOUT THE AUTHOR

John M. Paschal, Jr., P.E., is a Senior Technical Electrical Engineering Specialist with Bechtel, and has held a master electrician's license for three decades. He has worked in all facets of the electrical industry. He has worked "with the tools" in the electrical construction trade, as an electrical construction project manager, as chief electrical estimator for national and international construction firms, as a college assistant professor, and as an electrical engineer of some of the largest and most demanding commercial, health care, and industrial electrical systems throughout the world. Mr. Paschal is the author of many books on practical electrical engineering and construction methods, and books dealing with the *National Electrical Code*. He is the technical editor of EC&M Books and an author for *Electrical Construction & Maintenance* magazine.

Electrical Construction & Maintenance magazine, established in 1901, is the electrical industry's premier magazine for electrical design, construction, and maintenance.